Ruin Probabilities

Series Editor
Nikolaos Limnios

Ruin Probabilities

Smoothness, Bounds, Supermartingale Approach

Yuliya Mishura
Olena Ragulina

ISTE PRESS

ELSEVIER

First published 2016 in Great Britain and the United States by ISTE Press Ltd and Elsevier Ltd

ISTE Press Ltd
27-37 St George's Road
London SW19 4EU
UK

www.iste.co.uk

Elsevier Ltd
The Boulevard, Langford Lane
Kidlington, Oxford, OX5 1GB
UK

www.elsevier.com

British Library Cataloguing-in-Publication Data
A CIP record for this book is available from the British Library
Library of Congress Cataloging in Publication Data
A catalog record for this book is available from the Library of Congress
ISBN 978-1-78548-218-2

Contents

Preface

Risk theory is traditionally considered as a branch of insurance mathematics and its active area of research. Since Filip Lundberg [LUN 03] introduced the collective risk model in 1903, the estimation of ruin probabilities has been one of the central directions for investigations in risk theory. The infinite- and finite-horizon ruin probabilities are the probabilities that the surplus of an insurance company becomes negative on infinite and finite time intervals, respectively. The infinite-horizon ruin probability is usually considered as a function of the initial surplus, and the finite-horizon ruin probability is assumed to be a function of the initial surplus and time interval. Instead of ruin probabilities, one sometimes deals with the corresponding survival probabilities, which are the probabilities that the surplus of the insurance company does not become negative. So, the survival of the insurance company at least implies that it is able to reimburse claims and meet its obligations. Although other characteristics have also been proposed and investigated (e.g. distribution of the ruin time, surplus prior to ruin, deficit at ruin and Gerber–Shiu expected discounted penalty functions), the ruin or survival probabilities are still the main objects of interest in risk theory.

In this book, we will consider continuous-time risk models. Due to the obvious relationship between ruin and survival probabilities, from the mathematical viewpoint it makes no difference which of these two characteristics is studied. Since sometimes it is more convenient to deal with the ruin probabilities and sometimes with the survival probabilities, in what follows we consider both of them. To be more precise, we investigate both survival probabilities in Chapters 2–6 and ruin probabilities in Chapters 7–9.

This monograph mainly provides the results obtained by the authors recently in this field. It is a revised and extended version of the material published in [AND 07a, AND 07b, BON 12, MIS 15, RAG 10, RAG 11, RAG 12, RAG 14, RAG 15].

The book deals with several aspects of risk theory. The first of them is the smoothness of the survival probabilities. Particularly, we investigate the continuity and differentiability of the infinite- and finite-horizon survival probabilities in detail for different risk models. The initial motivation for this investigation is the reasonable derivation of integro-differential equations for these functions. These equations turn out to be very useful for calculation of the survival probabilities. In some cases, one can find their closed-form solutions. In other cases, the integro-differential equations provide some analytic information concerning the survival probabilities. In particular, they are used for obtaining bounds and approximations for the survival probabilities or investigation of their asymptotics. Note that it is not a problem to write down the integro-differential equations for the survival probabilities and most of the equations we derive have already been obtained and used for further investigation. Furthermore, one can easily write down all these equations in terms of infinitesimal generators for stopped surplus processes, which are Markov processes (see, e.g. [ASM 10, p. 248] or [SCH 08b, pp. 220, 225]). Nevertheless, usually one has to make the additional assumption concerning the differentiability of the survival probabilities or conditions when these functions that are really differentiable are not studied. Thus, investigation of the continuity and differentiability of the survival probabilities is of great interest. As we will see, these functions are not always differentiable in the classical sense. Moreover, sometimes points of discontinuity are also possible. So, the continuity and differentiability of the survival probabilities must be investigated separately for each risk model, and we study these properties for different models in this book.

Next, we concentrate on some possible applications of the results concerning the smoothness of the survival probabilities. Firstly, we find analytic expressions for the infinite-horizon survival probabilities in a few cases when the c.d.f. of the insurance compensation is a sum of absolutely continuous and discrete components. Secondly, we solve the problems of optimal control by franchise and deductible amounts in the classical risk model from the viewpoint of survival probability maximization. Thirdly, we obtain relationships connecting accuracy and reliability of uniform approximations of the survival probabilities from their statistical estimates.

Finally, we introduce and apply the supermartingale approach to obtain upper exponential bounds for the infinite-horizon ruin probabilities in different generalizations of the classical risk model with risky investments. This approach generalizes the martingale one introduced by Gerber [GER 73] to obtain the famous exponential bound for the infinite-horizon ruin probability in the classical risk model, which is known as the Lundberg inequality.

The book is not a textbook in risk theory and does not provide general results. Nevertheless, we give some basic results concerning ruin probabilities in the classical risk model and the risk model with stochastic premiums in Chapter 1. The reader can find the general results in risk theory in [ASM 10, BEA 84, BÜH 70, DIC 10, GER 79, GRA 91, ROL 99]. Furthermore, we refer the reader to [SCH 08b] for a systematic description of optimal control methods applied to insurance problems and [AZC 14] for the viscosity approach to optimal control problems in insurance.

This work is intended for researchers in probability theory, actuarial sciences and financial mathematics, as well as for graduate and postgraduate students.

Though the book is mainly theoretical, it is also accessible to practitioners who want to extend their knowledge in insurance mathematics.

The book is self-contained and does not require any complementary material. All its main results are provided with proofs. It is assumed that the reader has a basic knowledge in probability theory, stochastic processes, stochastic calculus, ordinary and stochastic differential equations and calculus. We provide some well-known mathematical results which are often used in the main text in the Appendix. Moreover, we refer the reader to [DOO 53, ELL 82, JAC 06, KAR 91, ØKS 07, ROG 00, ROL 99, SHR 04] for basic notions of the theory of stochastic processes and stochastic calculus and to [CHE 05, IKE 89, KAR 91, ØKS 03] for basic facts concerning stochastic differential equations.

The main material of the book is divided into two parts and organized as follows. In Part 1, we investigate the continuity and differentiability of survival probabilities for different risk models and concentrate on some possible applications of these results. In Chapter 1, we give some basic results concerning ruin probabilities in the classical risk model and the risk model with stochastic premiums. In Chapter 2, we deal with a generalization of the classical risk model where an insurance company invests all surplus in a risk-free asset. We investigate the continuity and differentiability of infinite- and finite-horizon survival probabilities in detail. In Chapter 3, we obtain the corresponding results for the risk model with stochastic premiums and investments in a risk-free asset. In Chapter 4, we consider the classical risk model under the additional assumption that an insurance company applies a franchise and a liability limit. Assuming that claim sizes are exponentially distributed we find analytic expressions for the infinite-horizon survival probabilities and investigate how a franchise and a liability limit change the survival probability for small and large enough initial surpluses. In Chapter 5, we solve the problems of optimal control by franchise and deductible amounts in the classical risk model from the viewpoint of survival probability maximization. In Chapter 6, we extend the results of Chapters 2 and 3 to a case where an insurance company invests all surplus in risk-free and risky assets proportionally and the price of the risky asset follows a jump process. Moreover, we obtain relationships connecting accuracy and reliability of uniform approximations of the survival probabilities by their statistical estimates.

In Part 2, we apply the supermartingale approach to obtain upper exponential bounds for infinite-horizon ruin probabilities in different generalizations of the classical risk model where a premium intensity depends on a current surplus of an insurance company. To this end, in Chapter 7 we assume that all surplus is invested in one risky asset, the price of which follows a geometric Brownian motion, and the premium intensity grows rapidly with increasing surplus. Furthermore, we investigate the question concerning the probability of explosion of the surplus process between claim arrivals in detail. In Chapter 8, we suppose that all surplus is also invested in one risky asset, the price of which follows a geometric Brownian motion, but an insurance company stops its investment activity when the price of the risky asset goes below some fixed level or outside some fixed interval. In Chapter 9, we assume that the surplus is invested in one risk-free and a few risky assets, the prices of which follow geometric Brownian motions, and borrowing is also possible.

We conclude the preface by thanking all those without whose help and encouragement this book would never have been written. First of all, it is our pleasure to thank Hanspeter Schmidli, who is a great authority in risk theory, for his hospitality during our visits to the University of Cologne and numerous helpful discussions of the problems considered in the book. Moreover, we are grateful to the co-authors of our papers, in particular Maryna Lundgren (Androshchuk), for their contribution to our joint work.

Yuliya MISHURA
Olena RAGULINA
Kyiv
May, 2016

PART 1

SMOOTHNESS OF THE SURVIVAL PROBABILITIES WITH APPLICATIONS

1

Classical Results on the Ruin Probabilities

In this chapter, we formulate some basic results concerning ruin probabilities in the classical risk model and the risk model with stochastic premiums. These results are used in Chapters 2–9.

Let $(\Omega, \mathfrak{F}, \mathbb{P})$ be a probability space satisfying the usual conditions and let all the stochastic objects that we use in Chapters 1–9 be defined on it.

1.1. Classical risk model

We now deal with the *classical risk model*, which is also called the *Cramér–Lundberg model*. For details and more information, see, for example [ASM 10, BEA 84, BÜH 70, DIC 10, GER 79, GRA 91, KAA 08, ROL 99, SCH 08b].

1.1.1. *Description of the model*

In the classical risk model, claim sizes form a sequence $(Y_i)_{i\geq 1}$ of non-negative i.i.d. r.v.'s with c.d.f. $F(y) = \mathbb{P}[Y_i \leq y]$ and finite expectation $\mu > 0$. Let τ_i be the time when the ith claim arrives. The number of claims on the time interval $[0, t]$ is a Poisson process $(N_t)_{t\geq 0}$ with constant intensity $\lambda > 0$. The r.v.'s $(Y_i)_{i\geq 1}$ and the process $(N_t)_{t\geq 0}$ are mutually independent. Thus, the total claims on $[0, t]$ equal $\sum_{i=1}^{N_t} Y_i$. We set $\sum_{i=1}^{0} Y_i = 0$ if $N_t = 0$. Let $(\mathfrak{F}_t)_{t\geq 0}$ be the filtration generated by $(Y_i)_{i\geq 1}$ and $(N_t)_{t\geq 0}$. The insurance company has a non-negative initial surplus x and receives premiums with constant intensity $c > 0$.

Let $X_t(x)$ be the surplus of the insurance company at time t provided that its initial surplus is x. Then, the *surplus process* $\left(X_t(x)\right)_{t\geq 0}$ follows:

$$X_t(x) = x + ct - \sum_{i=1}^{N_t} Y_i, \quad t \geq 0. \qquad [1.1]$$

The *ruin time* is defined by

$$\tau(x) = \inf\{t \geq 0\colon X_t(x) < 0\}.$$

We suppose that $\tau(x) = \infty$ if $X_t(x) \geq 0$ for all $t \geq 0$.

The *infinite-horizon ruin probability* is given by

$$\psi(x) = \mathbb{P}\left[\inf_{s\geq 0} X_s(x) < 0\right],$$

which is equivalent to

$$\psi(x) = \mathbb{P}[\tau(x) < \infty].$$

The *finite-horizon ruin probability* is given by

$$\psi(x,t) = \mathbb{P}\left[\inf_{0\leq s\leq t} X_s(x) < 0\right],$$

which is equivalent to

$$\psi(x,t) = \mathbb{P}[\tau(x) \leq t].$$

The corresponding *infinite-horizon* and *finite-horizon survival probabilities* equal

$$\varphi(x) = 1 - \psi(x)$$

and

$$\varphi(x,t) = 1 - \psi(x,t).$$

The notions of ruin time as well as infinite- and finite-horizon ruin and survival probabilities introduced above are applicable to all risk models considered in this book, unless otherwise stated.

It is easily seen that $\varphi(x)$ and $\varphi(x,t)$ are non-decreasing functions w.r.t. x. Furthermore, $\varphi(x,t)$ is a non-increasing function w.r.t. t. But it is not clear whether these functions are continuous or differentiable. As we will see later, survival probabilities are not always differentiable. Moreover, sometimes points of discontinuity of survival probabilities are also possible (see Chapter 3). So, the continuity and differentiability of these functions must be investigated separately for each risk model.

Since there is elementary dependence between ruin probabilities and the corresponding survival probabilities, in what follows, we formulate results for either of them. Sometimes it is more convenient to deal with the ruin probabilities and sometimes with the survival probabilities.

1.1.2. *Net profit condition and the behavior of the infinite-horizon ruin probability*

It turns out that the behavior of $\psi(x)$ depends substantially on whether the condition $c > \lambda\mu$ is true or not.

LEMMA 1.1.– Let the surplus process $\left(X_t(x)\right)_{t\geq 0}$ follow [1.1].

i) If $c \leq \lambda\mu$, then $\psi(x) = 1$ for all $x \geq 0$.

ii) If $c > \lambda\mu$, then $\lim_{x\to+\infty}\psi(x) = 0$.

To prove lemma 1.1, we will need the following theorem from the theory of random walks.

THEOREM 1.1.– ([ROL 99, p. 233], theorem 6.3.1) Let $(Z_i)_{i\geq 1}$ be a sequence of i.i.d. r.v.'s, such that $\mathbb{P}[Z_i > 0] > 0$, $\mathbb{P}[Z_i < 0] > 0$ and $\mathbb{E}[Z_i] < \infty$. Moreover, let the r.v.'s $\tilde{Z}_n$, $n \geq 1$, be given by $\tilde{Z}_n = \sum_{i=1}^{n} Z_i$.

i) If $\mathbb{E}[Z_i] > 0$, then $\mathbb{P}\left[\lim_{n\to\infty}\tilde{Z}_n = +\infty\right] = 1$.

ii) If $\mathbb{E}[Z_i] < 0$, then $\mathbb{P}\left[\lim_{n\to\infty}\tilde{Z}_n = -\infty\right] = 1$.

iii) If $\mathbb{E}[Z_i] = 0$, then $\mathbb{P}\left[\limsup_{n\to\infty}\tilde{Z}_n = +\infty\right] = 1$ and $\mathbb{P}\left[\liminf_{n\to\infty}\tilde{Z}_n = -\infty\right] = 1$.

Note that the proof of theorem 1.1 is straightforward in the first two cases. Indeed, by the strong law of large numbers, we get

$\mathbb{P}\big[\lim_{n\to\infty} \tilde{Z}_n/n = \mathbb{E}[Z_i]\big] = 1$. Therefore, $\mathbb{P}\big[\lim_{n\to\infty} \tilde{Z}_n = +\infty\big] = 1$ if $\mathbb{E}[Z_i] > 0$ and $\mathbb{P}\big[\lim_{n\to\infty} \tilde{Z}_n = -\infty\big] = 1$ if $\mathbb{E}[Z_i] < 0$. The case $\mathbb{E}[Z_i] = 0$ is much more difficult (see, for example [ROL 99]).

PROOF (Proof of lemma 1.1).– Let $Z_i = Y_i - c(\tau_i - \tau_{i-1})$, $i \geq 1$. It is easily seen that the r.v.'s Z_i satisfy the requirements of theorem 1.1 and $\mathbb{E}[Z_i] = \mu - c/\lambda$.

Next, let $\tilde{Z}_n = \sum_{i=1}^{n} Z_i$, $n \geq 1$, and $M = \sup_{t\geq 0}\Big(\sum_{i=1}^{N_t} Y_i - ct\Big)$. It is evident that $M = \sup_{n\geq 1} \tilde{Z}_n$, which gives

$$\psi(x) = \mathbb{P}[M > x], \quad x \geq 0. \qquad [1.2]$$

If $c \leq \lambda\mu$, then $\mathbb{E}[Z_i] \geq 0$ and $\mathbb{P}\big[\limsup_{n\to\infty} \tilde{Z}_n = +\infty\big] = 1$ by theorem 1.1. Hence, from [1.2] we conclude that $\psi(x) = 1$ for all $x \geq 0$.

If $c > \lambda\mu$, then $\mathbb{E}[Z_i] < 0$ and $\mathbb{P}\big[\lim_{n\to\infty} \tilde{Z}_n = -\infty\big] = 1$ by theorem 1.1, which yields $\mathbb{P}[M < \infty] = 1$. Therefore, $\lim_{x\to+\infty} \psi(x) = 0$ by [1.2]. □

The condition $c > \lambda\mu$ is called the *net profit condition*. This name can be explained by the fact that $\mathbb{E}\big[X_t(x) - x\big] = (c - \lambda\mu)t$. In what follows, we concentrate on that case.

Moreover, we often assume that the insurance company uses the *expected value principle* for premium calculation, which means that $c = \lambda\mu(1+\theta)$, where $\theta > 0$ is a *safety loading*.

1.1.3. *Integro-differential equations for the survival probabilities*

THEOREM 1.2.– ([ROL 99, pp. 162–163], theorem 5.3.1) Let the surplus process $\big(X_t(x)\big)_{t\geq 0}$ follow [1.1]. Then, $\varphi(x)$ is continuous on $\mathbb{R}_+$ with right and left derivatives $\varphi'_+(x)$ and $\varphi'_-(x)$, respectively. Moreover, we have

$$c\varphi'_+(x) = \lambda\varphi(x) - \lambda \int_0^x \varphi(x-y)\,\mathrm{d}F(y), \quad x \geq 0, \qquad [1.3]$$

and

$$c\varphi'_-(x) = \lambda\varphi(x) - \lambda \int_0^{x-} \varphi(x-y)\,\mathrm{d}F(y), \quad x > 0. \qquad [1.4]$$

PROOF.– It is easily seen that ruin does not occur up to the time τ_1 and it will not take place after τ_1 if and only if $Y_1 \leq x + c\tau_1$ and ruin does not happen on the time interval $(\tau_1, +\infty)$ with the initial surplus $x + c\tau_1 - Y_1$. Moreover, note that τ_1 is exponentially distributed with mean $1/\lambda$. So, we have $\mathbb{P}[\tau_1 > t] = \mathrm{e}^{-\lambda t}$ for all $t \geq 0$. Thus, by the law of total probability, we get

$$\varphi(x) = \mathrm{e}^{-\lambda t}\,\varphi(x+ct) + \int_0^t \lambda \mathrm{e}^{-\lambda s} \int_0^{x+cs} \varphi(x+cs-y)\,\mathrm{d}F(y)\,\mathrm{d}s \quad [1.5]$$

for all $x \geq 0$ and $t \geq 0$. Letting $t \downarrow 0$ in [1.5] we conclude that $\varphi(x)$ is right-continuous on $\mathbb{R}_+$.

Furthermore, we can rewrite [1.5] as

$$\begin{aligned} c\,\frac{\varphi(x+ct)-\varphi(x)}{ct} &= \frac{1-\mathrm{e}^{-\lambda t}}{t}\,\varphi(x+ct) \\ &- \frac{1}{t}\int_0^t \lambda \mathrm{e}^{-\lambda s} \int_0^{x+cs} \varphi(x+cs-y)\,\mathrm{d}F(y)\,\mathrm{d}s. \end{aligned} \quad [1.6]$$

This shows that $\varphi(x)$ is differentiable from the right on $\mathbb{R}_+$. Letting $t \downarrow 0$ in [1.6] yields [1.3].

Next, for $x > 0$ and $t \leq x/c$, we can rewrite [1.5] in the form

$$\begin{aligned} \varphi(x-ct) &= \mathrm{e}^{-\lambda t}\,\varphi(x) \\ &+ \int_0^t \lambda \mathrm{e}^{-\lambda s} \int_0^{x-c(t-s)} \varphi(x-c(t-s)-y)\,\mathrm{d}F(y)\,\mathrm{d}s. \end{aligned} \quad [1.7]$$

Letting $t \downarrow 0$ in [1.7] we deduce that $\varphi(x)$ is left-continuous on $(0, +\infty)$. Rewriting [1.7] in the form

$$\begin{aligned} c\,\frac{\varphi(x)-\varphi(x-ct)}{ct} &= \frac{1-\mathrm{e}^{-\lambda t}}{t}\,\varphi(x) \\ &- \frac{1}{t}\int_0^t \lambda \mathrm{e}^{-\lambda s} \int_0^{x-c(t-s)} \varphi(x-c(t-s)-y)\,\mathrm{d}F(y)\,\mathrm{d}s \end{aligned} \quad [1.8]$$

shows that $\varphi(x)$ is differentiable from the left on $(0, +\infty)$. Letting $t \downarrow 0$ in [1.8] yields [1.4]. □

REMARK 1.1.– It follows from [1.3] and [1.4] that if $c > \lambda\mu$, then $\varphi(x)$ is differentiable on $\mathbb{R}_+$, except at positive points y where $F(y)$ is not continuous.

Equation [1.3] involves both the derivative and an integral of $\varphi(x)$, so it is not easily solved in the general case.

Furthermore, we can easily show that $\varphi(x,t)$ satisfies the following partial integro-differential equation:

$$\frac{\partial\varphi(x,t)}{\partial t} - c\,\frac{\partial\varphi(x,t)}{\partial x} + \lambda\varphi(x,t) - \lambda\int_0^x \varphi(x-y,\, t)\,\mathrm{d}F(y) = 0.$$

The question regarding the existence of partial derivatives of $\varphi(x,t)$ is investigated in section 2.3.

1.1.4. *Integral equation for the infinite-horizon ruin probability*

THEOREM 1.3.– Let the surplus process $\big(X_t(x)\big)_{t\geq 0}$ follow [1.1] and $c > \lambda\mu$. Then, $\psi(x)$ satisfies the integral equation

$$c\psi(x) = \lambda\int_x^{+\infty}\big(1-F(y)\big)\,\mathrm{d}y + \lambda\int_0^x \psi(x-y)\big(1-F(y)\big)\,\mathrm{d}y, \quad x \geq 0. \quad [1.9]$$

PROOF.– Integrating [1.3] over $[0,x]$ yields

$$\begin{aligned}
c\big(\varphi(x)-\varphi(0)\big) &= \lambda\int_0^x \varphi(z)\,\mathrm{d}z - \lambda\int_0^x\int_0^z \varphi(z-y)\,\mathrm{d}F(y)\,\mathrm{d}z \\
&= \lambda\int_0^x \varphi(z)\,\mathrm{d}z - \lambda\int_0^x\int_y^x \varphi(z-y)\,\mathrm{d}z\,\mathrm{d}F(y) \\
&= \lambda\int_0^x \varphi(z)\,\mathrm{d}z - \lambda\int_0^x\int_0^{x-y} \varphi(z)\,\mathrm{d}z\,\mathrm{d}F(y) \\
&= \lambda\int_0^x \varphi(z)\,\mathrm{d}z - \lambda\int_0^x\int_0^{x-z} \mathrm{d}F(y)\,\varphi(z)\,\mathrm{d}z \\
&= \lambda\int_0^x \varphi(z)\big(1-F(x-z)\big)\,\mathrm{d}z \\
&= \lambda\int_0^x \varphi(x-y)\big(1-F(y)\big)\,\mathrm{d}y.
\end{aligned}$$

Thus, we get

$$c\big(\varphi(x) - \varphi(0)\big) = \lambda \int_0^x \varphi(x-y)\big(1 - F(y)\big)\,\mathrm{d}y, \quad x \geq 0. \qquad [1.10]$$

By monotone convergence, letting $x \to +\infty$ in [1.10] gives

$$c\left(\lim_{x\to+\infty} \varphi(x) - \varphi(0)\right) = \lambda\mu \lim_{x\to+\infty} \varphi(x).$$

Hence, we have

$$\varphi(0) = \left(1 - \frac{\lambda\mu}{c}\right) \lim_{x\to+\infty} \varphi(x). \qquad [1.11]$$

Since $c > \lambda\mu$, $\lim_{x\to+\infty} \varphi(x) = 1$ by lemma 1.1. Consequently, from [1.11] we get

$$\varphi(0) = 1 - \frac{\lambda\mu}{c}. \qquad [1.12]$$

Finally, substituting $\varphi(x) = 1 - \psi(x)$ into [1.10] and using [1.12] we obtain

$$c\psi(x) = \lambda\mu - \lambda \int_0^x \big(1 - F(y)\big)\,\mathrm{d}y + \lambda \int_0^x \psi(x-y)\big(1 - F(y)\big)\,\mathrm{d}y, \quad x \geq 0,$$

which is equivalent to [1.9]. □

Equation [1.9] is called a *defective renewal equation.*

REMARK 1.2.– Note that there is linear dependence [1.11] between $\varphi(0)$ and $\lim_{x\to+\infty} \varphi(x)$ in this risk model. In particular, [1.12] holds if and only if $\lim_{x\to+\infty} \varphi(x) = 1$. If the insurance company uses the expected value principle for premium calculation, then [1.12] can be rewritten as $\varphi(0) = \theta/(1+\theta)$.

1.1.5. *Laplace transform of the infinite-horizon survival probability*

Let $\hat{l}_F(z)$ and $\hat{L}_\varphi(z)$ be the *Laplace–Stieltjes transform* of $F(y)$ and the *Laplace transform* of $\varphi(x)$, respectively, i.e.

$$\hat{l}_F(z) = \int_0^{+\infty} \mathrm{e}^{-zy}\,\mathrm{d}F(y), \quad z \geq 0,$$

and

$$\hat{L}_\varphi(z) = \int_0^{+\infty} \mathrm{e}^{-zx}\,\varphi(x)\,\mathrm{d}x, \quad z > 0.$$

THEOREM 1.4.– Let the surplus process $\left(X_t(x)\right)_{t\geq 0}$ follow [1.1] and $c > \lambda\mu$. Then

$$\hat{L}_\varphi(z) = \frac{c - \lambda\mu}{cz - \lambda\left(1 - \hat{l}_F(z)\right)}, \quad z > 0. \tag{1.13}$$

PROOF.– Multiplying [1.3] by e^{-zx} and integrating over $[0, +\infty)$ gives

$$\begin{aligned} c\int_0^{+\infty} \varphi'_+(x)\,\mathrm{e}^{-zx}\,\mathrm{d}x &= \lambda\int_0^{+\infty} \varphi(x)\,\mathrm{e}^{-zx}\,\mathrm{d}x \\ &- \lambda\int_0^{+\infty}\int_0^{x} \varphi(x-y)\,\mathrm{d}F(y)\,\mathrm{e}^{-zx}\,\mathrm{d}x, \quad z > 0. \end{aligned} \tag{1.14}$$

It is easily seen that

$$\begin{aligned} \int_0^{+\infty} \varphi'_+(x)\,\mathrm{e}^{-zx}\,\mathrm{d}x &= -\varphi(0) + z\int_0^{+\infty} \varphi(x)\,\mathrm{e}^{-zx}\,\mathrm{d}x \\ &= -\varphi(0) + z\hat{L}_\varphi(z) \end{aligned} \tag{1.15}$$

and

$$\begin{aligned} &\int_0^{+\infty}\int_0^{x} \varphi(x-y)\,\mathrm{d}F(y)\,\mathrm{e}^{-zx}\,\mathrm{d}x \\ &\quad = \int_0^{+\infty}\int_y^{+\infty} \varphi(x-y)\,\mathrm{e}^{-zx}\,\mathrm{d}x\,\mathrm{d}F(y) \\ &\quad = \int_0^{+\infty}\int_0^{+\infty} \varphi(x)\,\mathrm{e}^{-z(x+y)}\,\mathrm{d}x\,\mathrm{d}F(y) \\ &\quad = \int_0^{+\infty} \varphi(x)\,\mathrm{e}^{-zx}\,\mathrm{d}x \int_0^{+\infty} \mathrm{e}^{-zy}\,\mathrm{d}F(y) = \hat{L}_\varphi(z)\,\hat{l}_F(z). \end{aligned} \tag{1.16}$$

Substituting [1.15] and [1.16] into [1.14] yields

$$c\left(z\hat{L}_{\varphi}(z) - \varphi(0)\right) = \lambda \hat{L}_{\varphi}(z)\left(1 - \hat{l}_F(z)\right), \quad z > 0. \qquad [1.17]$$

Since $c > \lambda\mu$, we have $\varphi(0) = 1 - \lambda\mu/c$. Hence, [1.17] can be rewritten as

$$cz\hat{L}_{\varphi}(z) - c + \lambda\mu = \lambda \hat{L}_{\varphi}(z)\left(1 - \hat{l}_F(z)\right), \quad z > 0,$$

which is equivalent to [1.13]. □

REMARK 1.3.– Since there is a one-to-one correspondence between a function and its Laplace transform, theorem 1.4 implies the existence and uniqueness of the solution to [1.3], such that $\lim_{x\to+\infty} \varphi(x) = 1$, provided that $c > \lambda\mu$.

1.1.6. *Analytic expressions for the infinite-horizon survival probability*

There are only a few cases where we can find analytic expressions for the infinite-horizon survival probability. We now consider two such cases.

THEOREM 1.5.– Let the surplus process $\left(X_t(x)\right)_{t\geq 0}$ follow [1.1], $c > \lambda\mu$ and the claim sizes be exponentially distributed with mean μ. Then

$$\varphi(x) = 1 - \frac{\lambda\mu}{c} \exp\left(-\frac{(c - \lambda\mu)\, x}{c\mu}\right), \quad x \geq 0. \qquad [1.18]$$

PROOF.– Since $F(y)$ is continuous, by theorem 1.2, $\varphi(x)$ is differentiable on $\mathbb{R}_+$ and satisfies the equation

$$c\varphi'(x) = \lambda\varphi(x) - \frac{\lambda}{\mu}\, e^{-x/\mu} \int_0^x \varphi(y)\, e^{y/\mu}\, dy. \qquad [1.19]$$

From [1.19] we see that $\varphi'(x)$ is differentiable on $\mathbb{R}_+$ and

$$\begin{aligned} c\varphi''(x) &= \lambda\varphi'(x) + \frac{\lambda}{\mu^2}\, e^{-x/\mu} \int_0^x \varphi(y)\, e^{y/\mu}\, dy - \frac{\lambda}{\mu}\, \varphi(x) \\ &= \lambda\varphi'(x) - \frac{1}{\mu}\left(c\varphi'(x) - \lambda\varphi(x)\right) - \frac{\lambda}{\mu}\, \varphi(x) \\ &= \left(\lambda - \frac{c}{\mu}\right) \varphi'(x). \end{aligned}$$

Thus, we get the differential equation

$$c\varphi''(x) + \left(\frac{c}{\mu} - \lambda\right)\varphi'(x) = 0. \qquad [1.20]$$

The general solution to [1.20] is

$$\varphi(x) = C_1 - C_2 \exp\left(-\frac{(c-\lambda\mu)x}{c\mu}\right),$$

where $C_1 \in \mathbb{R}$ and $C_2 \in \mathbb{R}$. Since $\lim_{x\to+\infty}\varphi(x) = 1$, we conclude that $C_1 = 1$. Since $\varphi(0) = 1 - \lambda\mu/c$, we get $C_2 = \lambda\mu/c$, which completes the proof. □

If the insurance company uses the expected value principle for premium calculation, then [1.18] can be rewritten as

$$\varphi(x) = 1 - \frac{1}{1+\theta}\exp\left(-\frac{\theta x}{\mu(1+\theta)}\right), \quad x \geq 0.$$

Using theorem 1.4, we can prove the following theorem for the case where the claim sizes have a degenerate distribution.

THEOREM 1.6.– Let the surplus process $(X_t(x))_{t\geq 0}$ follow [1.1], $c > \lambda\mu$ and $\mathbb{P}[Y_i = \mu] = 1$, $i \geq 1$. Then

$$\begin{aligned}\varphi(x) = {}& \left(1 - \frac{\lambda\mu}{c}\right)\sum_{k=0}^{\infty}\frac{1}{k!}\left(-\frac{\lambda}{c}\right)^k(\max\{x-k\mu,0\})^k \\ & \times \exp\left(\frac{\lambda}{c}\max\{x-k\mu,0\}\right), \quad x \geq 0.\end{aligned} \qquad [1.21]$$

REMARK 1.4.– It is easy to check that $\varphi(x)$ given by [1.21] is not differentiable at the point $x = \mu$ and

$$\varphi'_-(\mu) - \varphi'_+(\mu) = \frac{\lambda}{c}\left(1 - \frac{\lambda\mu}{c}\right) > 0. \qquad [1.22]$$

Indeed, from [1.21], we see that

$$\varphi(x) = \left(1 - \frac{\lambda\mu}{c}\right)\exp\left(\frac{\lambda x}{c}\right), \quad 0 \leq x \leq \mu,$$

$$\varphi(x) = \left(1 - \frac{\lambda\mu}{c}\right)\left(\exp\left(\frac{\lambda x}{c}\right)\right.$$
$$\left. - \frac{\lambda}{c}(x-\mu)\exp\left(\frac{\lambda(x-\mu)}{c}\right)\right), \quad \mu \leq x \leq 2\mu.$$

Hence, we have

$$\varphi'(x) = \frac{\lambda}{c}\left(1 - \frac{\lambda\mu}{c}\right)\exp\left(\frac{\lambda x}{c}\right), \quad 0 \leq x \leq \mu,$$

$$\varphi'(x) = \frac{\lambda}{c}\left(1 - \frac{\lambda\mu}{c}\right)\left(\exp\left(\frac{\lambda x}{c}\right) - \frac{\lambda}{c}(x-\mu)\exp\left(\frac{\lambda(x-\mu)}{c}\right)\right.$$
$$\left. - \exp\left(\frac{\lambda(x-\mu)}{c}\right)\right), \quad \mu \leq x \leq 2\mu.$$

Consequently,

$$\varphi'_-(\mu) = \frac{\lambda}{c}\left(1 - \frac{\lambda\mu}{c}\right)\exp\left(\frac{\lambda\mu}{c}\right)$$

and

$$\varphi'_+(\mu) = \frac{\lambda}{c}\left(1 - \frac{\lambda\mu}{c}\right)\left(\exp\left(\frac{\lambda\mu}{c}\right) - 1\right),$$

which gives [1.22].

Note that the same conclusion follows from theorem 1.2. □

1.1.7. *Cramér–Lundberg approximation for the infinite-horizon ruin probability*

We now assume that there is $\hat{R} > 0$, such that

$$\frac{\lambda}{c}\int_0^{+\infty} \mathrm{e}^{\hat{R}y}(1 - F(y))\,\mathrm{d}y = 1. \quad [1.23]$$

Let

$$\hat{\mu} = \frac{\lambda}{c}\int_0^{+\infty} y\mathrm{e}^{\hat{R}y}(1 - F(y))\,\mathrm{d}y.$$

Using the integral equation [1.9] and the key renewal theorem, we can prove the following theorem, which is called the *Cramér–Lundberg theorem.*

THEOREM 1.7.– Let the surplus process $\left(X_t(x)\right)_{t\geq 0}$ follow [1.1], $c > \lambda\mu$ and there be $\hat{R} > 0$, such that [1.23] holds.

i) If $\hat{\mu} < +\infty$, then

$$\lim_{x\to+\infty} \psi(x)\,\mathrm{e}^{\hat{R}x} = \frac{c-\lambda\mu}{c\hat{R}\hat{\mu}}. \qquad [1.24]$$

ii) If $\hat{\mu} = +\infty$, then

$$\lim_{x\to+\infty} \psi(x)\,\mathrm{e}^{\hat{R}x} = 0.$$

From [1.24] we conclude that

$$\psi(x) \sim \frac{c-\lambda\mu}{c\hat{R}\hat{\mu}}\,\mathrm{e}^{-\hat{R}x} \quad \text{as} \quad x \to +\infty. \qquad [1.25]$$

This approximation for $\psi(x)$ is called the *Cramér–Lundberg approximation.* Particularly, if the insurance company uses the expected value principle for premium calculation, then [1.25] can be rewritten as

$$\psi(x) \sim \frac{\theta}{(1+\theta)\hat{R}\hat{\mu}}\,\mathrm{e}^{-\hat{R}x} \quad \text{as} \quad x \to +\infty.$$

We now investigate the conditions when there is $\hat{R} > 0$, such that [1.23] holds.

Let $h\colon \mathbb{R}_+ \to \mathbb{R}_+$ be the shifted *moment generating function* of Y_i, such that $h(0) = 0$, i.e.

$$h(R) = \mathbb{E}\left[\mathrm{e}^{RY_i}\right] - 1.$$

We make the following classical assumption concerning $h(R)$: there is $R_\infty \in (0, +\infty]$, such that $h(R) < +\infty$ for all $R \in [0, R_\infty)$ and $\lim_{R\uparrow R_\infty} h(R) = +\infty$ (see [GRA 91, p. 2] and [ROL 99, p. 170]). It is easily seen that

$h(R)$ is increasing, convex and continuous on $[0, R_\infty)$. Moreover, we can show that $h(R)$ is infinitely differentiable on $[0, R_\infty)$.

Integrating by parts gives

$$\int_0^{+\infty} e^{\hat{R}y} (1 - F(y))\, dy = \frac{1}{\hat{R}} \int_0^{+\infty} (1 - F(y))\, d(e^{\hat{R}y})$$

$$= \frac{1}{\hat{R}} \left(\int_0^{+\infty} e^{\hat{R}y}\, dF(y) - 1 \right) = \frac{h(\hat{R})}{\hat{R}}.$$

Hence, we can rewrite [1.23] in the form

$$\lambda h(\hat{R}) = c\hat{R}. \qquad [1.26]$$

LEMMA 1.2.– Let $c > \lambda\mu$ and there be $R_\infty \in (0, +\infty]$, such that $h(R) < +\infty$ for all $R \in [0, R_\infty)$ and $\lim_{R \uparrow R_\infty} h(R) = +\infty$. Then, there is a unique solution $\hat{R} \in (0, R_\infty)$ to [1.26].

The proof of lemma 1.2 follows easily from the properties of $h(R)$.

The unique solution $\hat{R} \in (0, R_\infty)$ to [1.26], if it exists, is called the *adjustment coefficient* or the *Lundberg exponent*. If the conditions of lemma 1.2 hold, then

$$\hat{\mu} = \frac{\lambda}{c} \int_0^{+\infty} y e^{\hat{R}y} (1 - F(y))\, dy$$

$$= \frac{\lambda}{c} \int_0^{+\infty} (1 - F(y))\, d\left(\left(\frac{y}{\hat{R}} - \frac{1}{\hat{R}^2} \right) e^{\hat{R}y} \right)$$

$$= \frac{\lambda}{c\hat{R}^2} + \frac{\lambda}{c} \int_0^{+\infty} \left(\frac{y}{\hat{R}} - \frac{1}{\hat{R}^2} \right) e^{\hat{R}y}\, dF(y)$$

$$= \frac{\lambda}{c} \left(\frac{1}{\hat{R}^2} + \frac{1}{\hat{R}} \int_0^{+\infty} y e^{\hat{R}y}\, dF(y) - \frac{1}{\hat{R}^2} \int_0^{+\infty} e^{\hat{R}y}\, dF(y) \right)$$

$$= \frac{\lambda}{c} \left(\frac{1}{\hat{R}^2} + \frac{h'(\hat{R})}{\hat{R}} - \frac{h(\hat{R}) + 1}{\hat{R}^2} \right)$$

$$= \frac{\lambda}{c} \left(\frac{h'(\hat{R})}{\hat{R}} - \frac{c\hat{R}/\lambda}{\hat{R}^2} \right) = \frac{\lambda h'(\hat{R}) - c}{c\hat{R}} < +\infty.$$

Therefore, in this case from [1.25] we get

$$\psi(x) \sim \frac{c - \lambda\mu}{\lambda h'(\hat{R}) - c}\,\mathrm{e}^{-\hat{R}x} \quad \text{as} \quad x \to +\infty,$$

where $\hat{R} \in (0, R_\infty)$ is the unique solution to [1.26].

1.1.8. *Lundberg inequality for the infinite-horizon ruin probability*

We now give an upper exponential bound for the infinite-horizon ruin probability. To this end, we use the martingale approach introduced by Gerber [GER 73].

For all $R \geq 0$, we define the exponential process $\big(V_t(R)\big)_{t\geq 0}$ by

$$V_t(R) = \exp\Bigg(-R\left(ct - \sum_{i=1}^{N_t} Y_i\right)\Bigg).$$

We now establish the martingale property for $\big(V_t(R)\big)_{t\geq 0}$.

LEMMA 1.3.– Let the surplus process $\big(X_t(x)\big)_{t\geq 0}$ follow [1.1]. If there is $\hat{R} > 0$, such that [1.26] holds, then $\big(V_t(\hat{R})\big)_{t\geq 0}$ is an $(\mathfrak{F}_t)$-martingale.

PROOF.– For all $R > 0$, such that $\mathbb{E}\left[\mathrm{e}^{RY_i}\right] < +\infty$, if any, we have

$$\begin{aligned}
\mathbb{E}[V_t(R)] &= \mathrm{e}^{-cRt}\,\mathbb{E}\left[\exp\Bigg(R\sum_{i=1}^{N_t} Y_i\Bigg)\right] \\
&= \mathrm{e}^{-cRt}\sum_{j=0}^{\infty} \mathrm{e}^{-\lambda t}\frac{(\lambda t)^j}{j!}\left(\mathbb{E}\left[\mathrm{e}^{RY_i}\right]\right)^j \qquad [1.27] \\
&= \exp\!\big(t\left(\lambda\,\mathbb{E}\left[\mathrm{e}^{RY_i}\right] - \lambda - cR\right)\big) \\
&= \exp(t\left(\lambda h(R) - cR\right)).
\end{aligned}$$

If there is $\hat{R} > 0$, such that [1.26] holds, then $\mathbb{E}\left[e^{\hat{R}Y_i}\right] < +\infty$, so for all $t_2 \geq t_1 \geq 0$, we have

$$\mathbb{E}\left[V_{t_2}(\hat{R}) \,/\, \mathfrak{F}_{t_1}\right] = \mathbb{E}\left[\exp\left(-\hat{R}\left(ct_2 - \sum_{i=1}^{N_{t_2}} Y_i\right)\right) \,/\, \mathfrak{F}_{t_1}\right]$$

$$= \mathbb{E}\left[\exp\left(-\hat{R}\left(ct_1 - \sum_{i=1}^{N_{t_1}} Y_i\right)\right)\right]$$

$$\times\, \mathbb{E}\left[\exp\left(-\hat{R}\left(c(t_2 - t_1) - \sum_{i=N_{t_1}}^{N_{t_2}} Y_i\right)\right) \,/\, \mathfrak{F}_{t_1}\right]$$

$$= \mathbb{E}\left[V_{t_1}(\hat{R})\right] \mathbb{E}\left[\exp\left(-\hat{R}\left(c(t_2 - t_1) - \sum_{i=1}^{N_{t_2 - t_1}} Y_i\right)\right)\right] = \mathbb{E}\left[V_{t_1}(\hat{R})\right],$$

which shows that $\left(V_t(\hat{R})\right)_{t\geq 0}$ is an $(\mathfrak{F}_t)$-martingale.

Here we used the fact that

$$\mathbb{E}\left[\exp\left(-\hat{R}\left(c(t_2-t_1) - \sum_{i=1}^{N_{t_2-t_1}} Y_i\right)\right)\right] = \exp\left((t_2-t_1)(\lambda h(\hat{R}) - c\hat{R})\right) = 1$$

by [1.26] and [1.27]. □

THEOREM 1.8.– Let the surplus process $\left(X_t(x)\right)_{t\geq 0}$ follow [1.1]. If there is $\hat{R} > 0$, such that [1.26] holds, then for all $x \geq 0$, we have

$$\psi(x) \leq e^{-\hat{R}x}. \qquad [1.28]$$

PROOF.– It is easily seen that the ruin time $\tau(x)$ is an $(\mathfrak{F}_t)$-stopping time. Hence, $\tau(x) \wedge t$ is a bounded $(\mathfrak{F}_t)$-stopping time for any $t \geq 0$. The process $\left(V_t(\hat{R})\right)_{t\geq 0}$ is an $(\mathfrak{F}_t)$-martingale by lemma 1.3. Moreover, $\left(V_t(\hat{R})\right)_{t\geq 0}$ is positive a.s. by its definition. Consequently, applying the optional stopping theorem yields

$$1 = V_0(\hat{R}) = \mathbb{E}\left[V_{\tau(x)\wedge t}(\hat{R})\right]$$

$$= \mathbb{E}\left[V_{\tau(x)}(\hat{R})\, \mathbb{1}_{\{\tau(x)<t\}}\right] + \mathbb{E}\left[V_t(\hat{R})\, \mathbb{1}_{\{\tau(x)\geq t\}}\right]$$

$$\geq \mathbb{E}\left[\exp\left(-\hat{R}\left(c\tau(x) - \sum_{i=1}^{N_{\tau(x)}} Y_i\right)\right) \mathbb{1}_{\{\tau(x)<t\}}\right]$$

$$\geq \mathrm{e}^{\hat{R}x}\, \mathbb{P}[\tau(x) < t].$$

This gives

$$\mathbb{P}[\tau(x) < t] \leq \mathrm{e}^{-\hat{R}x} \qquad [1.29]$$

for all $t > 0$. Letting $t \to \infty$ in [1.29] yields

$$\mathbb{P}[\tau(x) < \infty] \leq \mathrm{e}^{-\hat{R}x},$$

which is our assertion. □

Inequality [1.28] is called the *Lundberg inequality* for the infinite-horizon ruin probability.

1.1.9. *Bibliographical notes*

The classical risk model was introduced by Lundberg [LUN 03, LUN 26], who first considered the problem of finding the ruin probability and gave the so-called Lundberg inequality. Since these works appeared before the development of the general theory of stochastic processes, their results were not quite mathematically rigorous. Next, the classical risk model was studied by Cramér [CRA 30, CRA 55]. In particular, he first gave the Cramér–Lundberg approximation. That is why the classical risk model is also called the Cramér–Lundberg model. Equation [1.3] for the infinite-horizon survival probability also goes back to these pioneering works, but the differentiability of $\varphi(x)$ was not investigated. The fact that $\varphi(x)$ may not be differentiable at some points was first mentioned in [ROL 99].

In modern risk theory, there are some classical proofs of the Lundberg inequality and the Cramér–Lundberg approximation (see, for example [ASM 10, GRA 91, ROL 99] and references therein). In particular, the approach to the proof of the Cramér–Lundberg approximation using the key renewal theorem goes back to [FEL 71]. The martingale approach to the proof of the Lundberg inequality was introduced by Gerber [GER 73]. Furthermore,

Taylor [TAY 76] obtained a sharpening of the Lundberg inequality and a lower exponential bound for $\psi(x)$. Later, Vinogradov [VIN 99] showed that condition [1.23] is necessary for the asymptotical exponentiality of $\psi(x)$.

Mammitzsch [MAM 86] obtained necessary and sufficient conditions for the existence of the adjustment coefficient. Bounds for the adjustment coefficient are given in [GER 79]. Different approaches to its estimation in the classical risk model were proposed by Csörgő and Steinebach [CSÖ 91], Csörgő and Teugels [CSÖ 90], Deheuvels and Steinebach [DEH 90], Embrechts and Mikosch [EMB 91], Grandell [GRA 79] and Herkenrath [HER 86]. Pitts, Grübel and Embrechts [PIT 96] and Schmidli [SCH 97] got estimates for the adjustment coefficient in more general risk models.

There are only a few cases where we can obtain analytic expressions for ruin or survival probabilities. The simplest one is when the claim sizes are exponentially distributed. Next, analytic expressions for $\psi(x)$ were obtained by Thorin [THO 73a, THO 86] for gamma-distributed claim sizes and by Dufresne and Gerber [DUF 88, DUF 89] for the case where claim sizes are combinations of translated exponential distributions. The case where the set of possible values of the claim sizes is finite was considered in [KAA 91, SCH 90, SHI 89].

Practical approximations for $\psi(x)$ can be divided conventionally into two classes. They are heuristic approximations and approximations that are based on limit theorems of probability theory. The most famous heuristic ones are the Beekman-Bowers and De Vylder approximations introduced in [BEE 69] and [DEV 78], respectively. The De Vylder approximation is simple enough and in order to apply it, we need to compute only the first three moments of the distribution of the claim sizes. Nevertheless, it gives surprisingly good results when the initial surplus is not too small, especially for light-tailed distributions of the claim sizes. An explanation for this fact was given later by Grandell [GRA 00]. Some heuristic approximations, which improve the Cramér–Lundberg one and are acceptable for small initial surpluses, were suggested in [CHO 10].

If the claim sizes are heavy-tailed, the asymptotics of $\psi(x)$ considerably differs from that given by [1.25]. Von Bahr [VON 75] obtained the asymptotic approximation for $\psi(x)$ when the claim sizes are Pareto distributed, while Thorin and Wikstad [THO 77b] dealt with the case where the claim sizes have a lognormal distribution. In the more general case, namely if the integrated tail distribution of the claim sizes is subexponential, the first-order asymptotic approximation for the infinite-horizon ruin probability was given by Embrechts and Veraverbeke [EMB 82] and Teugels and Veraverbeke [TEU 73]; see also [EMB 88, KLÜ 89]. Asmussen and Binswanger [ASM 97] suggested an asymptotic approximation, which is considerably better when the initial surplus is not too large. For higher-order asymptotic approximations for $\psi(x)$, we refer the reader to [ALB 10b, BAL 99, BAL 05, MIK 01]. Upper and lower bounds for $\psi(x)$ in the heavy-tailed case are proposed by Cai and Garrido [CAI 99] and Kalashnikov [KAL 99]. De Vylder and Goovaerts [DEV 84, GOO 84a] also obtained upper and lower bounds for $\psi(x)$ using integral equation [1.9] and suggested a recursive algorithm for its evaluation.

The infinite-horizon ruin probability can be computed by numerical inversion of its Laplace transform. To this end, we can apply, for instance, inversion techniques proposed by Abate and Whitt [ABA 92, ABA 95] and Jagerman [JAG 78, JAG 82]. Moreover, numerical inversion of the Laplace transform is one of the classical approaches to computation of the ruin probabilities in different risk models (see, for example [ALB 10a, EMB 93, THO 73b]). For some numerical schemes for calculation of $\psi(x)$, see [DUF 89, MEY 87, PAN 86, SHI 88, THO 77a].

Seal's formulas, which express the finite-horizon survival probability in terms of the distribution of the total claims, were derived in [PRA 61, SEA 69, SEA 74]. Under the assumption that the claim sizes are integer-valued, Picard and Lefèvre [PIC 97] showed that $\psi(x,t)$ can be expressed in terms of generalized Appell polynomials. Rullière and Loisel [RUL 04] proposed some alternatives to this expression. Asmussen [ASM 84] and Pervozvansky [PER 98] found analytic expressions for $\psi(x,t)$ by different techniques in the case of exponentially distributed claim sizes.

Garcia [GAR 05] showed that analytic expressions for $\psi(x,t)$ in terms of infinite series can be obtained through the inversion of the double Laplace transform of the distribution of time to ruin when the claim sizes have a mixed exponential or Erlang distribution. Dickson and Willmot [DIC 05] used a similar approach to get an analytic expression for $\psi(x,t)$ when the claim sizes have a mixed Erlang distribution. Some recursive schemes for evaluation of $\psi(x,t)$ were given in [DIC 91, KLI 91].

The classical normal-type approximation for $\psi(x,t)$, which can be considered as a time-dependent version of the Cramér–Lundberg approximation, was proposed by Segerdahl [SEG 55]. For refinements of this approximation, we refer the reader to [ASM 84, MAL 94]. A time-dependent version of the Lundberg inequality was obtained by Gerber [GER 73]. Arfwedson [ARF 55] and Barndorff-Nielsen and Schmidli [BAR 95] applied saddle-point techniques to obtain approximations for the finite-horizon ruin probability.

Iglehart [IGL 69] and Grandell [GRA 77, GRA 78] suggested diffusion approximations, the idea of which is to replace the surplus process by a Brownian motion by fitting the first two moments and use the Brownian first passage probabilities as approximations for the ruin probabilities. Some corrected diffusion approximations were proposed in [ASM 84, ASM 97].

Some results for ruin probabilities can be obtained using the equivalence between characteristics in risk and queuing theories. In particular, in the classical risk model, $\psi(x)$ can be represented in the form of a series of convolutions. This representation is called the Pollaczek–Khinchin formula by analogy with the celebrated formula in queuing theory (see, for example [ASM 03, ASM 10, FEL 71, FRA 82, ROL 99]). In risk theory, it is also referred to as Beekman's convolution formula [BEE 85]. Moreover, the martingale approach to the proof of the Lundberg inequality was given by Kingman [KIN 64] in a setting of queuing theory. For the heavy and light traffic approximations for $\psi(x)$, which also came from queuing theory, we refer the reader to [ASM 10].

In general, the finite-horizon ruin probability can be evaluated by the crude Monte Carlo method for a fixed initial surplus. To this end, we simulate paths of the surplus process and compute the part of the paths which lead to ruin. Such arguments are not applicable for the evaluation of the infinite-horizon ruin probability because the surplus process cannot be simulated in infinite time. To get around this difficulty, some alternative representations for $\psi(x)$ were suggested (see, for example [ASM 10]). Simulation via the Pollaczek–Khinchin formula allows us to evaluate $\psi(x)$, but to maintain the efficiency of this approach for large initial surpluses, we have to combine it with some variance reduction methods (e.g. conditional Monte Carlo or importance sampling). For some such algorithms, see [ASM 97, ASM 06, ASM 00, JUN 07, JUN 02]. For some importance sampling algorithms, we refer the reader to [ASM 85, BLA 08, DUP 07, GLY 89, LEH 92].

For the distributions that are used in insurance, see, for example [ASM 10, BEA 84, HOG 84, ROL 99]. For more information on premium calculation principles and their properties, see, for example [ASM 10, BÜH 84, DEN 90, GOO 82, GOO 84b, KAA 08, REI 86, ROL 99, SCH 08b].

1.2. Risk model with stochastic premiums

We now consider the *risk model with stochastic premiums* introduced in [BOI 03, LIV 99]. For details and proofs of the theorems given below, see [BOI 03].

1.2.1. *Description of the model*

In the risk model with stochastic premiums, claim sizes form a sequence $(Y_i)_{i\geq 1}$ of non-negative i.i.d. r.v.'s with c.d.f. $F(y) = \mathbb{P}[Y_i \leq y]$ and finite expectation $\mu > 0$. The number of claims on the time interval $[0, t]$ is a Poisson process $(N_t)_{t\geq 0}$ with constant intensity $\lambda > 0$. Next, premium sizes form a sequence $(\bar{Y}_i)_{i\geq 1}$ of non-negative i.i.d. r.v.'s with c.d.f. $\bar{F}(y) = \mathbb{P}[\bar{Y}_i \leq y]$ and finite expectation $\bar{\mu} > 0$. The number of premiums on the time interval $[0, t]$ is a Poisson process $(\bar{N}_t)_{t\geq 0}$ with constant intensity $\bar{\lambda} > 0$. The r.v.'s $(Y_i)_{i\geq 1}$ and $(\bar{Y}_i)_{i\geq 1}$ and the processes $(N_t)_{t\geq 0}$ and $(\bar{N}_t)_{t\geq 0}$ are mutually independent. Thus, the total claims and premiums on $[0, t]$ equal $\sum_{i=1}^{N_t} Y_i$ and $\sum_{i=1}^{\bar{N}_t} \bar{Y}_i$, respectively. We set $\sum_{i=1}^{0} Y_i = 0$ if $N_t = 0$ and $\sum_{i=1}^{0} \bar{Y}_i = 0$ if $\bar{N}_t = 0$. Let $(\mathfrak{F}_t)_{t\geq 0}$ be the filtration generated by $(Y_i)_{i\geq 1}$, $(\bar{Y}_i)_{i\geq 1}$, $(N_t)_{t\geq 0}$, $(\bar{N}_t)_{t\geq 0}$. Moreover, the insurance company has a non-negative initial surplus x.

Let $X_t(x)$ be the surplus of the insurance company at time t provided that its initial surplus is x. Then, the *surplus process* $\big(X_t(x)\big)_{t\geq 0}$ follows the equation

$$X_t(x) = x + \sum_{i=1}^{\bar{N}_t} \bar{Y}_i - \sum_{i=1}^{N_t} Y_i, \quad t \geq 0. \qquad [1.30]$$

The ruin time $\tau(x)$, the infinite-horizon ruin probability $\psi(x)$, the finite-horizon ruin probability $\psi(x, t)$, the infinite-horizon survival probability $\varphi(x)$ and the finite-horizon survival probability $\varphi(x,t)$ are defined as in section 1.1.1 for the classical risk model. As in section 1.1, we formulate some results for the ruin probabilities and other results for the survival probabilities.

1.2.2. *Basic results*

Applying theorem 1.1 and arguments similar to those in the proof of lemma 1.1 one can prove the following lemma.

LEMMA 1.4.– Let the surplus process $\left(X_t(x)\right)_{t\geq 0}$ follow [1.30].

i) If $\bar{\lambda}\bar{\mu} \leq \lambda\mu$, then $\psi(x) = 1$ for all $x \geq 0$.

ii) If $\bar{\lambda}\bar{\mu} > \lambda\mu$, then $\lim_{x\to+\infty} \psi(x) = 0$.

In this model, the condition $\bar{\lambda}\bar{\mu} > \lambda\mu$ is the analogue of the net profit condition in the classical risk model.

THEOREM 1.9.– Let the surplus process $\left(X_t(x)\right)_{t\geq 0}$ follow [1.30]. Then, $\varphi(x)$ satisfies the integral equation

$$(\bar{\lambda}+\lambda)\varphi(x) = \bar{\lambda}\int_0^{+\infty} \varphi(x+y)\,\mathrm{d}\bar{F}(y) + \lambda\int_0^{x} \varphi(x-y)\,\mathrm{d}F(y). \quad [1.31]$$

Equation [1.31] can be derived in a standard way applying the law of total probability. Note that this equation is the analogue of the integro-differential equation [1.3] in the classical risk model, but it does not involve the derivative of $\varphi(x)$. That is why one does not need to investigate the differentiability of this function.

If $\bar{\lambda}\bar{\mu} > \lambda\mu$, then $\varphi(0)$ cannot be found explicitly, but it is easy to show that $\varphi(0) > 0$ in this case. Indeed, from [1.31], we get

$$\varphi(0) = \frac{\bar{\lambda}}{\bar{\lambda}+\lambda}\int_0^{+\infty} \varphi(y)\,\mathrm{d}\bar{F}(y). \quad [1.32]$$

Moreover, $\lim_{x\to+\infty} \varphi(x) = 1$ by lemma 1.4 and the function $\varphi(x)$ is non-decreasing by its definition. Assume that the equality $\varphi(0) = 0$ is true. Then by [1.32], there is a point $x_0 > 0$, such that $\varphi(x) = 0$ for all $x \in [0, x_0]$. Substituting $x = x_0$ into [1.31] and applying the same arguments give that

$\varphi(x) = 0$ for all $x \in [0, 2x_0]$. Thus, by induction, one can show that $\varphi(x) = 0$ for all $x \geq 0$, which contradicts the fact that $\lim_{x\to+\infty} \varphi(x) = 1$.

Furthermore, the function $\varphi(x,t)$ satisfies the partial integro-differential equation

$$\frac{\partial \varphi(x,t)}{\partial t} + (\bar{\lambda} + \lambda)\, \varphi(x,t) \\ - \bar{\lambda} \int_0^{+\infty} \varphi(x+y,\, t)\, \mathrm{d}\bar{F}(y) - \lambda \int_0^{x} \varphi(x-y,\, t)\, \mathrm{d}F(y) = 0.$$

The question concerning the existence of a partial derivative of $\varphi(x,t)$ w.r.t. t is investigated in section 3.3.

Theorems 1.10 and 1.11 below give analytic expressions for the infinite-horizon survival probability in two particular cases.

THEOREM 1.10.– Let the surplus process $\big(X_t(x)\big)_{t\geq 0}$ follow [1.30], $\bar{\lambda}\bar{\mu} > \lambda\mu$, the premium sizes be exponentially distributed with mean $\bar{\mu}$ and the claim sizes be exponentially distributed with mean μ. Then

$$\varphi(x) = 1 - \frac{\lambda(\mu + \bar{\mu})}{\bar{\mu}(\lambda + \bar{\lambda})} \exp\left(- \frac{(\bar{\lambda}\bar{\mu} - \lambda\mu)\, x}{\mu\bar{\mu}(\lambda + \bar{\lambda})} \right), \quad x \geq 0.$$

THEOREM 1.11.– Let the surplus process $\big(X_t(x)\big)_{t\geq 0}$ follow [1.1], $\bar{\lambda} > \lambda$ and $\mathbb{P}[\bar{Y}_i = 1] = \mathbb{P}[Y_i = 1] = 1$, $i \geq 1$. Then

$$\varphi(x) = 1 - \left(\frac{\lambda}{\bar{\lambda}}\right)^{\lfloor x \rfloor + 1}, \quad x \geq 0.$$

The proof of theorem 1.10 is similar to the proof of theorem 1.5. To prove theorem 1.11, it is sufficient to notice that $\varphi(x) = \varphi(\lfloor x \rfloor)$ for all $x \geq 0$, and for integer $x \geq 1$, integral equation [1.31] becomes the difference equation

$$\bar{\lambda}\varphi(x+1) - (\lambda + \bar{\lambda})\varphi(x) + \lambda\varphi(x-1) = 0,$$

which can be solved in a standard way.

In particular, it can be easily seen from theorem 1.11 that $\varphi(x)$ may not be continuous on $\mathbb{R}_+$ in this risk model. In theorem 1.11, $\varphi(x)$ is not continuous at all integer points $x \geq 1$.

Theorem 1.12 below gives an upper exponential bound for the infinite-horizon ruin probability.

THEOREM 1.12.– Let the surplus process $\left(X_t(x)\right)_{t\geq 0}$ follow [1.30]. If there is $\hat{R} > 0$, such that

$$\bar{\lambda}\left(\mathbb{E}\left[\mathrm{e}^{-\hat{R}\bar{Y}_i}\right] - 1\right) + \lambda\left(\mathbb{E}\left[\mathrm{e}^{\hat{R}Y_i}\right] - 1\right) = 0, \qquad [1.33]$$

then for all $x \geq 0$, we have

$$\psi(x) \leq \mathrm{e}^{-\hat{R}x}.$$

The proof of theorem 1.12 is similar to the proof of theorem 1.8 and based on the martingale approach. In particular, to prove theorem 1.12, we need to show that the exponential process $\left(V_t(\hat{R})\right)_{t\geq 0}$ given by

$$V_t(\hat{R}) = \exp\left(-\hat{R}\left(\sum_{i=1}^{\bar{N}_t}\bar{Y}_i - \sum_{i=1}^{N_t}Y_i\right)\right)$$

is an $(\mathfrak{F}_t)$-martingale, where $\hat{R} > 0$ is such that [1.33] holds.

1.2.3. *Bibliographical notes*

The risk model with stochastic premiums was introduced independently by Boikov [BOI 03] and Livshits [LIV 99] as a natural generalization of the classical risk model. All the basic results given in section 1.2.2 can be found in [BOI 03]. Some similar results were obtained in [LIV 99] by other techniques.

Zinchenko and Andrusiv [ZIN 08] suggested an analogue of the De Vylder approximation for $\psi(x)$, got diffusion approximations for $\psi(x)$ and $\psi(x,t)$ and considered the strong invariance principle for the risk process with stochastic premiums.

Bening and Korolev [BEN 03] dealt with a generalized risk model where the premium rate and the claim intensity are stochastic. They constructed a statistical estimate for the infinite-horizon ruin probability and investigated its asymptotic properties.

Norkin [NOR 06] derived integral equations for $\varphi(x)$ in the risk model where the premium intensity depends on the current surplus of the insurance company, and stochastic premium and claim arrivals are not necessarily Poisson. The method of successive approximations was proposed for numerical solution of these equations. In particular, necessary and sufficient conditions for the existence and uniqueness of solutions of the equations were obtained.

Some distributions of functionals for the risk processes with stochastic premiums were studied by Gusak [GUS 05a, HUS 07] and Gusak and Karnaukh [GUS 05b].

2

Classical Risk Model with Investments in a Risk-Free Asset

In this chapter, we deal with a generalization of the classical risk model where an insurance company invests all surplus in a risk-free asset. We investigate the continuity and differentiability of the infinite- and finite-horizon survival probabilities in detail. Moreover, we derive integro-differential equations for these functions and get bounds for their derivatives w.r.t. the initial surplus.

2.1. Description of the model

We consider a generalization of the *classical risk model* described in section 1.1.1: an insurance company invests all surplus in a *risk-free asset*, the price of which equals $B_t = B_0\,\mathrm{e}^{rt}$ at time t, where B_0 is the price of the asset at the time $t = 0$ and $r > 0$ is a risk-free interest rate.

Then, the *surplus process* $\big(X_t(x)\big)_{t\geq 0}$ follows the equation

$$X_t(x) = x + \int_0^t \big(rX_s(x) + c\big)\,\mathrm{d}s - \sum_{i=1}^{N_t} Y_i, \quad t \geq 0. \tag{2.1}$$

The rest of the chapter is organized as follows. In section 2.2, we investigate the continuity and differentiability of the infinite-horizon survival probability. Section 2.3 is devoted to investigation of these properties for the finite-horizon survival probability. Some bibliographical notes concerning the classical risk model with investments in a risk-free asset are given in section 2.4.

2.2. Continuity and differentiability of the infinite-horizon survival probability

THEOREM 2.1.–

1) Let the surplus process $(X_t(x))_{t\geq 0}$ follow [2.1] under the above assumptions. Then, the function $\varphi(x)$ is continuous on $\mathbb{R}_+$.

2) Moreover, let $c > \lambda\mu$. Then, the following assertions hold:

i) The function $\varphi(x)$ is continuously differentiable on $\mathbb{R}_+$, except at positive points of discontinuity of $F(y)$.

ii) If $x > 0$ is a point of discontinuity of $F(y)$ and $F(x) - F(x_-) = p$, then $\varphi(x)$ has the left and right derivatives $\varphi'_-(x)$ and $\varphi'_+(x)$, respectively, and

$$\varphi'_-(x) - \varphi'_+(x) = \frac{\lambda p \varphi(0)}{rx + c} > 0. \qquad [2.2]$$

iii) The function $\varphi(x)$ satisfies the integro-differential equation

$$(rx + c)\varphi'_+(x) = \lambda\varphi(x) - \lambda \int_0^x \varphi(x - y)\, \mathrm{d}F(y) \qquad [2.3]$$

on $\mathbb{R}_+$ with the boundary condition $\lim_{x\to+\infty} \varphi(x) = 1$.

iv) We have

$$\sup_{x\in[0,+\infty)} \left|\varphi'_+(x)\right| \leq \frac{\lambda}{c}. \qquad [2.4]$$

PROOF.– It follows immediately from [2.1] that $X_t(x) = (x + c/r)\,\mathrm{e}^{rt} - c/r$ up to the time τ_1. Let $\tau_1 = s$ and $Y_1 = y$. It is easily seen that ruin does not occur up to the time τ_1 and it will not take place after τ_1 if and only if $y \leq (x + c/r)\,\mathrm{e}^{rs} - c/r$ and ruin does not occur on the time interval $(s, +\infty)$ with the initial surplus $(x + c/r)\,\mathrm{e}^{rs} - c/r - y$. Note that τ_1 is exponentially distributed with mean $1/\lambda$. Therefore, by the law of total probability, for all $x \geq 0$, we get

$$\begin{aligned}\varphi(x) = &\int_0^{+\infty} \lambda \mathrm{e}^{-\lambda s} \\ &\times \int_0^{(x+c/r)\,\mathrm{e}^{rs}-c/r} \varphi\big((x + c/r)\,\mathrm{e}^{rs} - c/r - y\big)\, \mathrm{d}F(y)\, \mathrm{d}s.\end{aligned} \qquad [2.5]$$

Here and subsequently, all integrals w.r.t. a c.d.f. are assumed to be the Lebesgue–Stieltjes integrals.

Changing the variable $(x + c/r)\,\mathrm{e}^{rs} - c/r = u$ in the outer integral in the right-hand side of [2.5] we have

$$\varphi(x) = \lambda(rx+c)^{\lambda/r} \int_x^{+\infty} \left(\frac{1}{ru+c}\right)^{\lambda/r+1} \int_0^u \varphi(u-y)\,\mathrm{d}F(y)\,\mathrm{d}u. \quad [2.6]$$

Since $\varphi(x)$ is non-decreasing and bounded, the integral $\int_0^u \varphi(u-y)\,\mathrm{d}F(y)$ is also a non-decreasing and bounded function of u. Hence, the function $\left(\frac{1}{ru+c}\right)^{\lambda/r+1} \int_0^u \varphi(u-y)\,\mathrm{d}F(y)$ is Riemann integrable on any finite interval in $[x, +\infty)$ and the improper integral in the right-hand side of [2.6] is a continuous function of x by theorem A.1 and remark A.1. Thus, $\varphi(x)$ is continuous in $\mathbb{R}_+$, which proves the first assertion of the theorem.

By Lebesgue's decomposition theorem, we can write

$$F(y) = F^{\mathrm{c}}(y) + F^{\mathrm{d}}(y),$$

where $F^{\mathrm{c}}(y)$ and $F^{\mathrm{d}}(y)$ are continuous and discrete components of $F(y)$, respectively.

Let $(y_k)_{k\geq 1}$ be a sequence of all points of discontinuity of $F(y)$, if any. Set $p_k = F(y_k) - \lim_{y\uparrow y_k} F(y)$. Then, we can rewrite [2.6] as

$$\varphi(x) = \varphi_1(x) + \varphi_2(x), \quad [2.7]$$

where

$$\varphi_1(x) = \lambda(rx+c)^{\lambda/r} \int_x^{+\infty} \left(\frac{1}{ru+c}\right)^{\lambda/r+1} I(u)\,\mathrm{d}u, \quad [2.8]$$

$$\varphi_2(x) = \lambda(rx+c)^{\lambda/r} \times \int_x^{+\infty} \left(\frac{1}{ru+c}\right)^{\lambda/r+1} \Bigg(\sum_{k\colon y_k\leq u} p_k\,\varphi(u-y_k)\Bigg)\,\mathrm{d}u, \quad [2.9]$$

$$I(u) = \int_0^u \varphi(u-y)\,\mathrm{d}F^{\mathrm{c}}(y).$$

We now prove that $I(u)$ is left-continuous on $(0,+\infty)$. Fix any $u_0 > 0$. For $u \le u_0$, we have

$$\begin{aligned} I(u_0) - I(u) = {} & \int_0^u \big(\varphi(u_0-y) - \varphi(u-y)\big)\,\mathrm{d}F^{\mathrm{c}}(y) \\ & + \int_u^{u_0} \varphi(u_0-y)\,\mathrm{d}F^{\mathrm{c}}(y). \end{aligned} \qquad [2.10]$$

It is obvious that

$$0 \le \lim_{u \uparrow u_0} \int_u^{u_0} \varphi(u_0-y)\,\mathrm{d}F^{\mathrm{c}}(y) \le \lim_{u \uparrow u_0} \big(F^{\mathrm{c}}(u_0) - F^{\mathrm{c}}(u)\big) = 0. \qquad [2.11]$$

Since $\varphi(x)$ is continuous on $[0,u_0]$, it is uniformly continuous on this interval by the Heine-Cantor theorem. Hence, for any $\varepsilon > 0$, there is $\delta > 0$, such that for $\big|(u_0-y) - (u-y)\big| < \delta$, $u \in [0,u_0]$, $y \in [0,u]$, we have $\big|\varphi(u_0-y) - \varphi(u-y)\big| < \varepsilon$. Therefore, for $|u_0 - u| < \delta$, we get

$$\left|\int_0^u \big(\varphi(u_0-y) - \varphi(u-y)\big)\,\mathrm{d}F^{\mathrm{c}}(y)\right| < \varepsilon F(u_0) \le \varepsilon. \qquad [2.12]$$

Thus, [2.10]–[2.12] imply the left continuity of $I(u)$ at any $u_0 > 0$. The right continuity of $I(u)$ on $\mathbb{R}_+$ is proved similarly. We only note that for $u \ge u_0$, we have

$$\begin{aligned} I(u) - I(u_0) = {} & \int_0^{u_0} \big(\varphi(u-y) - \varphi(u_0-y)\big)\,\mathrm{d}F^{\mathrm{c}}(y) \\ & + \int_{u_0}^{u} \varphi(u-y)\,\mathrm{d}F^{\mathrm{c}}(y). \end{aligned}$$

Thus, the integrand in the right-hand side of [2.8] is continuous in $\mathbb{R}_+$ w.r.t. u. Hence, by [2.8], theorem A.1 and remark A.1, $\varphi_1(x)$ is differentiable on $\mathbb{R}_+$ and

$$\varphi_1'(x) = \frac{\lambda}{rx+c}\varphi_1(x) - \frac{\lambda}{rx+c} I(x). \qquad [2.13]$$

Since the integrand in the right-hand side of [2.9] is right-continuous on $\mathbb{R}_+$ w.r.t. u, by theorem A.1 and remarks A.1 and A.2, there is a right derivative of $\varphi_2(x)$ on $\mathbb{R}_+$ (we denote it by $(\varphi_2(x))'_+$) and

$$(\varphi_2(x))'_+ = \frac{\lambda}{rx+c}\,\varphi_2(x) - \frac{\lambda}{rx+c}\sum_{k:\, y_k \le x} p_k\,\varphi(x-y_k). \qquad [2.14]$$

Since the Riemann integral does not depend on values of the integrand at a countable number of points, we can rewrite [2.9] as

$$\varphi_2(x) = \lambda(rx+c)^{\lambda/r} \times \int_x^{+\infty}\left(\frac{1}{ru+c}\right)^{\lambda/r+1}\left(\sum_{k:\, y_k < u} p_k\,\varphi(u-y_k)\right)\mathrm{d}u. \qquad [2.15]$$

The integrand in the right-hand side of [2.15] is left-continuous on $(0,+\infty)$ w.r.t. u. Hence, by theorem A.1 and remarks A.1 and A.2, there is a left derivative of $\varphi_2(x)$ on $(0,+\infty)$ (we denote it by $(\varphi_2(x))'_-$) and

$$(\varphi_2(x))'_- = \frac{\lambda}{rx+c}\,\varphi_2(x) - \frac{\lambda}{rx+c}\sum_{k:\, y_k < x} p_k\,\varphi(x-y_k). \qquad [2.16]$$

From [2.14] and [2.16] we conclude that

$$(\varphi_2(x))'_- - (\varphi_2(x))'_+ = \frac{\lambda p_k\,\varphi(0)\,\mathbb{1}_{\{x=y_k\}}}{rx+c} \qquad [2.17]$$

for $x > 0$. If $c > \lambda\mu$, then $\lim_{x\to+\infty}\varphi(x) = 1$ and $\varphi(0) = 1 - \lambda\mu/c > 0$ in the classical risk model without any investments (see sections 1.1.2 and 1.1.4). Moreover, it is easily seen that at any time the surplus in the risk model with investments in the risk-free asset is not less than the surplus in the corresponding risk model without investments. Therefore, investments in the risk-free asset do not decrease the survival probability. From this we get $\lim_{x\to+\infty}\varphi(x) = 1$ and $\varphi(0) > 0$ for the model with investments.

Hence, from [2.17] it follows that $\varphi_2(x)$ is differentiable on $\mathbb{R}_+$, except at positive points y_k, $k \ge 1$. Consequently, taking into account [2.7], we deduce that $\varphi(x)$ is also differentiable on $\mathbb{R}_+$, except at positive points of discontinuity of $F(y)$, and [2.2] holds. Moreover, from [2.7], [2.13] and [2.14], we obtain

$$\varphi'_+(x) = \frac{\lambda}{rx+c}\left(\varphi_1(x) + \varphi_2(x) - I(x) - \sum_{k:\, y_k \le x} p_k\, \varphi(x - y_k)\right)$$
$$= \frac{\lambda}{rx+c}\left(\varphi(x) - \int_0^x \varphi(x-y)\,\mathrm{d}F(y)\right),$$

which yields [2.3]. Note that the right-hand side of [2.3] is continuous on $\mathbb{R}_+$, except at positive points of discontinuity of $F(y)$. Therefore, $\varphi'_+(x)$ is also continuous on $\mathbb{R}_+$, except at these points. Bound [2.4] follows immediately from [2.3], which completes the proof. □

REMARK 2.1.– Slight changes in the proof show that theorem 2.1 is also true for $r = 0$. Another proof of this theorem for the case $r = 0$ is given in [ROL 99, pp. 162–163] (see also section 1.1.3).

REMARK 2.2.– Bound [2.4], as well as all similar bounds that will be obtained later, can be used for derivation of a relation connecting accuracy and reliability of a uniform approximation of the survival probability by its statistical estimate (see Chapter 6).

2.3. Continuity of the finite-horizon survival probability and existence of its partial derivatives

2.3.1. *Main theorem*

THEOREM 2.2.– Let the surplus process $\left(X_t(x)\right)_{t\ge 0}$ follow [2.1] under the above assumptions.

1) If Y_i, $i \ge 1$, have a p.d.f. $f(y)$, which is continuous on $\mathbb{R}_+$, then $\varphi(x,t)$ is continuous on $\mathbb{R}_+^2$ as a function of two variables.

2) If the p.d.f. $f(y)$ of Y_i, $i \ge 1$, has the derivative $f'(y)$ on $\mathbb{R}_+$, such that $|f'(y)|$ is integrable and bounded on $\mathbb{R}_+$, then:

i) $\varphi(x,t)$ has partial derivatives w.r.t. x and t on $\mathbb{R}_+^2$, which are continuous as functions of two variables;

ii) $\varphi(x,t)$ satisfies the partial integro-differential equation

$$\frac{\partial \varphi(x,t)}{\partial t} - (rx+c)\frac{\partial \varphi(x,t)}{\partial x} + \lambda\varphi(x,t) - \lambda\int_0^x \varphi(x-y,\, t)\,\mathrm{d}F(y) = 0 \qquad [2.18]$$

on $\mathbb{R}^2_+$ with the boundary conditions $\varphi(x,0)=1$ and $\lim_{x\to+\infty}\varphi(x,t)=1$;

iii) for all $T>0$, we have

$$\sup_{\substack{x\in[0,+\infty),\\ t\in[0,T]}} \left|\frac{\partial \varphi(x,t)}{\partial x}\right| \le C_1C_2, \qquad [2.19]$$

where

$$C_1 = \begin{cases} \dfrac{1-\mathrm{e}^{(r-\lambda)T}}{\lambda - r} & \text{if} \quad \lambda \ne r, \\ T & \text{if} \quad \lambda = r, \end{cases}$$

$$C_2 = \lambda\Big(f(0) + \int_0^{+\infty} |f'(y)|\,\mathrm{d}y\Big).$$

PROOF.– As in the proof of theorem 2.1, note that $X_t(x) = (x+c/r)\,\mathrm{e}^{rt} - c/r$ up to the time τ_1 and let $\tau_1 = s$ and $Y_1 = y$. Note also that $\mathbb{P}\big[\tau_1 > t\big] = \mathrm{e}^{-\lambda t}$ for all $t \ge 0$. Moreover, if $\tau_1 \le t$, ruin does not occur up to the time τ_1 and it will not take place after τ_1 if and only if $y \le (x+c/r)\,\mathrm{e}^{rs} - c/r$ and ruin does not occur on the time interval $(s,t]$ with the initial surplus $(x+c/r)\,\mathrm{e}^{rs} - c/r - y$. Hence, by the law of total probability, for all $x \ge 0$ and $t \ge 0$, we obtain

$$\varphi(x,t) = \int_0^t \lambda \mathrm{e}^{-\lambda s}\int_0^{(x+c/r)\,\mathrm{e}^{rs}-c/r} \varphi\big((x+c/r)\,\mathrm{e}^{rs} - c/r - y,\, t-s\big)\,\mathrm{d}F(y)\,\mathrm{d}s + \mathrm{e}^{-\lambda t}. \qquad [2.20]$$

Changing the variable $t-s=v$ in the outer integral in the right-hand side of [2.20] we get

$$\varphi(x,t) = \lambda\mathrm{e}^{-\lambda t}\int_0^t \mathrm{e}^{\lambda v}\int_0^{(x+c/r)\,\mathrm{e}^{r(t-v)}-c/r} \varphi\big((x+c/r)\,\mathrm{e}^{r(t-v)} - c/r - y,\, v\big)\,\mathrm{d}F(y)\,\mathrm{d}v + \mathrm{e}^{-\lambda t}. \qquad [2.21]$$

Note that $\varphi(x,t)$ is bounded, non-decreasing w.r.t. x and non-increasing w.r.t. t. Hence, the inner integral in the right-hand side of [2.21] is a non-increasing function w.r.t. v. Consequently, it is Riemann integrable on $[0,t]$ and the right-hand sides of [2.20] and [2.21] are well-defined.

We now use that Y_i, $i \geq 1$, have a p.d.f. $f(y)$. Changing the variable $(x + c/r)\,\mathrm{e}^{r(t-v)} - c/r - y = u$ in the inner integral in the right-hand side of [2.21] we have

$$\begin{aligned}\varphi(x,t) = \lambda \mathrm{e}^{-\lambda t} \int_0^t \mathrm{e}^{\lambda v} \int_0^{(x+c/r)\,\mathrm{e}^{r(t-v)}-c/r} & \varphi(u,v) \\ & \times f\big((x+c/r)\,\mathrm{e}^{r(t-v)} - c/r - u\big)\,\mathrm{d}u\,\mathrm{d}v + \mathrm{e}^{-\lambda t}.\end{aligned} \qquad [2.22]$$

Let

$$\begin{aligned}I(x,t) = \int_0^t \mathrm{e}^{\lambda v} \int_0^{(x+c/r)\,\mathrm{e}^{r(t-v)}-c/r} & \varphi(u,v) \\ & \times f\big((x+c/r)\,\mathrm{e}^{r(t-v)} - c/r - u\big)\,\mathrm{d}u\,\mathrm{d}v.\end{aligned}$$

We now use that $f(y)$ is continuous on $\mathbb{R}_+$. By [2.22], in order to prove that $\varphi(x,t)$ is continuous on $\mathbb{R}_+^2$ as a function of two variables, it is sufficient to show that the function $I(x,t)$ is so, that is, for any fixed $(x_0,t_0) \in \mathbb{R}_+^2$ and any $\varepsilon > 0$, there is a $\delta > 0$, such that

$$\big|I(x,t) - I(x_0,t_0)\big| < \varepsilon \qquad [2.23]$$

whenever $|x - x_0| < \delta$, $|t - t_0| < \delta$ and $(x,t) \in \mathbb{R}_+^2$.

We now fix $(x_0,t_0) \in (0,+\infty) \times (0,+\infty)$ and assume that $0 \leq x \leq x_0$ and $0 \leq t \leq t_0$ (similar arguments can be applied to other cases). We have

$$\begin{aligned}&\big|I(x,t) - I(x_0,t_0)\big| \\ &\leq \int_0^t \mathrm{e}^{\lambda v} \int_0^{(x+c/r)\,\mathrm{e}^{r(t-v)}-c/r} \big|f\big((x_0+c/r)\,\mathrm{e}^{r(t_0-v)} - c/r - u\big) \\ &- f\big((x+c/r)\,\mathrm{e}^{r(t-v)} - c/r - u\big)\big|\,\mathrm{d}u\,\mathrm{d}v + \int_0^t \mathrm{e}^{\lambda v} \\ &\times \int_{(x+c/r)\,\mathrm{e}^{r(t-v)}-c/r}^{(x_0+c/r)\,\mathrm{e}^{r(t_0-v)}-c/r} f\big((x_0+c/r)\,\mathrm{e}^{r(t_0-v)} - c/r - u\big)\,\mathrm{d}u\,\mathrm{d}v\end{aligned}$$

$$+\int_t^{t_0} \mathrm{e}^{\lambda v} \times \int_0^{(x_0+c/r)\,\mathrm{e}^{r(t_0-v)}-c/r} f\big((x_0+c/r)\,\mathrm{e}^{r(t_0-v)}-c/r-u\big)\,\mathrm{d}u\,\mathrm{d}v. \qquad [2.24]$$

Consider now the first summand in the right-hand side of [2.24]. The function $f\big((x+c/r)\,\mathrm{e}^{r(t-v)}-c/r-u\big)$ is continuous w.r.t. $(x,t)\in\mathbb{R}_+^2$, so it is uniformly continuous on any compact set in $\mathbb{R}_+^2$, that is, for any $\varepsilon_1>0$, there is a $\delta_1>0$, such that for all $v\in[0,t_0]$ and $u\in[0,(x_0+c/r)\,\mathrm{e}^{r(t_0-v)}-c/r]$, we have

$$\big|f\big((x_0+c/r)\,\mathrm{e}^{r(t_0-v)}-c/r-u\big)-f\big((x+c/r)\,\mathrm{e}^{r(t-v)}-c/r-u\big)\big|<\varepsilon_1$$

whenever $0\le x_0-x<\delta_1$, $0\le t_0-t<\delta_1$ and $(x,t)\in[0,x_0]\times[0,t_0]$.

Therefore, for $0\le x_0-x<\delta_1$, $0\le t_0-t<\delta_1$ and $(x,t)\in[0,x_0]\times[0,t_0]$, we get

$$\begin{aligned}&\int_0^t \mathrm{e}^{\lambda v}\int_0^{(x+c/r)\,\mathrm{e}^{r(t-v)}-c/r} \big|f\big((x_0+c/r)\,\mathrm{e}^{r(t_0-v)}-c/r-u\big)\\&\quad-f\big((x+c/r)\,\mathrm{e}^{r(t-v)}-c/r-u\big)\big|\,\mathrm{d}u\,\mathrm{d}v\\&\le\varepsilon_1\int_0^{t_0}\mathrm{e}^{\lambda v}\big((x_0+c/r)\,\mathrm{e}^{r(t_0-v)}-c/r\big)\,\mathrm{d}v\\&\le\varepsilon_1\bigg(\frac{(x_0+c/r)(\mathrm{e}^{\lambda t_0}-\mathrm{e}^{rt_0})}{\lambda-r}-\frac{c(\mathrm{e}^{\lambda t_0}-1)}{\lambda r}\bigg).\end{aligned} \qquad [2.25]$$

Consider now the second summand in the right-hand side of [2.24]. Since $f(y)$ is continuous, it is bounded and attains its maximum on any finite interval in $\mathbb{R}_+$. Set $M=\max_{y\in[0,\,(x_0+c/r)\,\mathrm{e}^{rt_0}-c/r]} f(y)$. The function $(x+c/r)\,\mathrm{e}^{r(t-v)}$ is continuous as a function of two variables x and t. Hence, it is continuous on any compact set in $\mathbb{R}_+^2$, that is, for any $\varepsilon_2>0$, there is a $\delta_2>0$, such that for all $v\in[0,t_0]$, we have

$$\big|(x_0+c/r)\,\mathrm{e}^{r(t_0-v)}-(x+c/r)\,\mathrm{e}^{r(t-v)}\big|<\varepsilon_2$$

whenever $0\le x_0-x<\delta_2$, $0\le t_0-t<\delta_2$ and $(x,t)\in[0,x_0]\times[0,t_0]$.

Therefore, we get

$$\int_0^t \mathrm{e}^{\lambda v} \int_{(x+c/r)\,\mathrm{e}^{r(t-v)}-c/r}^{(x_0+c/r)\,\mathrm{e}^{r(t_0-v)}-c/r} f\big((x_0+c/r)\,\mathrm{e}^{r(t_0-v)} - c/r - u\big)\,\mathrm{d}u\,\mathrm{d}v$$
$$\le \varepsilon_2 M \int_0^{t_0} \mathrm{e}^{\lambda v}\,\mathrm{d}v = \frac{\varepsilon_2 M(\mathrm{e}^{\lambda t_0}-1)}{\lambda}. \qquad [2.26]$$

Consider now the third summand in the right-hand side of [2.24]. Since

$$\int_t^{t_0} \mathrm{e}^{\lambda v} \int_0^{(x_0+c/r)\,\mathrm{e}^{r(t_0-v)}-c/r} f\big((x_0+c/r)\,\mathrm{e}^{r(t_0-v)} - c/r - u\big)\,\mathrm{d}u\,\mathrm{d}v$$
$$\le \int_t^{t_0} \mathrm{e}^{\lambda v}\,\mathrm{d}v \le \mathrm{e}^{\lambda t_0}(t_0-t),$$

for any $\varepsilon_3 > 0$, there is a $\delta_3 > 0$, such that

$$\int_t^{t_0} \mathrm{e}^{\lambda v} \int_0^{(x_0+c/r)\,\mathrm{e}^{r(t_0-v)}-c/r} f\big((x_0+c/r)\,\mathrm{e}^{r(t_0-v)} - c/r - u\big)\,\mathrm{d}u\,\mathrm{d}v$$
$$< \varepsilon_3 \qquad [2.27]$$

whenever $0 \le t_0 - t < \delta_3$ and $t \in [0, t_0]$.

We now fix any $\varepsilon > 0$. Let

$$\varepsilon_1 = \frac{\varepsilon}{3\left(\dfrac{(x_0+c/r)(\mathrm{e}^{\lambda t_0}-\mathrm{e}^{r t_0})}{\lambda - r} - \dfrac{c(\mathrm{e}^{\lambda t_0}-1)}{\lambda r}\right)},$$

$$\varepsilon_2 = \frac{\varepsilon\lambda}{3M(\mathrm{e}^{\lambda t_0}-1)},$$

$$\varepsilon_3 = \frac{\varepsilon}{3}.$$

By [2.24]–[2.27], we conclude that [2.23] holds whenever $0 \le x_0 - x < \delta$, $0 \le t_0 - t < \delta$ and $(x,t) \in [0,x_0] \times [0,t_0]$, where $\delta = \delta_1 \wedge \delta_2 \wedge \delta_3$.

We now use that the p.d.f. $f(y)$ of Y_i, $i \ge 1$, has the derivative $f'(y)$ on $\mathbb{R}_+$, such that $|f'(y)|$ is integrable and bounded on $\mathbb{R}_+$. By [2.22], in order to prove the existence of partial derivatives of $\varphi(x,t)$, it is sufficient to show the existence of partial derivatives of $I(x,t)$.

Let

$$I(x,t,v) = \int_0^{(x+c/r)\,\mathrm{e}^{r(t-v)}-c/r} \varphi(u,v)\, f\big((x+c/r)\,\mathrm{e}^{r(t-v)}-c/r-u\big)\,\mathrm{d}u.$$

By theorem A.2, to prove the existence of a partial derivative of $I(x,t)$ w.r.t. x for any fixed t, it suffices to show that $I(x,t,v)$ is continuous w.r.t. v on $[0,t]$ for any fixed $(x,t) \in \mathbb{R}_+^2$ and has a partial derivative w.r.t. x, which is continuous as a function of two variables x and v, where $(x,v) \in [0,+\infty) \times [0,t]$.

The continuity of $I(x,t,v)$ w.r.t. v on $[0,t]$ is evident. By theorems A.3 and A.5, to prove the existence of a partial derivative of $I(x,t,v)$ w.r.t. x, it suffices to show that for any fixed $t \in [0,+\infty)$ and $v \in [0,t]$, the following conditions hold:

– the function $(x+c/r)\,\mathrm{e}^{r(t-v)} - c/r$ has a derivative w.r.t. x;

– the function $\varphi(u,v)\, f\big((x+c/r)\,\mathrm{e}^{r(t-v)} - c/r - u\big)$ is continuous as a function of two variables x and u, where $(x,u) \in [0,+\infty) \times [0,(x+c/r)\,\mathrm{e}^{r(t-v)} - c/r]$;

– the function $f\big((x+c/r)\,\mathrm{e}^{r(t-v)} - c/r - u\big)$ has a derivative w.r.t. x, which is integrable w.r.t. u on $[0,(x+c/r)\,\mathrm{e}^{r(t-v)} - c/r]$ and bounded for all $x \in [0,+\infty)$ and $u \in [0,(x+c/r)\,\mathrm{e}^{r(t-v)} - c/r]$.

It is easily seen that all these conditions hold. Hence, there is a partial derivative of $I(x,t,v)$ w.r.t. x and

$$\frac{\partial I(x,t,v)}{\partial x} = \mathrm{e}^{r(t-v)}\Bigg(f(0)\,\varphi\big((x+c/r)\,\mathrm{e}^{r(t-v)} - c/r,\, v\big) + \int_0^{(x+c/r)\,\mathrm{e}^{r(t-v)}-c/r} \varphi(u,v)\, f'\big((x+c/r)\,\mathrm{e}^{r(t-v)} - c/r - u\big)\,\mathrm{d}u\Bigg).$$

Arguments similar to those in the proof of the continuity of $I(x,t)$ show that the partial derivative of $I(x,t,v)$ w.r.t. x is continuous as a function of two variables x and v, where $(x,v) \in [0,+\infty) \times [0,t]$, for all fixed $t \in \mathbb{R}_+$. We only note that to this end, we use the equality

$$\int_0^{(x+c/r)\,\mathrm{e}^{r(t-v)}-c/r} \varphi(u,v)\, f'\big((x+c/r)\,\mathrm{e}^{r(t-v)} - c/r - u\big)\,\mathrm{d}u$$

$$= \int_0^{(x+c/r)\,\mathrm{e}^{r(t-v)}-c/r} \varphi\big((x+c/r)\,\mathrm{e}^{r(t-v)} - c/r - u,\, v\big)\, f'(u)\,\mathrm{d}u.$$

Thus, there is a partial derivative of $\varphi(x,t)$ w.r.t. x and

$$\frac{\partial \varphi(x,t)}{\partial x} = \lambda \mathrm{e}^{(r-\lambda)t} \int_0^t \mathrm{e}^{(\lambda-r)v} \Big(f(0)\, \varphi\big((x+c/r)\,\mathrm{e}^{r(t-v)} - c/r,\, v\big)$$

$$+ \int_0^{(x+c/r)\,\mathrm{e}^{r(t-v)}-c/r} \varphi(u,v) \qquad [2.28]$$

$$\times f'\big((x+c/r)\,\mathrm{e}^{r(t-v)} - c/r - u\big)\,\mathrm{d}u \Big)\mathrm{d}v.$$

We conclude from [2.28] that

$$\left|\frac{\partial \varphi(x,t)}{\partial x}\right| \le \lambda \Big(f(0) + \int_0^{+\infty} |f'(y)|\,\mathrm{d}y \Big)\, \mathrm{e}^{(r-\lambda)t} \int_0^t \mathrm{e}^{(\lambda-r)v}\,\mathrm{d}v$$

for all $(x,t) \in \mathbb{R}_+^2$. Hence, we get [2.19].

Next, by theorem A.2, to prove the existence of a partial derivative of $I(x,t)$ w.r.t. t for any fixed x, it suffices to show that $I(x,t,v)$ is continuous as a function of two variables t and v, where $t \in [0,+\infty)$ and $v \in [0,t]$, for all fixed $x \in \mathbb{R}_+$ and $I(x,t,v)$ has a partial derivative w.r.t. t, which is continuous as a function of two variables t and v, where $t \in [0,+\infty)$ and $v \in [0,t]$, for all fixed $x \in \mathbb{R}_+$.

Arguments similar to those in the proof of the continuity of $I(x,t)$ show that $I(x,t,v)$ is continuous. By theorems A.3 and A.5, to prove the existence of a partial derivative of $I(x,t,v)$ w.r.t. t, it suffices to show that for any fixed $(x,v) \in [0,+\infty) \times [0,t]$, the following conditions hold:

– the function $(x+c/r)\,\mathrm{e}^{r(t-v)} - c/r$ has a derivative w.r.t. t;

– the function $\varphi(u,v)\, f\big((x+c/r)\,\mathrm{e}^{r(t-v)} - c/r - u\big)$ is continuous as a function of two variables t and u, where $(t,u) \in [0,+\infty) \times [0,(x+c/r)\,\mathrm{e}^{r(t-v)} - c/r]$;

– the function $f\big((x+c/r)\,\mathrm{e}^{r(t-v)} - c/r - u\big)$ has a derivative w.r.t. t, which is integrable w.r.t. u on $[0,(x+c/r)\,\mathrm{e}^{r(t-v)} - c/r]$ and bounded for all $t \in [0,+\infty)$ and $u \in [0,(x+c/r)\,\mathrm{e}^{r(t-v)} - c/r]$.

It is easily seen that the first two conditions hold. In the third condition, the derivative of $f\big((x+c/r)\,\mathrm{e}^{r(t-v)} - c/r - u\big)$ w.r.t. t may not be bounded for all $t \in [0,+\infty)$ and $u \in [0,(x+c/r)\,\mathrm{e}^{r(t-v)} - c/r]$, but it is bounded if $t \in [0,T]$ for any $T > 0$. So, we must prove the existence of a partial derivative of $I(x,t,v)$ w.r.t. t when $t \in [0,T]$ for any $T > 0$. This implies the existence of the partial derivative for all $t \in [0,+\infty)$.

Hence, there is a partial derivative of $I(x,t,v)$ w.r.t. t and

$$\frac{\partial I(x,t,v)}{\partial t} = (rx+c)\mathrm{e}^{r(t-v)}\Big(f(0)\,\varphi\big((x+c/r)\,\mathrm{e}^{r(t-v)} - c/r,\, v\big) + \int_0^{(x+c/r)\,\mathrm{e}^{r(t-v)}-c/r} \varphi(u,v)\, f'\big((x+c/r)\,\mathrm{e}^{r(t-v)} - c/r - u\big)\,\mathrm{d}u\Big).$$

Arguments similar to those in the proof of the continuity of $I(x,t)$ show that the partial derivative of $I(x,t,v)$ w.r.t. t is continuous as a function of two variables t and v, where $t \in [0,+\infty)$ and $v \in [0,t]$, for all fixed $x \in \mathbb{R}_+$.

Thus, there exists a partial derivative of $\varphi(x,t)$ w.r.t. t and

$$\begin{aligned}\frac{\partial \varphi(x,t)}{\partial t} &= -\lambda^2\,\mathrm{e}^{-\lambda t}\int_0^t \mathrm{e}^{\lambda v}\int_0^{(x+c/r)\,\mathrm{e}^{r(t-v)}-c/r} \varphi(u,v)\\ &\quad\times f\big((x+c/r)\,\mathrm{e}^{r(t-v)} - c/r - u\big)\,\mathrm{d}u\,\mathrm{d}v\\ &\quad+\lambda\int_0^x \varphi(u,t)\,f(x-u)\,\mathrm{d}u + \lambda(rx+c)\,\mathrm{e}^{(r-\lambda)t}\\ &\quad\times\int_0^t \mathrm{e}^{(\lambda-r)v}\Big(f(0)\,\varphi\big((x+c/r)\,\mathrm{e}^{r(t-v)} - c/r,\, v\big)\\ &\quad+\int_0^{(x+c/r)\,\mathrm{e}^{r(t-v)}-c/r}\varphi(u,v)\\ &\quad\times f'\big((x+c/r)\,\mathrm{e}^{r(t-v)} - c/r - u\big)\,\mathrm{d}u\Big)\mathrm{d}v - \lambda\mathrm{e}^{-\lambda t}.\end{aligned} \qquad [2.29]$$

Combining [2.22], [2.28] and [2.29], we have

$$\frac{\partial\varphi(x,t)}{\partial t} = (rx+c)\,\frac{\partial\varphi(x,t)}{\partial x} - \lambda\varphi(x,t) + \lambda\int_0^x \varphi(x-y,\,t)\,\mathrm{d}F(y),$$

which yields [2.18]. Arguments similar to those in the proof of the continuity of $I(x,t)$ show that the partial derivatives of $\varphi(x,t)$ are continuous.

The boundary condition $\varphi(x,0) = 1$ is evident. The condition $\lim_{x\to+\infty}\varphi(x,t) = 1$ follows immediately from the fact that there can be only a finite number of claims on any finite time interval a.s. This completes the proof. □

REMARK 2.3.– Slight changes in the proof show that theorem 2.2 is also true for $r = 0$.

REMARK 2.4.– The requirements that we impose on the distribution of the claim sizes in theorem 2.2 are not necessary for the continuity of $\varphi(x,t)$ or the existence of its partial derivatives. This is verified by example 2.1 below. Furthermore, it is clear that the second assertion of the theorem is true if $f(y)$ is continuous and has a derivative everywhere except at a finite number of points, where it has finite one-sided derivatives, and $|f'(y)|$ is integrable and bounded on $\mathbb{R}_+$. On the other hand, $\varphi(x,t)$ may not have partial derivatives in some cases. This can be seen from example 2.2 below.

2.3.2. *Examples*

EXAMPLE 2.1.– Let Y_i, $i \geq 1$, be uniformly distributed on $[0,1]$. It is evident that the p.d.f. of Y_i is not continuous at the points $y = 0$ and $y = 1$. Hence, it is not differentiable at these points. Nevertheless, we now show that $\varphi(x,t)$ has partial derivatives w.r.t. x and t on $\mathbb{R}_+^2$, which are continuous as functions of two variables.

We can rewrite [2.22] as

$$\varphi(x,t) = \lambda\mathrm{e}^{-\lambda t}\int_0^t \mathrm{e}^{\lambda v}\int_{0\,\vee\,((x+c/r)\,\mathrm{e}^{r(t-v)}-c/r-1)}^{(x+c/r)\,\mathrm{e}^{r(t-v)}-c/r} \varphi(u,v)\,\mathrm{d}u\,\mathrm{d}v + \mathrm{e}^{-\lambda t}.$$

[2.30]

Consider now three cases.

1) If $(x,t) \in [0,1] \times \left[0, \frac{1}{r}\ln\frac{r+c}{rx+c}\right]$, then for all $v \in [0,t]$, we have

$$(x+c/r)\,\mathrm{e}^{r(t-v)} - c/r - 1 \leq (x+c/r)(\mathrm{e}^{rt}-1) \leq 0.$$

Hence, we can rewrite [2.30] as

$$\varphi(x,t) = \lambda \mathrm{e}^{-\lambda t}\int_0^t \mathrm{e}^{\lambda v}\int_0^{(x+c/r)\,\mathrm{e}^{r(t-v)}-c/r} \varphi(u,v)\,\mathrm{d}u\,\mathrm{d}v + \mathrm{e}^{-\lambda t}. \quad [2.31]$$

2) If $(x,t) \in [0,1] \times \left(\frac{1}{r}\ln\frac{r+c}{rx+c}, +\infty\right)$, then

$$(x+c/r)\,\mathrm{e}^{r(t-v)} - c/r - 1 \geq 0 \quad \text{for} \quad v \in \left[0,\, t - \frac{1}{r}\ln\frac{r+c}{rx+c}\right],$$

$$(x+c/r)\,\mathrm{e}^{r(t-v)} - c/r - 1 \leq 0 \quad \text{for} \quad v \in \left[t - \frac{1}{r}\ln\frac{r+c}{rx+c},\, t\right].$$

Hence, we can rewrite [2.30] as

$$\varphi(x,t) = \lambda \mathrm{e}^{-\lambda t}\left(\int_0^{t-\frac{1}{r}\ln\frac{r+c}{rx+c}} \mathrm{e}^{\lambda v}\int_{(x+c/r)\,\mathrm{e}^{r(t-v)}-c/r-1}^{(x+c/r)\,\mathrm{e}^{r(t-v)}-c/r} \varphi(u,v)\,\mathrm{d}u\,\mathrm{d}v\right.$$

$$\left. + \int_{t-\frac{1}{r}\ln\frac{r+c}{rx+c}}^{t} \mathrm{e}^{\lambda v}\int_0^{(x+c/r)\,\mathrm{e}^{r(t-v)}-c/r} \varphi(u,v)\,\mathrm{d}u\,\mathrm{d}v\right) + \mathrm{e}^{-\lambda t}. \quad [2.32]$$

3) If $(x,t) \in (1,+\infty) \times [0,+\infty)$, then for all $v \in [0,t]$, we have

$$(x+c/r)\,\mathrm{e}^{r(t-v)} - c/r - 1 \geq x - 1 > 0.$$

Hence, we can rewrite [2.30] as

$$\varphi(x,t) = \lambda \mathrm{e}^{-\lambda t}\int_0^t \mathrm{e}^{\lambda v}\int_{(x+c/r)\,\mathrm{e}^{r(t-v)}-c/r-1}^{(x+c/r)\,\mathrm{e}^{r(t-v)}-c/r} \varphi(u,v)\,\mathrm{d}u\,\mathrm{d}v + \mathrm{e}^{-\lambda t}. \quad [2.33]$$

Thus, we can divide the domain of $\varphi(x,t)$ into three sets

$$S_1 = [0,1] \times \left[0, \frac{1}{r}\ln\frac{r+c}{rx+c}\right],$$

$$S_2 = [0,1] \times \left(\frac{1}{r} \ln \frac{r+c}{rx+c}, +\infty\right),$$

$$S_3 = (1, +\infty) \times [0, +\infty),$$

where it satisfies equations [2.31], [2.32] and [2.33], respectively.

Applying arguments similar to those in the proof of theorem 2.2 and using [2.31], [2.32] and [2.33], we can show that $\varphi(x,t)$ is continuous as a function of two variables on S_1, S_2, S_3. Moreover, we can show that $\varphi(x,t)$ is continuous as a function of two variables on $\mathbb{R}_+^2$. Then, it satisfies equations [2.31], [2.32] and [2.33] on

$$\bar{S}_1 = [0,1] \times \left[0, \frac{1}{r} \ln \frac{r+c}{rx+c}\right],$$

$$\bar{S}_2 = [0,1] \times \left[\frac{1}{r} \ln \frac{r+c}{rx+c}, +\infty\right),$$

$$\bar{S}_3 = [1, +\infty) \times [0, +\infty),$$

respectively.

Applying arguments similar to those in the proof of theorem 2.2 and using [2.31], [2.32] and [2.33], we can also show that there are continuous partial derivatives of $\varphi(x,t)$ w.r.t. x and t on $\bar{S}_1$, $\bar{S}_2$, $\bar{S}_3$. Note that we imply one-sided derivatives on the boundary of these sets.

If $(x,t) \in \bar{S}_1$, then

$$\frac{\partial \varphi(x,t)}{\partial x} = \lambda e^{(r-\lambda)t} \int_0^t e^{(\lambda-r)v} \varphi\big((x+c/r)\, e^{r(t-v)} - c/r,\, v\big)\, dv, \quad [2.34]$$

$$\frac{\partial \varphi(x,t)}{\partial t} = -\lambda^2 e^{-\lambda t} \int_0^t e^{\lambda v} \int_0^{(x+c/r)\, e^{r(t-v)} - c/r} \varphi(u,v)\, du\, dv$$
$$+\lambda \int_0^x \varphi(u,t)\, du + \lambda(rx+c)\, e^{(r-\lambda)t} \quad [2.35]$$
$$\times \int_0^t e^{(\lambda-r)v} \varphi\big((x+c/r)\, e^{r(t-v)} - c/r,\, v\big)\, dv - \lambda e^{-\lambda t}.$$

If $(x, t) \in \bar{S}_2$, then

$$\frac{\partial \varphi(x,t)}{\partial x} = \lambda e^{(r-\lambda)t} \left(\int_0^{t-\frac{1}{r}\ln\frac{r+c}{rx+c}} e^{(\lambda-r)v} \right.$$
$$\times \left(\varphi\big((x+c/r)\, e^{r(t-v)} - c/r,\, v\big)\right.$$
$$\left. - \varphi\big((x+c/r)\, e^{r(t-v)} - c/r - 1,\, v\big)\right) dv$$
$$\left. + \int_{t-\frac{1}{r}\ln\frac{r+c}{rx+c}}^{t} e^{(\lambda-r)v}\, \varphi\big((x+c/r)\, e^{r(t-v)} - c/r,\, v\big)\, dv \right), \qquad [2.36]$$

$$\frac{\partial \varphi(x,t)}{\partial t} = -\lambda^2 e^{-\lambda t} \left(\int_0^{t-\frac{1}{r}\ln\frac{r+c}{rx+c}} e^{\lambda v} \right.$$
$$\times \int_{(x+c/r)\, e^{r(t-v)}-c/r-1}^{(x+c/r)\, e^{r(t-v)}-c/r} \varphi(u,v)\, du\, dv$$
$$\left. + \int_{t-\frac{1}{r}\ln\frac{r+c}{rx+c}}^{t} e^{\lambda v} \int_0^{(x+c/r)\, e^{r(t-v)}-c/r} \varphi(u,v)\, du\, dv \right)$$
$$+ \lambda \int_0^x \varphi(u,t)\, du$$
$$+ \lambda(rx+c)\, e^{(r-\lambda)t} \left(\int_0^{t-\frac{1}{r}\ln\frac{r+c}{rx+c}} e^{(\lambda-r)v} \right.$$
$$\times \left(\varphi\big((x+c/r)\, e^{r(t-v)} - c/r,\, v\big)\right.$$
$$\left. - \varphi\big((x+c/r)\, e^{r(t-v)} - c/r - 1,\, v\big)\right) dv$$
$$\left. + \int_{t-\frac{1}{r}\ln\frac{r+c}{rx+c}}^{t} e^{(\lambda-r)v}\, \varphi\big((x+c/r)\, e^{r(t-v)} - c/r,\, v\big)\, dv \right)$$
$$- \lambda e^{-\lambda t}. \qquad [2.37]$$

If $(x,t) \in \bar{S}_3$, then

$$\frac{\partial \varphi(x,t)}{\partial x} = \lambda \mathrm{e}^{(r-\lambda)t} \int_0^t \mathrm{e}^{(\lambda-r)v} \big(\varphi\big((x+c/r)\,\mathrm{e}^{r(t-v)} - c/r,\, v\big) - \varphi\big((x+c/r)\,\mathrm{e}^{r(t-v)} - c/r - 1,\, v\big)\big)\,\mathrm{d}v, \quad [2.38]$$

$$\frac{\partial \varphi(x,t)}{\partial t} = -\lambda^2 \mathrm{e}^{-\lambda t} \int_0^t \mathrm{e}^{\lambda v} \int_{(x+c/r)\,\mathrm{e}^{r(t-v)}-c/r-1}^{(x+c/r)\,\mathrm{e}^{r(t-v)}-c/r} \varphi(u,v)\,\mathrm{d}u\,\mathrm{d}v + \lambda \int_{x-1}^{x} \varphi(u,t)\,\mathrm{d}u + \lambda(rx+c)\,\mathrm{e}^{(r-\lambda)t} \int_0^t \mathrm{e}^{(\lambda-r)v} \times \big(\varphi\big((x+c/r)\,\mathrm{e}^{r(t-v)} - c/r,\, v\big) - \varphi\big((x+c/r)\,\mathrm{e}^{r(t-v)} - c/r - 1,\, v\big)\big)\,\mathrm{d}v - \lambda \mathrm{e}^{-\lambda t}. \quad [2.39]$$

By [2.34]–[2.37], we conclude that there are continuous partial derivatives of $\varphi(x,t)$ on the boundary of $\bar{S}_1$ and $\bar{S}_2$, i.e. for $x \in [0,1]$ and $t = \frac{1}{r}\ln\frac{r+c}{rx+c}$. Furthermore, by [2.36]–[2.39], we see that there are continuous partial derivatives of $\varphi(x,t)$ on the boundary of $\bar{S}_2$ and $\bar{S}_3$, i.e. for $x = 1$ and $t \in \mathbb{R}_+$. Thus, $\varphi(x,t)$ has continuous partial derivatives on $\mathbb{R}_+^2$ and we can show that it satisfies [2.18].

EXAMPLE 2.2.– Let $\mathbb{P}[Y_i = 1] = 1$, $i \geq 1$. We now show that $\varphi(x,t)$ does not have partial derivatives on some sets on $\mathbb{R}_+^2$. Consider the following three cases:

1) If $(x,t) \in [0,1] \times \left[0, \frac{1}{r}\ln\frac{r+c}{rx+c}\right]$, then by [2.21], we have

$$\varphi(x,t) = \mathrm{e}^{-\lambda t}. \quad [2.40]$$

2) If $(x,t) \in [0,1] \times \left(\frac{1}{r}\ln\frac{r+c}{rx+c}, +\infty\right)$, then

$$(x+c/r)\,\mathrm{e}^{r(t-v)} - c/r > 1 \quad \text{for} \quad v \in \left[0,\, t - \frac{1}{r}\ln\frac{r+c}{rx+c}\right].$$

Hence, we can rewrite [2.21] as

$$\varphi(x,t) = \lambda \mathrm{e}^{-\lambda t} \int_0^{t-\frac{1}{r}\ln\frac{r+c}{rx+c}} \mathrm{e}^{\lambda v}\, \varphi\big((x+c/r)\,\mathrm{e}^{r(t-v)} - c/r - 1,\, v\big)\,\mathrm{d}v + \mathrm{e}^{-\lambda t}. \quad [2.41]$$

3) If $(x,t) \in (1,+\infty) \times [0,+\infty)$, then for all $v \in [0,t]$, we have

$$(x + c/r)\,\mathrm{e}^{r(t-v)} - c/r - 1 \geq x > 1.$$

Hence, we can rewrite [2.21] as

$$\varphi(x,t) = \lambda \mathrm{e}^{-\lambda t} \int_0^t \mathrm{e}^{\lambda v}\, \varphi\big((x+c/r)\,\mathrm{e}^{r(t-v)} - c/r - 1,\, v\big)\, \mathrm{d}v + \mathrm{e}^{-\lambda t}. \qquad [2.42]$$

Thus, as in example 2.1, we can divide the domain of $\varphi(x,t)$ into the same three sets S_1, S_2, S_3, where it is defined by [2.40], [2.41] and [2.42], respectively. The sets $\bar{S}_1$, $\bar{S}_2$ and $\bar{S}_3$ are also defined as in example 2.1.

If $(x,t) \in \bar{S}_1$, then it can easily be seen from [2.40] that $\varphi(x,t)$ is continuous as a function of two variables and has continuous partial derivatives on $\bar{S}_1$. Moreover,

$$\frac{\partial \varphi(x,t)}{\partial x} = 0 \quad \text{and} \quad \frac{\partial \varphi(x,t)}{\partial t} = -\lambda \mathrm{e}^{-\lambda t}. \qquad [2.43]$$

On the contrary, [2.41] and [2.42] do not imply these properties of $\varphi(x,t)$ on $\bar{S}_2$ and $\bar{S}_3$. Nevertheless, we make the additional assumption, which is intuitively natural, that $\varphi(x,t)$ is continuous as a function of two variables on $\mathbb{R}_+^2$ and has continuous partial derivatives on $\bar{S}_2$ and $\bar{S}_3$. Let us denote by $\varphi_1'(\cdot,\cdot)$ the partial derivative of $\varphi(x,t)$ w.r.t. the first argument.

If $(x,t) \in \bar{S}_2$, then by [2.41], we get

$$\frac{\partial \varphi(x,t)}{\partial x} = \frac{\lambda}{rx+c}\left(\frac{rx+c}{r+c}\right)^{\lambda/r} \varphi\left(0,\, t - \frac{1}{r}\ln\frac{r+c}{rx+c}\right) + \lambda \mathrm{e}^{(r-\lambda)t}$$

$$\times \int_0^{t-\frac{1}{r}\ln\frac{r+c}{rx+c}} \mathrm{e}^{(\lambda-r)v}\, \varphi_1'\big((x+c/r)\,\mathrm{e}^{r(t-v)} - c/r - 1,\, v\big)\, \mathrm{d}v, \qquad [2.44]$$

$$\frac{\partial\varphi(x,t)}{\partial t} = -\lambda^2 \mathrm{e}^{-\lambda t}\int_0^{t-\frac{1}{r}\ln\frac{r+c}{rx+c}} \mathrm{e}^{\lambda v}$$
$$\times\, \varphi\big((x+c/r)\,\mathrm{e}^{r(t-v)} - c/r - 1,\, v\big)\,\mathrm{d}v$$
$$+\lambda\left(\frac{rx+c}{r+c}\right)^{\lambda/r}\varphi\left(0,\, t-\frac{1}{r}\ln\frac{r+c}{rx+c}\right) \qquad [2.45]$$
$$+\lambda(rx+c)\,\mathrm{e}^{(r-\lambda)t}\int_0^{t-\frac{1}{r}\ln\frac{r+c}{rx+c}} \mathrm{e}^{(\lambda-r)v}$$
$$\times\, \varphi_1'\big((x+c/r)\,\mathrm{e}^{r(t-v)} - c/r - 1,\, v\big)\,\mathrm{d}v - \lambda \mathrm{e}^{-\lambda t}.$$

If $(x,t)\in\bar{S}_3$, then by [2.42], we get

$$\frac{\partial\varphi(x,t)}{\partial x} = \lambda \mathrm{e}^{(r-\lambda)t}\int_0^t \mathrm{e}^{(\lambda-r)v}\,\varphi_1'\big((x+c/r)\,\mathrm{e}^{r(t-v)} - c/r - 1,\, v\big)\,\mathrm{d}v, \qquad [2.46]$$

$$\frac{\partial\varphi(x,t)}{\partial t} = -\lambda^2 \mathrm{e}^{-\lambda t}\int_0^t \mathrm{e}^{\lambda v}\,\varphi\big((x+c/r)\,\mathrm{e}^{r(t-v)} - c/r - 1,\, v\big)\,\mathrm{d}v$$
$$+\lambda\varphi(x-1,t) + \lambda(rx+c)\,\mathrm{e}^{(r-\lambda)t}\int_0^t \mathrm{e}^{(\lambda-r)v}$$
$$\times\, \varphi_1'\big((x+c/r)\,\mathrm{e}^{r(t-v)} - c/r - 1,\, v\big)\,\mathrm{d}v - \lambda \mathrm{e}^{-\lambda t}.$$

However, the partial derivatives of $\varphi(x,t)$ w.r.t. x and t do not exist on the boundary of $\bar{S}_1$ and $\bar{S}_2$ even under these additional assumptions. Moreover, the partial derivative of $\varphi(x,t)$ w.r.t. x does not exist on the boundary of $\bar{S}_2$ and $\bar{S}_3$.

Indeed, if $x\in[0,1]$ and $t=\frac{1}{r}\ln\frac{r+c}{rx+c}$, then by [2.43], we have

$$\frac{\partial\varphi(x,t)}{\partial x} = 0 \quad\text{and}\quad \frac{\partial\varphi(x,t)}{\partial t} = -\lambda \mathrm{e}^{-\lambda t}$$

on $\bar{S}_1$ and by [2.44] and [2.45], we have

$$\frac{\partial\varphi(x,t)}{\partial x} = \frac{\lambda}{rx+c}\left(\frac{rx+c}{r+c}\right)^{\lambda/r}$$

and

$$\frac{\partial\varphi(x,t)}{\partial t} = \lambda\left(\frac{rx+c}{r+c}\right)^{\lambda/r} - \lambda \mathrm{e}^{-\lambda t}$$

on $\bar{S}_2$.

Furthermore, if $x = 1$ and $t \in \mathbb{R}_+$, then by [2.44], we have

$$\frac{\partial\varphi(x,t)}{\partial x} = \frac{\lambda}{r+c}\,\varphi(0,t)$$
$$+ \lambda \mathrm{e}^{(r-\lambda)t} \int_0^t \mathrm{e}^{(\lambda-r)v}\,\varphi_1'\big((1+c/r)\,\mathrm{e}^{r(t-v)} - c/r - 1,\, v\big)\,\mathrm{d}v$$

on $\bar{S}_2$ and by [2.46], we have

$$\frac{\partial\varphi(x,t)}{\partial x} = \lambda \mathrm{e}^{(r-\lambda)t} \int_0^t \mathrm{e}^{(\lambda-r)v}\,\varphi_1'\big((1+c/r)\,\mathrm{e}^{r(t-v)} - c/r - 1,\, v\big)\,\mathrm{d}v$$

on $\bar{S}_3$.

It only remains to note that $\varphi(0,t) > 0$ for all $t \in \mathbb{R}_+$ by [2.40] and [2.41]. □

2.4. Bibliographical notes

For some early results concerning the classical risk model with investments in a risk-free asset, we refer the reader to [BOO 87, BOO 88, DEL 87, EMA 75, GER 71, HAR 77, VIT 87]. In particular, Delbaen and Haezendonck [DEL 87] studied the influence of macro-economic factors such as interest and inflation on the classical surplus process. Boogaert and Crijns [BOO 87] investigated the influence of a negative safety loading on the upper bounds for the ruin probabilities when macro-economic factors are taken into account. Later, Embrechts and Schmidli [EMB 94] and Schmidli [SCH 94] studied risk models with borrowing, investment and inflation.

In the classical risk model with investments in a risk-free asset, Sundt and Teugels [SUN 95, SUN 97] derived integro-differential and integral equations for $\varphi(x)$ and obtained some approximations and upper and lower bounds for it. Moreover, they found the closed-form solution to the integro-differential equation when the claim sizes are exponentially distributed. Pervozvansky [PER 98] derived an integro-differential equation for $\varphi(x, t)$ and found the closed-form solution to it when the claim sizes are exponentially distributed and the risk-free interest rate is zero.

Asmussen [ASM 98], Kalashnikov and Konstantinides [KAL 00] and Klüppelberg and Stadtmüller [KLÜ 98] obtained asymptotic approximations for $\psi(x)$ when the claim sizes are heavy-tailed. These approximations differ considerably from those in the classical risk model without any investments. Thus, investments in a risk-free asset change the asymptotics of the infinite-horizon ruin probability. Tang [TAN 05] established an asymptotic formula for $\psi(x, t)$ in the case of subexponential claims.

Two-sided bounds for the infinite-horizon ruin probability in the case of heavy-tailed claim sizes were obtained by Kalashnikov [KAL 96] and Konstantinides, Tang and Tsitsiashvili [KON 02, KON 04]. Cai and Dickson [CAI 03] derived upper exponential bounds for $\psi(x)$ by martingale and recursive techniques in the more general Sparre Andersen model and gave applications of the results to the classical risk model.

Kasozi and Paulsen [KAS 05] proposed a numerical procedure for estimation of $\psi(x)$. Albrecher, Teugels and Tichy [ALB 01] got explicit formulas for the finite-horizon ruin probability in the case of exponentially distributed claim sizes in terms of finite gamma series. Knessl and Peter [KNE 96] studied detailed asymptotics of $\psi(x, t)$ for small interest rate and exponentially distributed claim sizes. Brekelmans and De Waegenaere [BRE 01] presented an algorithm that allows us to determine lower and upper bounds for the finite-horizon ruin probability. Cardoso and Waters [CAR 03, CAR 05] suggested numerical algorithms for calculation of $\psi(x, t)$ based on discrete time Markov chains and obtained upper and lower bounds for $\psi(x, t)$.

For bibliographical notes concerning the ruin or survival probabilities in the classical risk model without investments, see section 1.1.9.

3

Risk Model with Stochastic Premiums and Investments in a Risk-Free Asset

In this chapter, we consider a generalization of the risk model with stochastic premiums where an insurance company invests all surplus in a risk-free asset. We investigate the continuity and differentiability of the infinite- and finite-horizon survival probabilities and compare these results with those in the case of zero interest rate. Furthermore, we derive integro-differential equations for the survival probabilities and get bounds for their derivatives w.r.t. an initial surplus.

3.1. Description of the model

We consider a generalization of the *risk model with stochastic premiums* described in section 1.2.1: an insurance company invests all surplus in a *risk-free asset,* the price of which equals $B_t = B_0\, \mathrm{e}^{rt}$ at time t, where B_0 is the price of the asset at the time $t = 0$ and r is a risk-free interest rate. In what follows, we assume that $r > 0$, but deal with the case $r = 0$ in section 3.2.2.

Then, the *surplus process* $\big(X_t(x)\big)_{t\geq 0}$ follows the equation

$$X_t(x) = x + r\int_0^t X_s(x)\,\mathrm{d}s + \sum_{i=1}^{\bar{N}_t}\bar{Y}_i - \sum_{i=1}^{N_t} Y_i, \quad t \geq 0. \qquad [3.1]$$

In this chapter, we discuss the continuity and differentiability of the survival probabilities in this risk model. Moreover, we compare these properties of the survival probabilities with those in the risk model with stochastic premiums and without any investments. They turn out to be quite different in contrast to the classical risk model where investments in a risk-free asset do not have an influence on the continuity and differentiability of these functions. In particular, we will show that the infinite-horizon survival probability may not be differentiable at some points. Moreover, points of discontinuity of this function are also possible.

The rest of the chapter is organized as follows. Section 3.2 is devoted to investigation of the continuity and differentiability of the infinite-horizon survival probability. In section 3.3, we investigate these properties for the finite-horizon survival probability. The bibliographical notes regarding the risk model with stochastic premiums are given in section 1.2.3.

3.2. Continuity and differentiability of the infinite-horizon survival probability

3.2.1. *Main results*

THEOREM 3.1.–

1) Let the surplus process $\left(X_t(x)\right)_{t\geq 0}$ follow [3.1] under the above assumptions. Then, the function $\varphi(x)$ is continuous on $(0, +\infty)$.

2) Moreover, let $\bar{\lambda}\bar{\mu} > \lambda\mu$. Then the following assertions hold:

i) The function $\varphi(x)$ is continuous on $\mathbb{R}_+$.

ii) The function $\varphi(x)$ is continuously differentiable on $(0, +\infty)$, except at positive points of discontinuity of $F(y)$.

iii) If $x > 0$ is a point of discontinuity of $F(y)$ and $F(x) - F(x_-) = p$, then $\varphi(x)$ has the left and right derivatives $\varphi'_-(x)$ and $\varphi'_+(x)$, respectively, and

$$\varphi'_-(x) - \varphi'_+(x) = \frac{\lambda p \varphi(0)}{rx} > 0. \qquad [3.2]$$

iv) The function $\varphi(x)$ satisfies the integro-differential equation

$$rx\varphi'_+(x) = (\bar{\lambda} + \lambda)\varphi(x) - \bar{\lambda}\int_0^{+\infty} \varphi(x+y)\,\mathrm{d}\bar{F}(y) - \lambda\int_0^{x} \varphi(x-y)\,\mathrm{d}F(y) \qquad [3.3]$$

on $(0, +\infty)$ with the boundary condition $\lim_{x\to+\infty} \varphi(x) = 1$ and its value at the point $x = 0$ is given by

$$\varphi(0) = \frac{\bar{\lambda}}{\bar{\lambda} + \lambda} \int_0^{+\infty} \varphi(y)\, \mathrm{d}\bar{F}(y). \qquad [3.4]$$

v) For all $x_* > 0$, we have

$$\sup_{x\in[x_*,+\infty)} \left|\varphi'_+(x)\right| \le \frac{\bar{\lambda} + \lambda}{r x_*}. \qquad [3.5]$$

PROOF.– It follows easily from [3.1] that $X_t(x) = x\mathrm{e}^{rt}$ up to the time $\bar{\tau}_1 \wedge \tau_1$. Let $\bar{\tau}_1 \wedge \tau_1 = s$. Furthermore, let $\bar{Y}_1 = y$ if $\bar{\tau}_1 \wedge \tau_1 = \bar{\tau}_1$ and $Y_1 = y$ if $\bar{\tau}_1 \wedge \tau_1 = \tau_1$. It is evident that ruin does not occur up to the time $\bar{\tau}_1 \wedge \tau_1$ and it will not take place after that time if and only if one of the following two conditions holds:

– $\bar{\tau}_1 \wedge \tau_1 = \bar{\tau}_1$ and ruin does not occur on the time interval $(s, +\infty)$ with the initial surplus $x\mathrm{e}^{rs} + y$;

– $\bar{\tau}_1 \wedge \tau_1 = \tau_1$, $y \le x\mathrm{e}^{rs}$ and ruin does not occur on the time interval $(s, +\infty)$ with the initial surplus $x\mathrm{e}^{rs} - y$.

Next, $\mathbb{P}\left[\bar{\tau}_1 \wedge \tau_1 = \bar{\tau}_1\right] = \bar{\lambda}/(\bar{\lambda} + \lambda)$, $\mathbb{P}\left[\bar{\tau}_1 \wedge \tau_1 = \tau_1\right] = \lambda/(\bar{\lambda} + \lambda)$ and $\bar{\tau}_1 \wedge \tau_1$ is exponentially distributed with mean $1/(\bar{\lambda} + \lambda)$. Hence, by the law of total probability, for all $x \ge 0$, we get

$$\begin{aligned}\varphi(x) = \int_0^{+\infty} \mathrm{e}^{-(\bar{\lambda}+\lambda)s} \bigg(\bar{\lambda} \int_0^{+\infty} \varphi\big(x\mathrm{e}^{rs} + y\big)\, \mathrm{d}\bar{F}(y) \\ + \lambda \int_0^{x\mathrm{e}^{rs}} \varphi\big(x\mathrm{e}^{rs} - y\big)\, \mathrm{d}F(y)\bigg)\, \mathrm{d}s.\end{aligned} \qquad [3.6]$$

We see at once that [3.4] follows immediately from [3.6] for $x = 0$. Changing the variable $x\mathrm{e}^{rs} = u$ in the outer integral in the right-hand side of [3.6] for $x > 0$ we obtain

$$\begin{aligned}\varphi(x) = \frac{x^{(\bar{\lambda}+\lambda)/r}}{r} \int_x^{+\infty} \left(\frac{1}{u}\right)^{(\bar{\lambda}+\lambda)/r+1} \\ \times \bigg(\bar{\lambda} \int_0^{+\infty} \varphi(u + y)\, \mathrm{d}\bar{F}(y) + \lambda \int_0^{u} \varphi(u - y)\, \mathrm{d}F(y)\bigg)\, \mathrm{d}u.\end{aligned} \qquad [3.7]$$

Since the inner integrals in the right-hand side of [3.7] are non-decreasing and bounded functions of u, the integrand in the outer integral is integrable on any $[x, +\infty)$, where $x > 0$. Therefore, $\varphi(x)$ is continuous on $(0, +\infty)$, which proves the first assertion of the theorem.

By Lebesgue's decomposition theorem, we have

$$F(y) = F^{\mathrm{c}}(y) + F^{\mathrm{d}}(y),$$

where $F^{\mathrm{c}}(y)$ and $F^{\mathrm{d}}(y)$ are continuous and discrete components of $F(y)$, respectively.

Denote by $(y_k)_{k\geq 1}$ a sequence of all points of discontinuity of $F(y)$, if any. Set $p_k = F(y_k) - \lim_{y\uparrow y_k} F(y)$. Then, we can rewrite [3.7] as

$$\varphi(x) = \varphi_1(x) + \varphi_2(x), \tag{3.8}$$

where

$$\varphi_1(x) = \frac{x^{(\bar{\lambda}+\lambda)/r}}{r} \int_x^{+\infty} \left(\frac{1}{u}\right)^{(\bar{\lambda}+\lambda)/r+1} \left(\bar{\lambda}\bar{I}(u) + \lambda I(u)\right) \mathrm{d}u, \tag{3.9}$$

$$\begin{aligned} \varphi_2(x) = \frac{\lambda x^{(\bar{\lambda}+\lambda)/r}}{r} \int_x^{+\infty} \left(\frac{1}{u}\right)^{(\bar{\lambda}+\lambda)/r+1} \\ \times \left(\sum_{k:\, y_k \leq u} p_k\, \varphi(u - y_k) \right) \mathrm{d}u, \end{aligned} \tag{3.10}$$

$$\bar{I}(u) = \int_0^{+\infty} \varphi(u + y)\, \mathrm{d}\bar{F}(y), \quad I(u) = \int_0^u \varphi(u - y)\, \mathrm{d}F^{\mathrm{c}}(y).$$

The continuity of $I(u)$ is established in the proof of theorem 2.1. Next, for all $u > 0$, $u_0 > 0$ and $M > 0$, we have

$$\begin{aligned} \left|\bar{I}(u) - \bar{I}(u_0)\right| &\leq \int_0^{+\infty} \left|\varphi(u + y) - \varphi(u_0 + y)\right| \mathrm{d}\bar{F}(y) \\ &\leq \left(1 - \bar{F}(M)\right) + \sup_{y\in[0,\, M]} \left|\varphi(u + y) - \varphi(u_0 + y)\right|. \end{aligned} \tag{3.11}$$

Choosing a large enough value of M, we can first make $1 - \bar{F}(M)$ small enough and then can make the second summand in [3.11] small enough when u is close to u_0. This proves the continuity of $\bar{I}(u)$ on $(0, +\infty)$.

Therefore, the integrand in the right-hand side of [3.9] is continuous on $(0,+\infty)$ w.r.t. u. By [3.9], theorem A.1 and remark A.1, $\varphi_1(x)$ is differentiable on $(0,+\infty)$ and

$$\varphi_1'(x) = \frac{\bar{\lambda}+\lambda}{rx}\,\varphi_1(x) - \frac{\bar{\lambda}\bar{I}(x) + \lambda I(x)}{rx}. \qquad [3.12]$$

Since the integrand in the right-hand side of [3.10] is right-continuous on $(0,+\infty)$ w.r.t. u, there is a right derivative of $\varphi_2(x)$ on $(0,+\infty)$ (we denote it by $\big(\varphi_2(x)\big)'_+$) and

$$\big(\varphi_2(x)\big)'_+ = \frac{\bar{\lambda}+\lambda}{rx}\,\varphi_2(x) - \frac{\lambda}{rx}\sum_{k\colon y_k\le x} p_k\,\varphi(x-y_k). \qquad [3.13]$$

Since the Riemann integral does not depend on values of the integrand at a countable number of points, we can rewrite [3.10] as

$$\begin{aligned}\varphi_2(x) = {} & \frac{\lambda x^{(\bar{\lambda}+\lambda)/r}}{r}\int_x^{+\infty}\left(\frac{1}{u}\right)^{(\bar{\lambda}+\lambda)/r+1} \\ & \times\left(\sum_{k\colon y_k<u} p_k\,\varphi(u-y_k)\right)du.\end{aligned} \qquad [3.14]$$

The integrand in the right-hand side of [3.14] is left-continuous on $(0,+\infty)$ w.r.t. u. Hence, there is a left derivative of $\varphi_2(x)$ on $(0,+\infty)$ (we denote it by $\big(\varphi_2(x)\big)'_-$) and

$$\big(\varphi_2(x)\big)'_- = \frac{\bar{\lambda}+\lambda}{rx}\,\varphi_2(x) - \frac{\lambda}{rx}\sum_{k\colon y_k<x} p_k\,\varphi(x-y_k). \qquad [3.15]$$

From [3.13] and [3.15] we deduce that

$$\big(\varphi_2(x)\big)'_- - \big(\varphi_2(x)\big)'_+ = \frac{\lambda p_k\,\varphi(0)\,\mathbb{1}_{\{x=y_k\}}}{rx} \qquad [3.16]$$

for $x>0$. If $\bar{\lambda}\bar{\mu} > \lambda\mu$, then $\lim_{x\to+\infty}\varphi(x) = 1$ and $\varphi(0) > 0$ in the risk model with stochastic premiums and without any investments (see section 1.2.2). Moreover, it is easily seen that at any time the surplus in the risk model with investments in the risk-free asset is not less than the surplus in the corresponding risk model without investments. Therefore, investments in the risk-free asset do not decrease the survival probability. From this we get $\lim_{x\to+\infty}\varphi(x) = 1$ and $\varphi(0) > 0$ for the model with investments.

Therefore, from [3.16] we see that $\varphi_2(x)$ is differentiable on $(0, +\infty)$, except at positive points y_k, $k \geq 1$. Hence, taking into account [3.8] we conclude that $\varphi(x)$ is also differentiable on $(0, +\infty)$, except at positive points of discontinuity of $F(y)$, and [3.2] holds. Furthermore, from [3.8], [3.12] and [3.13] we get

$$\begin{aligned}\varphi'_+(x) &= \frac{\bar{\lambda}+\lambda}{rx}\,(\varphi_1(x)+\varphi_2(x)) - \frac{\bar{\lambda}}{rx}\,\bar{I}(x) \\ &\quad - \frac{\lambda}{rx}\left(I(x) + \sum_{k:\,y_k\leq x} p_k\,\varphi(x-y_k)\right) \\ &= \frac{\bar{\lambda}+\lambda}{rx}\,\varphi(x) - \frac{\bar{\lambda}}{rx}\int_0^{+\infty}\varphi(x+y)\,\mathrm{d}\bar{F}(y) - \frac{\lambda}{rx}\int_0^{x}\varphi(x-y)\,\mathrm{d}F(y),\end{aligned}$$

which gives [3.3]. Since the right-hand side of [3.3] is continuous on $(0, +\infty)$, except at positive points of discontinuity of $F(y)$, we deduce that $\varphi'_+(x)$ is also continuous on $(0, +\infty)$, except at these points. Bound [3.5] follows easily from [3.3].

What is left is to show that $\varphi(x)$ is right-continuous at $x = 0$. Indeed, taking into account [3.4] and [3.7], the differentiability of $\varphi(x)$ in the right semi-neighborhood of the point $x = 0$ and L'Hopital's rule we have

$$\begin{aligned}\lim_{x\downarrow 0}\varphi(x) &= \lim_{x\downarrow 0}\frac{1}{rx^{-(\bar{\lambda}+\lambda)/r}}\int_x^{+\infty}\left(\frac{1}{u}\right)^{(\bar{\lambda}+\lambda)/r+1} \\ &\quad \times\left(\bar{\lambda}\int_0^{+\infty}\varphi(u+y)\,\mathrm{d}\bar{F}(y) + \lambda\int_0^{u}\varphi(u-y)\,\mathrm{d}F(y)\right)\mathrm{d}u \\ &= \lim_{x\downarrow 0}\frac{1}{(\bar{\lambda}+\lambda)\,x^{-(\bar{\lambda}+\lambda)/r-1}}\left(\frac{1}{x}\right)^{(\bar{\lambda}+\lambda)/r+1} \\ &\quad \times\left(\bar{\lambda}\int_0^{+\infty}\varphi(x+y)\,\mathrm{d}\bar{F}(y) + \lambda\int_0^{x}\varphi(x-y)\,\mathrm{d}F(y)\right) \\ &= \frac{\bar{\lambda}}{\bar{\lambda}+\lambda}\int_0^{+\infty}\varphi(y)\,\mathrm{d}\bar{F}(y) = \varphi(0),\end{aligned}$$

which completes the proof. □

REMARK 3.1.– In contrast to the classical risk model, theorem 3.1 is not true for $r = 0$. The continuity and differentiability of the infinite-horizon survival probability in the risk model with stochastic premiums and without investments is considered in section 3.2.2.

REMARK 3.2.– Theorem 3.1 does not indicate whether the right derivative of $\varphi(x)$ exists at $x = 0$ and $\varphi'(x)$ is bounded in the right semi-neighborhood of this point. Proposition 3.1 below yields sufficient conditions for existence of the right derivative of $\varphi(x)$ at $x = 0$ and boundedness of $\varphi'(x)$ for all $x \in \mathbb{R}_+$.

PROPOSITION 3.1.– Let the surplus process $\left(X_t(x)\right)_{t\geq 0}$ follow [3.1] under the above assumptions with $r < \bar{\lambda} + \lambda$. If $\bar{Y}_i$ and Y_i, $i \geq 1$, have p.d.f.'s $\bar{f}(y)$ and $f(y)$, respectively, which have the derivatives $\bar{f}'(y)$ and $f'(y)$ on $\mathbb{R}_+$, such that $|\bar{f}'(y)|$ and $|f'(y)|$ are integrable and bounded on $\mathbb{R}_+$, then $\varphi(x)$ is continuously differentiable on $\mathbb{R}_+$ and

$$\sup_{x\in[0,+\infty)} |\varphi'(x)| \leq \frac{|\bar{\lambda}\bar{f}(0) - \lambda f(0)| + \int_0^{+\infty} (\bar{\lambda}|\bar{f}'(y)| + \lambda|f'(y)|)\, dy}{\bar{\lambda} + \lambda - r}. \qquad [3.17]$$

PROOF.– If $\bar{Y}_i$ and Y_i, $i \geq 1$, have p.d.f.'s $\bar{f}(y)$ and $f(y)$, respectively, then for all $x \geq 0$, we can rewrite [3.6] as

$$\begin{aligned}\varphi(x) = \int_0^{+\infty} \mathrm{e}^{-(\bar{\lambda}+\lambda)s} \Bigg(\bar{\lambda} \int_{x\mathrm{e}^{rs}}^{+\infty} \varphi(y)\, \bar{f}\left(y - x\mathrm{e}^{rs}\right) \mathrm{d}y \\ + \lambda \int_0^{x\mathrm{e}^{rs}} \varphi(y)\, f\left(x\mathrm{e}^{rs} - y\right) \mathrm{d}y \Bigg)\, \mathrm{d}s.\end{aligned} \qquad [3.18]$$

If the assumptions of the proposition hold, then by [3.18] and theorems A.3, A.4 and A.5, it follows that $\varphi(x)$ is continuously differentiable on $\mathbb{R}_+$ and

$$\begin{aligned}\varphi'(x) = \int_0^{+\infty} & \mathrm{e}^{-(\bar{\lambda}+\lambda-r)s} \\ & \times\Bigg(-\bar{\lambda}\Bigg(\bar{f}(0)\, \varphi\left(x\mathrm{e}^{rs}\right) + \int_{x\mathrm{e}^{rs}}^{+\infty} \varphi(y)\, \bar{f}'\left(y - x\mathrm{e}^{rs}\right) \mathrm{d}y\Bigg) \\ & +\lambda\Bigg(f(0)\, \varphi\left(x\mathrm{e}^{rs}\right) + \int_0^{x\mathrm{e}^{rs}} \varphi(y)\, f'\left(x\mathrm{e}^{rs} - y\right) \mathrm{d}y\Bigg)\Bigg)\, \mathrm{d}s.\end{aligned} \qquad [3.19]$$

From [3.19] we conclude that

$$|\varphi'(x)| \leq \int_0^{+\infty} \mathrm{e}^{-(\bar{\lambda}+\lambda-r)s}$$
$$\times \left(|\bar{\lambda}\bar{f}(0) - \lambda f(0)| + \int_0^{+\infty} (\bar{\lambda}|\bar{f}'(y)| + \lambda|f'(y)|)\,\mathrm{d}y \right)$$

for all $x \geq 0$, which yields [3.17]. This proves the proposition. □

3.2.2. *Zero interest rate*

Let the *surplus process* $\left(X_t(x)\right)_{t\geq 0}$ be given by

$$X_t(x) = x + \sum_{i=1}^{\bar{N}_t} \bar{Y}_i - \sum_{i=1}^{N_t} Y_i, \quad t \geq 0. \qquad [3.20]$$

In [BOI 03], it was shown that $\varphi(x)$ satisfies the following integral equation:

$$(\bar{\lambda} + \lambda)\varphi(x) = \bar{\lambda} \int_0^{+\infty} \varphi(x+y)\,\mathrm{d}\bar{F}(y) + \lambda \int_0^x \varphi(x-y)\,\mathrm{d}F(y) \qquad [3.21]$$

on $\mathbb{R}_+$, which is an analogue of integro-differential equation [3.3] (see also section 1.2.2).

The analytic properties of the infinite-horizon survival probability in this risk model are quite different from those established in section 3.2.1. Firstly, $\varphi(x)$ may have points of discontinuity if the distributions of premium and claim sizes are discrete (see section 1.2.2 and [BOI 03]). Secondly, $\varphi(x)$ may not be differentiable at some points if the distributions of premium and claim sizes are absolutely continuous, which can be seen from example 3.1 below.

Note that if $\bar{Y}_i$ and Y_i, $i \geq 1$, have p.d.f.'s $\bar{f}(y)$ and $f(y)$, respectively, then we can rewrite [3.21] as

$$\varphi(x) = \frac{\bar{\lambda}}{\bar{\lambda}+\lambda} \int_x^{+\infty} \varphi(y)\,\bar{f}(y-x)\,\mathrm{d}y + \frac{\lambda}{\bar{\lambda}+\lambda} \int_0^x \varphi(y)\,f(x-y)\,\mathrm{d}y. \quad [3.22]$$

EXAMPLE 3.1.– Let the premium and claim sizes be uniformly distributed on $[\bar{y}_*, \bar{y}^*]$ and $[y_*, y^*]$, respectively, where $0 < \bar{y}_* < \bar{y}^* < y_* < y^*$ and $\bar{\lambda}(\bar{y}_* + \bar{y}^*) > \lambda(y_* + y^*)$.

Then from [3.22] we have

$$\varphi(x) = \frac{\bar{\lambda}}{(\bar{\lambda}+\lambda)(\bar{y}^* - \bar{y}_*)} \int_{x+\bar{y}_*}^{x+\bar{y}^*} \varphi(y)\,\mathrm{d}y, \quad 0 \le x \le y_*,$$

$$\varphi(x) = \frac{\bar{\lambda}}{(\bar{\lambda}+\lambda)(\bar{y}^* - \bar{y}_*)} \int_{x+\bar{y}_*}^{x+\bar{y}^*} \varphi(y)\,\mathrm{d}y + \frac{\lambda}{(\bar{\lambda}+\lambda)(y^* - y_*)} \int_{0}^{x-y_*} \varphi(y)\,\mathrm{d}y, \quad y_* \le x \le y^*,$$

$$\varphi(x) = \frac{\bar{\lambda}}{(\bar{\lambda}+\lambda)(\bar{y}^* - \bar{y}_*)} \int_{x+\bar{y}_*}^{x+\bar{y}^*} \varphi(y)\,\mathrm{d}y + \frac{\lambda}{(\bar{\lambda}+\lambda)(y^* - y_*)} \int_{x-y^*}^{x-y_*} \varphi(y)\,\mathrm{d}y, \quad x \ge y^*.$$

From this we conclude that $\varphi(x)$ is continuous on $\mathbb{R}_+$ and differentiable on $[0, y_*]$, $[y_*, y^*]$ and $[y^*, +\infty]$. Moreover, we have

$$\varphi'(x) = \frac{\bar{\lambda}}{(\bar{\lambda}+\lambda)(\bar{y}^* - \bar{y}_*)} \left(\varphi(x+\bar{y}^*) - \varphi(x+\bar{y}_*)\right), \quad 0 \le x \le y_*,$$

$$\varphi'(x) = \frac{\bar{\lambda}}{(\bar{\lambda}+\lambda)(\bar{y}^* - \bar{y}_*)} \left(\varphi(x+\bar{y}^*) - \varphi(x+\bar{y}_*)\right) + \frac{\lambda}{(\bar{\lambda}+\lambda)(y^* - y_*)} \varphi(x - y_*), \quad y_* \le x \le y^*,$$

$$\varphi'(x) = \frac{\bar{\lambda}}{(\bar{\lambda}+\lambda)(\bar{y}^* - \bar{y}_*)} \left(\varphi(x+\bar{y}^*) - \varphi(x+\bar{y}_*)\right) + \frac{\lambda}{(\bar{\lambda}+\lambda)(y^* - y_*)} \left(\varphi(x - y_*) - \varphi(x - y^*)\right), \quad x \ge y^*.$$

Nevertheless, $\varphi(x)$ is not differentiable at the points $x = y_*$ and $x = y^*$. Indeed, from the above, we have

$$\varphi'_+(y_*) - \varphi'_-(y_*) = \frac{\lambda\varphi(0)}{(\bar{\lambda}+\lambda)(y^* - y_*)}$$

and

$$\varphi'_-(y^*) - \varphi'_+(y^*) = \frac{\lambda\varphi(0)}{(\bar{\lambda} + \lambda)(y^* - y_*)}.$$

Furthermore, $\varphi(0) > 0$ since $\bar{\lambda}(\bar{y}_* + \bar{y}^*) > \lambda(y_* + y^*)$ (see section 1.2.2). □

The following proposition gives sufficient conditions for the continuity and differentiability of $\varphi(x)$ in the risk model with stochastic premiums and without any investments.

PROPOSITION 3.2.– Let the surplus process $\big(X_t(x)\big)_{t\geq 0}$ follow [3.20] under the above assumptions.

1) If $\bar{Y}_i$ and Y_i, $i \geq 1$, have p.d.f.'s $\bar{f}(y)$ and $f(y)$, respectively, which are continuous on $\mathbb{R}_+$, then $\varphi(x)$ is continuous on $\mathbb{R}_+$.

2) If the p.d.f.'s $\bar{f}(y)$ and $f(y)$ of $\bar{Y}_i$ and Y_i, $i \geq 1$, respectively, have the derivatives $\bar{f}'(y)$ and $f'(y)$ on $\mathbb{R}_+$, such that $|\bar{f}'(y)|$ and $|f'(y)|$ are integrable and bounded on $\mathbb{R}_+$, then $\varphi(x)$ is continuously differentiable on $\mathbb{R}_+$ and

$$\sup_{x\in[0,+\infty)} |\varphi'(x)| \leq \frac{\big|\bar{\lambda}\bar{f}(0) - \lambda f(0)\big| + \int_0^{+\infty}\big(\bar{\lambda}|\bar{f}'(y)| + \lambda|f'(y)|\big)\,\mathrm{d}y}{\bar{\lambda} + \lambda}. \quad [3.23]$$

PROOF.– Fix any $x_0 > 0$. We now prove that $\varphi(x)$ is left-continuous at this point. By [3.22], for all $x < x_0$ we have

$$\begin{aligned}\big|\varphi(x_0) - \varphi(x)\big| \leq{}& \frac{\bar{\lambda}}{\bar{\lambda}+\lambda}\bigg(\int_{x_0}^{M} \big|\bar{f}(y - x_0) - \bar{f}(y - x)\big|\,\mathrm{d}y \\ &+ 2\big(1 - \bar{F}(M - x_0)\big) + \bar{F}(x_0 - x)\bigg) \\ &+ \frac{\lambda}{\bar{\lambda}+\lambda}\bigg(\int_0^{x} \big|f(x_0 - y) - f(x - y)\big|\,\mathrm{d}y + F(x_0 - x)\bigg),\end{aligned}$$

where $M > x_0$ is arbitrary. Hence, choosing a large enough value of M, we can first make $2\big(1 - \bar{F}(M - x_0)\big)$ small enough, and then by the uniform continuity of $\bar{f}(y)$ and $f(y)$ on $[0, M]$ and $[0, x_0]$, respectively, we can make $\int_{x_0}^{M}\big|\bar{f}(y - x_0) - \bar{f}(y - x)\big|\,\mathrm{d}y$ and $\int_0^{x}\big|f(x_0 - y) - f(x - y)\big|\,\mathrm{d}y$ small enough when x is close to x_0. Note that we can also make $\bar{F}(x_0 - x)$ and $F(x_0 - x)$

small enough when x is close to x_0. Consequently, $\varphi(x)$ is left-continuous at $x = x_0$.

Similarly, we can show that $\varphi(x)$ is right-continuous at any point $x_0 \geq 0$. Thus, $\varphi(x)$ is continuous on $\mathbb{R}_+$.

If $\bar{f}(y)$ and $f(y)$ have derivatives, which satisfy the conditions of the proposition, then by [3.22], the continuity of $\varphi(x)$ and theorems A.3 and A.4, we conclude that $\varphi(x)$ is continuously differentiable on $\mathbb{R}_+$ and

$$\begin{aligned}\varphi'(x) = &- \frac{\bar{\lambda}}{\bar{\lambda}+\lambda}\left(\bar{f}(0)\,\varphi(x) + \int_x^{+\infty} \varphi(y)\,\bar{f}'(y-x)\,\mathrm{d}y\right) \\ &+ \frac{\lambda}{\bar{\lambda}+\lambda}\left(f(0)\,\varphi(x) + \int_0^{x} \varphi(y)\,f'(x-y)\,\mathrm{d}y\right).\end{aligned} \qquad [3.24]$$

Bound [3.23] follows immediately from [3.24]. This completes the proof. □

3.3. Continuity of the finite-horizon survival probability and existence of its partial derivatives

THEOREM 3.2.– Let the surplus process $(X_t(x))_{t\geq 0}$ follow [3.1] under the above assumptions.

1) If $\bar{Y}_i$ and Y_i, $i \geq 1$, have p.d.f.'s $\bar{f}(y)$ and $f(y)$, respectively, which are continuous on $\mathbb{R}_+$, then $\varphi(x,t)$ is continuous on $\mathbb{R}^2_+$ as a function of two variables.

2) If the p.d.f.'s $\bar{f}(y)$ and $f(y)$ of $\bar{Y}_i$ and Y_i, $i \geq 1$, respectively, have the derivatives $\bar{f}'(y)$ and $f'(y)$ on $\mathbb{R}_+$, such that $|\bar{f}'(y)|$ and $|f'(y)|$ are integrable and bounded on $\mathbb{R}_+$, then:

i) $\varphi(x,t)$ has partial derivatives w.r.t. x and t on $\mathbb{R}^2_+$, which are continuous as functions of two variables;

ii) $\varphi(x,t)$ satisfies the partial integro-differential equation

$$\begin{aligned}&\frac{\partial\varphi(x,t)}{\partial t} - rx\,\frac{\partial\varphi(x,t)}{\partial x} + (\bar{\lambda}+\lambda)\,\varphi(x,t) \\ &- \bar{\lambda}\int_0^{+\infty} \varphi(x+y,\,t)\,\mathrm{d}\bar{F}(y) - \lambda\int_0^{x} \varphi(x-y,\,t)\,\mathrm{d}F(y) = 0\end{aligned} \qquad [3.25]$$

on $\mathbb{R}^2_+$ with the boundary conditions $\varphi(x,0) = 1$ and $\lim_{x\to+\infty} \varphi(x,t) = 1$;

iii) for all $T > 0$, we have

$$\sup_{\substack{x\in[0,+\infty),\\ t\in[0,T]}} \left|\frac{\partial\varphi(x,t)}{\partial x}\right| \leq C_1 C_2, \qquad [3.26]$$

where

$$C_1 = \begin{cases} \dfrac{1-\mathrm{e}^{(r-\bar{\lambda}-\lambda)T}}{\bar{\lambda}+\lambda-r} & \text{if } \ \bar{\lambda}+\lambda \neq r, \\ T & \text{if } \ \bar{\lambda}+\lambda = r, \end{cases}$$

$$C_2 = \left|\bar{\lambda}\bar{f}(0) - \lambda f(0)\right| + \bar{\lambda}\int_0^{+\infty} |\bar{f}'(y)|\,\mathrm{d}y + \lambda\int_0^{+\infty} |f'(y)|\,\mathrm{d}y.$$

PROOF.– Applying arguments similar to those in the proofs of theorems 2.2 and 3.1, for all $x \geq 0$ and $t \geq 0$, we get

$$\begin{aligned}\varphi(x,t) = \int_0^t \mathrm{e}^{-(\bar{\lambda}+\lambda)s}\Big(\bar{\lambda}\int_0^{+\infty} \varphi\big(x\mathrm{e}^{rs}+y,\, t-s\big)\,\mathrm{d}\bar{F}(y) \\ + \lambda\int_0^{x\mathrm{e}^{rs}} \varphi\big(x\mathrm{e}^{rs}-y,\, t-s\big)\,\mathrm{d}F(y)\Big)\,\mathrm{d}s + \mathrm{e}^{-(\bar{\lambda}+\lambda)t}.\end{aligned} \qquad [3.27]$$

Changing the variable $t - s = v$ in the outer integral in the right-hand side of [3.27], we have

$$\begin{aligned}\varphi(x,t) &= \mathrm{e}^{-(\bar{\lambda}+\lambda)t}\int_0^t \mathrm{e}^{(\bar{\lambda}+\lambda)v}\Big(\bar{\lambda}\int_0^{+\infty} \varphi\big(x\mathrm{e}^{r(t-v)}+y,\, v\big)\,\mathrm{d}\bar{F}(y) \\ &+\lambda\int_0^{x\mathrm{e}^{r(t-v)}} \varphi\big(x\mathrm{e}^{r(t-v)}-y,\, v\big)\,\mathrm{d}F(y)\Big)\,\mathrm{d}v + \mathrm{e}^{-(\bar{\lambda}+\lambda)t}. \end{aligned} \qquad [3.28]$$

We now use that $\bar{Y}_i$ and Y_i, $i \geq 1$, have p.d.f.'s $\bar{f}(y)$ and $f(y)$, respectively. Changing the variables $x\mathrm{e}^{r(t-v)} + y = u$ and $x\mathrm{e}^{r(t-v)} - y = u$ in the first and second inner integrals, respectively, in the right-hand side of [3.28] we get

$$\varphi(x,t) = \mathrm{e}^{-(\bar\lambda+\lambda)t} \int_0^t \mathrm{e}^{(\bar\lambda+\lambda)v} \Big(\bar\lambda \int_{x\mathrm{e}^{r(t-v)}}^{+\infty} \varphi(u,v)\, \bar f\big(u - x\mathrm{e}^{r(t-v)}\big)\, \mathrm{d}u$$

$$+\lambda \int_0^{x\mathrm{e}^{r(t-v)}} \varphi(u,v)\, f\big(x\mathrm{e}^{r(t-v)} - u\big)\, \mathrm{d}u \Big)\, \mathrm{d}v + \mathrm{e}^{-(\bar\lambda+\lambda)t}. \qquad [3.29]$$

Let

$$I(x,t) = \int_0^t \mathrm{e}^{(\bar\lambda+\lambda)v} \Big(\bar\lambda \int_{x\mathrm{e}^{r(t-v)}}^{+\infty} \varphi(u,v)\, \bar f\big(u - x\mathrm{e}^{r(t-v)}\big)\, \mathrm{d}u$$

$$+ \lambda \int_0^{x\mathrm{e}^{r(t-v)}} \varphi(u,v)\, f\big(x\mathrm{e}^{r(t-v)} - u\big)\, \mathrm{d}u \Big)\, \mathrm{d}v.$$

If $\bar f(y)$ and $f(y)$ are continuous on $\mathbb{R}_+$, then we can show the continuity of $\varphi(x,t)$ as a function of two variables applying arguments similar to those in the proof of theorem 2.2. Indeed, for any fixed $(x_0,t_0) \in (0,+\infty) \times (0,+\infty)$ and all $0 \le x \le x_0$ and $0 \le t \le t_0$, we have

$$\big|I(x,t) - I(x_0,t_0)\big| \le \bar\lambda \int_0^t \mathrm{e}^{(\bar\lambda+\lambda)v} \Big(\int_{x_0\,\mathrm{e}^{r(t_0-v)}}^{M} \big|\bar f\big(u - x_0\,\mathrm{e}^{r(t_0-v)}\big)$$

$$- \bar f\big(u - x\mathrm{e}^{r(t-v)}\big)\big|\, \mathrm{d}u + 2\big(1 - \bar F\big(M - x_0\,\mathrm{e}^{rt_0}\big)\big) \Big)\, \mathrm{d}v$$

$$+ \bar\lambda \int_0^t \mathrm{e}^{(\bar\lambda+\lambda)v} \int_{x\mathrm{e}^{r(t-v)}}^{x_0\,\mathrm{e}^{r(t_0-v)}} \bar f\big(u - x\mathrm{e}^{r(t-v)}\big)\, \mathrm{d}u\, \mathrm{d}v$$

$$+ \lambda \int_0^t \mathrm{e}^{(\bar\lambda+\lambda)v} \int_{x\mathrm{e}^{r(t-v)}}^{x_0\,\mathrm{e}^{r(t_0-v)}} f\big(x_0\,\mathrm{e}^{r(t_0-v)} - u\big)\, \mathrm{d}u\, \mathrm{d}v$$

$$+ \lambda \int_0^t \mathrm{e}^{(\bar\lambda+\lambda)v} \int_0^{x\mathrm{e}^{r(t-v)}} \big|f\big(x_0\,\mathrm{e}^{r(t_0-v)} - u\big) \qquad [3.30]$$

$$- f\big(x\mathrm{e}^{r(t-v)} - u\big)\big|\, \mathrm{d}u\, \mathrm{d}v$$

$$+ \int_t^{t_0} \mathrm{e}^{(\bar\lambda+\lambda)v} \Big(\bar\lambda \int_{x_0\,\mathrm{e}^{r(t_0-v)}}^{+\infty} \bar f\big(u - x_0\,\mathrm{e}^{r(t_0-v)}\big)\, \mathrm{d}u$$

$$+ \lambda \int_0^{x_0\,\mathrm{e}^{r(t_0-v)}} f\big(x_0\,\mathrm{e}^{r(t_0-v)-u}\big)\, \mathrm{d}u \Big) \mathrm{d}v,$$

where $M > x_0\,\mathrm{e}^{rt_0}$ is arbitrary. Consequently, choosing a large enough value of M we can first make $2\big(1-\bar{F}\big(M - x_0\,\mathrm{e}^{rt_0}\big)\big)$ small enough and then by the continuity of $\bar{f}(y)$ and $f(y)$, we can make the right-hand side of [3.30] small enough when x is close to x_0 and t is close to t_0. Thus, $\varphi(x,t)$ is continuous at (x_0,t_0) for $x \le x_0$ and $t \le t_0$. Similarly, we can consider other cases.

If $\bar{f}(y)$ and $f(y)$ have the derivatives $\bar{f}'(y)$ and $f'(y)$ on $\mathbb{R}_+$, such that $|\bar{f}'(y)|$ and $|f'(y)|$ are integrable and bounded on $\mathbb{R}_+$, then we can show the existence of partial derivatives of $\varphi(x,t)$ as in the proof of theorem 2.2. Moreover, from [3.29] we conclude that

$$\begin{aligned}\frac{\partial\varphi(x,t)}{\partial x} &= \mathrm{e}^{(r-\bar{\lambda}-\lambda)t}\int_0^t \mathrm{e}^{(\bar{\lambda}+\lambda-r)v}\bigg(-\bar{\lambda}\bigg(\bar{f}(0)\,\varphi\big(x\mathrm{e}^{r(t-v)},\,v\big)\\ &\quad+\int_{x\mathrm{e}^{r(t-v)}}^{+\infty}\varphi(u,v)\,\bar{f}'\big(u - x\mathrm{e}^{r(t-v)}\big)\,\mathrm{d}u\bigg)\\ &\quad+\lambda\bigg(f(0)\,\varphi\big(x\mathrm{e}^{r(t-v)},\,v\big)\\ &\quad+\int_0^{x\mathrm{e}^{r(t-v)}}\varphi(u,v)\,f'\big(x\mathrm{e}^{r(t-v)} - u\big)\,\mathrm{d}u\bigg)\bigg)\mathrm{d}v\end{aligned} \qquad [3.31]$$

and

$$\begin{aligned}\frac{\partial\varphi(x,t)}{\partial t} &= -\,(\bar{\lambda}+\lambda)\,\varphi(x,t) + \bar{\lambda}\int_x^{+\infty}\varphi(u,t)\,\bar{f}(u-x)\,\mathrm{d}u\\ &\quad+\lambda\int_0^x \varphi(u,t)\,f(x-u)\,\mathrm{d}u\\ &\quad+rx\,\mathrm{e}^{(r-\bar{\lambda}-\lambda)t}\int_0^t \mathrm{e}^{(\bar{\lambda}+\lambda-r)v}\bigg(-\bar{\lambda}\bigg(\bar{f}(0)\,\varphi\big(x\mathrm{e}^{r(t-v)},\,v\big)\\ &\quad+\int_{x\mathrm{e}^{r(t-v)}}^{+\infty}\varphi(u,v)\,\bar{f}'\big(u - x\mathrm{e}^{r(t-v)}\big)\,\mathrm{d}u\bigg)\\ &\quad+\lambda\bigg(f(0)\,\varphi\big(x\mathrm{e}^{r(t-v)},\,v\big)\\ &\quad+\int_0^{x\mathrm{e}^{r(t-v)}}\varphi(u,v)\,\bar{f}'\big(x\mathrm{e}^{r(t-v)} - u\big)\,\mathrm{d}u\bigg)\bigg)\,\mathrm{d}v.\end{aligned} \qquad [3.32]$$

Combining [3.31] and [3.32], we have

$$\begin{aligned}\frac{\partial \varphi(x,t)}{\partial t} &= -(\bar{\lambda}+\lambda)\,\varphi(x,t) + \bar{\lambda}\int_x^{+\infty} \varphi(u,t)\,\bar{f}(u-x)\,\mathrm{d}u \\ &\quad + \lambda\int_0^x \varphi(u,t)\,f(x-u)\,\mathrm{d}u + rx\,\frac{\partial \varphi(x,t)}{\partial x} \\ &= rx\,\frac{\partial \varphi(x,t)}{\partial x} - (\bar{\lambda}+\lambda)\,\varphi(x,t) \\ &\quad + \bar{\lambda}\int_0^{+\infty} \varphi(x+y,\,t)\,\mathrm{d}\bar{F}(y) + \lambda\int_0^x \varphi(x-y,\,t)\,\mathrm{d}F(y),\end{aligned}$$

which yields [3.25].

From [3.31], it follows that

$$\begin{aligned}\left|\frac{\partial \varphi(x,t)}{\partial x}\right| &\le \left(\left|\bar{\lambda}\bar{f}(0) - \lambda f(0)\right| + \bar{\lambda}\int_0^{+\infty} |\bar{f}'(y)|\,\mathrm{d}y + \lambda\int_0^{+\infty} |f'(y)|\,\mathrm{d}y\right) \\ &\quad \times \mathrm{e}^{(r-\bar{\lambda}-\lambda)t}\int_0^t \mathrm{e}^{(\bar{\lambda}+\lambda-r)v}\,\mathrm{d}v\end{aligned}$$

for all $x \ge 0$ and $t \ge 0$, which gives [3.26]. As in theorem 2.2, the boundary condition $\varphi(x,0) = 1$ is evident and the condition $\lim_{x\to+\infty}\varphi(x,t) = 1$ follows immediately from the fact that there can be only a finite number of claims on any finite time interval a.s. This completes the proof. □

REMARK 3.3.– Let $r = 0$. If $\bar{Y}_i$ and Y_i, $i \ge 1$, have p.d.f.'s $\bar{f}(y)$ and $f(y)$, respectively, then by the law of total probability, for all $x \ge 0$ and $t \ge 0$, we have

$$\begin{aligned}\varphi(x,t) &= \mathrm{e}^{-(\bar{\lambda}+\lambda)t}\int_0^t \mathrm{e}^{(\bar{\lambda}+\lambda)v}\left(\bar{\lambda}\int_0^{+\infty}\varphi(x+y,\,v)\,\mathrm{d}\bar{F}(y)\right. \\ &\quad \left. + \lambda\int_0^x \varphi(x-y,\,v)\,\mathrm{d}F(y)\right)\mathrm{d}v + \mathrm{e}^{-(\bar{\lambda}+\lambda)t}.\end{aligned} \qquad [3.33]$$

By [3.33], we can show that if $\bar{f}(y)$ and $f(y)$ are continuous on $\mathbb{R}_+$, then $\varphi(x,t)$ is continuous on $\mathbb{R}_+^2$ as a function of two variables, has a continuous partial derivative w.r.t. t and satisfies the partial integro-differential equation

$$\frac{\partial\varphi(x,t)}{\partial t} + (\bar{\lambda}+\lambda)\,\varphi(x,t) - \bar{\lambda}\int_0^{+\infty}\varphi(x+y,\,t)\,\mathrm{d}\bar{F}(y) - \lambda\int_0^{x}\varphi(x-y,\,t)\,\mathrm{d}F(y) = 0$$

on $\mathbb{R}_+^2$ with the boundary conditions $\varphi(x,0) = 1$ and $\lim_{x\to+\infty}\varphi(x,t) = 1$. Nevertheless, we can guarantee the existence of a partial derivative w.r.t. x provided that the conditions of theorem 3.2 hold. In this case, bound [3.26] is also true with $r = 0$ in the definition of C_1. □

REMARK 3.4.– As in theorem 2.2, the requirements that we impose on the distribution of the premium and claim sizes in theorem 3.2 are not necessary for the continuity of $\varphi(x,t)$ or existence of its partial derivatives (see also remark 2.4). On the other hand, $\varphi(x,t)$ may not have partial derivatives in some cases.

4

Classical Risk Model with a Franchise and a Liability Limit

In this chapter, we deal with the classical risk model under the additional assumption that an insurance company applies a franchise and a liability limit. To be more precise, we consider three cases: the insurance company establishes a franchise only, a liability limit only and both a franchise and a liability limit. Assuming that claim sizes are exponentially distributed we find analytic expressions for the infinite-horizon survival probabilities in these cases. The expressions turn out different on certain intervals. Moreover, we investigate how a franchise and a liability limit change the survival probability for small and large enough initial surpluses.

4.1. Introduction

We consider the *classical risk model* described in section 1.1.1. Recall that in this case the *surplus process* $\big(X_t(x)\big)_{t\geq 0}$ follows

$$X_t(x) = x + ct - \sum_{i=1}^{N_t} Y_i, \quad t \geq 0. \qquad [4.1]$$

Let the *net profit condition* hold, i.e. $c > \lambda\mu$, throughout this chapter. Moreover, we assume that the insurance company uses the *expected value principle* for premium calculation, which means that $c = \lambda\mu(1+\theta)$, where $\theta > 0$ is a *safety loading*.

It is well known that $\varphi(x)$ is a solution to the integro-differential equation

$$c\varphi'_+(x) = \lambda\varphi(x) - \lambda \int_0^x \varphi(x-y)\,\mathrm{d}F(y), \tag{4.2}$$

where $\varphi'_+(x)$ is the right derivative of $\varphi(x)$. The question concerning the differentiability of $\varphi(x)$ is investigated in [ROL 99, pp. 162–163] and section 2.3 in different ways (see also section 1.1.3). Equation [4.2] has a unique solution, such that $\lim_{x\to+\infty}\varphi(x) = 1$, provided that the net profit condition holds, otherwise ruin on an infinite horizon occurs with probability 1. Moreover, it is known that the condition $\lim_{x\to+\infty}\varphi(x) = 1$ is true for $\varphi(x)$ if and only if the condition $\varphi(0) = 1 - \lambda\mu/c$, which is equivalent to $\varphi(0) = \theta/(1+\theta)$, holds (see sections 1.1.2 and 1.1.4). So, if we find any solution to [4.2] with $\varphi(0) = \theta/(1+\theta)$, then we can be sure that this solution is the survival probability.

If the claim sizes are exponentially distributed, then the closed form solution to [4.2] can be found and

$$\varphi(x) = 1 - \frac{1}{1+\theta}\exp\left(-\frac{\theta x}{\mu(1+\theta)}\right). \tag{4.3}$$

In this chapter, we will deal with the classical risk model under the additional assumption that the insurance company uses a franchise and a liability limit. A *franchise* is a provision in an insurance policy whereby an insurer does not pay unless damage exceeds the franchise amount. It is applied to prevent a large number of trivial claims. A *liability limit* determines the maximum amount that is paid by an insurer. It is used to restrict the insurer's liability to the insured.

Let d and L be franchise and liability limit amounts, respectively. We choose them at the initial time and do not change them later. We make the following natural assumption concerning these amounts: $0 \le d < L \le +\infty$. In particular, if $d = 0$, then a franchise is not used and if $L = +\infty$, then a liability limit is not used. Let $Y_i^{(d,\,L)}$, $i \ge 1$, denote an insurance compensation for the ith claim provided that the franchise and liability limit amounts are d and L. We let $F^{(d,\,L)}(y)$ stand for the c.d.f. of $Y_i^{(d,\,L)}$.

Normally, a franchise and a liability limit also imply reduction of insurance premiums. We suppose that the safety loading $\theta > 0$ is constant. Thus, the premium intensity is given by

$$c^{(d,\,L)} = \lambda(1+\theta)\,\mathbb{E}\big[Y_i^{(d,\,L)}\big]$$

provided that the insurance company uses the expected value principle for premium calculation.

Let $X_t^{(d,\,L)}(x)$ be the surplus of the insurance company at time t provided that its initial surplus is x and the franchise and liability limit amounts are d and L, respectively. Then, [4.1] for the *surplus process* $\big(X_t^{(d,\,L)}(x)\big)_{t\geq 0}$ can be rewritten as follows:

$$X_t^{(d,\,L)}(x) = x + c^{(d,\,L)}t - \sum_{i=1}^{N_t} Y_i^{(d,\,L)}, \quad t \geq 0. \qquad [4.4]$$

Let $\varphi^{(d,\,L)}(x)$ denote the corresponding *infinite-horizon survival probability.*

In what follows, we deal with exponentially distributed claim sizes only. This is one of the few cases where we succeed in finding an analytic expression for the survival probability. It is easily seen that the c.d.f. of the insurance compensation is a sum of absolutely continuous and discrete components when the claim sizes are exponentially distributed and the insurance company uses the franchise and liability limit. That is why analytic expressions for the survival probability turn out different on certain intervals as we will show later. To get them, we apply the results of Chapter 2, namely theorem 2.1 and remark 2.1 (see also section 1.1.3 and [ROL 99, pp. 162–163]). Furthermore, we investigate how a franchise and a liability limit change the survival probability for small and large enough initial surpluses.

The rest of the chapter is organized as follows. In section 4.2, we consider the case where the insurance company establishes a franchise only. In section 4.3, we suppose that the insurance company applies a liability limit only. Section 4.4 deals with the case when both a franchise and a liability limit are established. The bibliographical notes concerning approaches to calculation of the ruin or survival probabilities in the classical risk model are given in section 1.1.9. For details concerning various benefits that can be provided by insurance policies, see, for example [PIT 14] and references therein.

4.2. Survival probability in the classical risk model with a franchise

If the insurance company establishes a franchise only and the claim sizes are exponentially distributed, then equation [4.2] for $\varphi^{(d,+\infty)}(x)$ can be written as

$$
\begin{aligned}
&c^{(d,+\infty)}\left(\varphi^{(d,+\infty)}(x)\right)'_+ \\
&\quad = \lambda\varphi^{(d,+\infty)}(x) - \lambda\int_0^x \varphi^{(d,+\infty)}(x-y)\,\mathrm{d}F^{(d,+\infty)}(y),
\end{aligned}
\tag{4.5}
$$

where

$$
F^{(d,+\infty)}(y) = \begin{cases} 0 & \text{if } y<0, \\ 1-\mathrm{e}^{-d/\mu} & \text{if } 0\le y<d, \\ 1-\mathrm{e}^{-y/\mu} & \text{if } y\ge d, \end{cases}
$$

and

$$
\begin{aligned}
c^{(d,+\infty)} &= \lambda(1+\theta)\,\mathbb{E}\big[Y_i^{(d,+\infty)}\big] \\
&= \lambda(1+\theta)\int_d^{+\infty} \frac{y\,\mathrm{e}^{-y/\mu}}{\mu}\,\mathrm{d}y = \lambda(1+\theta)(\mu+d)\,\mathrm{e}^{-d/\mu}.
\end{aligned}
$$

In section 4.2.1, we derive analytic expressions for $\varphi^{(d,+\infty)}(x)$ in the case of exponentially distributed claim sizes. In section 4.2.2, we investigate how a franchise changes the survival probability for small and large enough initial surpluses. Note that throughout this chapter all sums equal 0 provided that their lower summation indices are greater than the upper ones.

4.2.1. *Analytic expression for the survival probability*

To formulate the next theorem, introduce the constants

$$
\gamma = (1+\theta)(\mu+d),
$$

$$
C_{1,1} = \frac{\theta}{1+\theta},
$$

$$A_{2,0} = -\frac{\theta}{(1+\theta)(\gamma+\mu)}\,\mathrm{e}^{-d/\gamma},$$

$$C_{2,1} = \frac{\theta}{1+\theta}\left(1 + \frac{\gamma\mu + d(\gamma+\mu)}{(\gamma+\mu)^2}\,\mathrm{e}^{-d/\gamma}\right),$$

$$C_{2,2} = -\frac{\theta\gamma\mu}{(1+\theta)(\gamma+\mu)^2}\,\mathrm{e}^{d/\mu}.$$

Moreover, let the constants $A_{n+1,\,j}$, $0 \le j \le n-1$, be given in a recurrent way by formulas

$$A_{n+1,\,n-1} = -\frac{A_{n,\,n-2}}{n(\gamma+\mu)}\,\mathrm{e}^{-d/\gamma}, \quad n \ge 2, \tag{4.6}$$

$$\begin{aligned} A_{n+1,\,j} &= -\frac{1}{\gamma+\mu}\left[(j+2)\gamma\mu A_{n+1,\,j+1} + \frac{1}{j+1}\right. \\ &\quad \left.\times\left(\sum_{i=j-1}^{n-2} A_{n,\,i}\,(-d)^{i-j+1}\binom{i+1}{j}\right)\mathrm{e}^{-d/\gamma}\right], \\ &\quad 1 \le j \le n-2, \quad n \ge 3, \end{aligned} \tag{4.7}$$

$$\begin{aligned} A_{n+1,\,0} &= -\frac{1}{\gamma+\mu}\left[2\gamma\mu A_{n+1,\,1}\right. \\ &\quad \left. + \left(C_{n,\,1} + \sum_{i=0}^{n-2} A_{n,\,i}\,(-d)^{i+1}\right)\mathrm{e}^{-d/\gamma}\right], \quad n \ge 2. \end{aligned} \tag{4.8}$$

Next, let the constants $B_{n+1,\,j}$, $0 \le j \le n-2$, be given in a recurrent way by formulas

$$B_{3,0} = \frac{C_{2,2}}{\gamma+\mu}\,\mathrm{e}^{d/\mu},$$

$$B_{n+1,\,n-2} = \frac{B_{n,\,n-3}}{(n-1)(\gamma+\mu)}\,\mathrm{e}^{d/\mu}, \quad n \ge 3, \tag{4.9}$$

$$B_{n+1,\,j} = \frac{1}{\gamma+\mu}\left[(j+2)\gamma\mu B_{n+1,\,j+1} + \frac{1}{j+1}\right.$$
$$\left.\times\left(\sum_{i=j-1}^{n-3} B_{n,\,i}\,(-d)^{i-j+1}\binom{i+1}{j}\right)\mathrm{e}^{d/\mu}\right], \qquad [4.10]$$
$$1 \le j \le n-3, \quad n \ge 4,$$

$$B_{n+1,\,0} = \frac{1}{\gamma+\mu}\left[2\gamma\mu B_{n+1,\,1}\right.$$
$$\left.+\left(C_{n,\,2} + \sum_{i=0}^{n-3} B_{n,\,i}\,(-d)^{i+1}\right)\mathrm{e}^{d/\mu}\right], \quad n \ge 3. \qquad [4.11]$$

Finally, let the constants $C_{n+1,\,1}$ and $C_{n+1,\,2}$ be given by formulas

$$C_{n+1,\,1} = C_{n,\,1} + \frac{\gamma\mu(A_{n,\,0} - A_{n+1,\,0})}{\gamma+\mu} + \sum_{i=0}^{n-3}\Bigg(A_{n,\,i} - A_{n+1,\,i}$$
$$+\frac{(i+2)\gamma\mu(A_{n,\,i+1} - A_{n+1,\,i+1})}{\gamma+\mu}\Bigg)(nd)^{i+1}$$
$$+\left(A_{n,\,n-2} - A_{n+1,\,n-2} - \frac{n\gamma\mu A_{n+1,\,n-1}}{\gamma+\mu}\right)(nd)^{n-1}$$
$$- A_{n+1,\,n-1}\,(nd)^{n} \qquad [4.12]$$
$$+\frac{\gamma\mu}{\gamma+\mu}\Bigg(\sum_{i=0}^{n-3}(i+1)(B_{n,\,i} - B_{n+1,\,i})(nd)^{i}$$
$$-(n-1)B_{n+1,\,n-2}\,(nd)^{n-2}\Bigg)\exp\left(-nd\,\frac{\gamma+\mu}{\gamma\mu}\right),$$
$$n \ge 2,$$

$$C_{3,\,2} = C_{2,\,2} + \frac{\gamma\mu B_{3,\,0}}{\gamma+\mu} - 2dB_{3,\,0}$$
$$+\frac{\gamma\mu(A_{3,\,0} - A_{2,\,0} + 4dA_{3,\,1})}{\gamma+\mu}\exp\left(2d\,\frac{\gamma+\mu}{\gamma\mu}\right),$$

$$\begin{aligned} C_{n+1,2} = C_{n,2} &+ \frac{\gamma\mu(B_{n+1,0} - B_{n,0})}{\gamma+\mu} + \sum_{i=0}^{n-4}\Bigg(B_{n,i} - B_{n+1,i} \\ &+ \frac{(i+2)\gamma\mu(B_{n+1,i+1} - B_{n,i+1})}{\gamma+\mu}\Bigg)(nd)^{i+1} \\ &+ \left(B_{n,n-3} - B_{n+1,n-3} + \frac{(n-1)\gamma\mu B_{n+1,n-2}}{\gamma+\mu}\right) \\ &\times (nd)^{n-2} - B_{n+1,n-2}\,(nd)^{n-1} \\ &+ \frac{\gamma\mu}{\gamma+\mu}\Bigg(\sum_{i=0}^{n-2}(i+1)(A_{n+1,i} - A_{n,i})(nd)^i \\ &+ nA_{n+1,n-1}\,(nd)^{n-1}\Bigg)\exp\left(nd\,\frac{\gamma+\mu}{\gamma\mu}\right), \\ &n \geq 3. \end{aligned} \qquad [4.13]$$

Note that to compute the constants by the formulas above for any $n \geq 2$, we have to know all the constants for $n-1$. Moreover, for any fixed $n \geq 2$, we start from the computation of $A_{n+1,j}$ for j from $n-1$ to 0, $B_{n+1,j}$ for j from $n-2$ to 0, and then we can compute $C_{n+1,1}$ and $C_{n+1,2}$. We introduced all the constants only to formulate the next theorem. We will get them in the proof of this theorem.

THEOREM 4.1.– Let the surplus process $\big(X_t^{(d,+\infty)}(x)\big)_{t\geq 0}$ follow [4.4] under the above assumptions with $0 < d < +\infty$ and $L = +\infty$, and the claim sizes be exponentially distributed with mean μ. Then,

$$\varphi^{(d,+\infty)}(x) = \varphi_{n+1}^{(d,+\infty)}(x) \quad \text{for all} \quad x \in [nd, (n+1)d), \quad n \geq 0,$$

where

$$\varphi_1^{(d,+\infty)}(x) = C_{1,1}\,e^{x/\gamma}, \qquad [4.14]$$

$$\varphi_2^{(d,+\infty)}(x) = \big(C_{2,1} + A_{2,0}\,x\big)\,e^{x/\gamma} + C_{2,2}\,e^{-x/\mu}, \qquad [4.15]$$

$$\begin{aligned} \varphi_{n+1}^{(d,+\infty)}(x) = &\left(C_{n+1,1} + \sum_{j=0}^{n-1} A_{n+1,j}\,x^{j+1}\right)e^{x/\gamma} \\ &+ \left(C_{n+1,2} + \sum_{j=0}^{n-2} B_{n+1,j}\,x^{j+1}\right)e^{-x/\mu}, \quad n \geq 2. \end{aligned} \qquad [4.16]$$

PROOF.– Substituting $c^{(d,+\infty)}$ and $F^{(d,+\infty)}(y)$ into [4.5] yields

$$(1+\theta)(\mu+d)\big(\varphi^{(d,+\infty)}(x)\big)'_+ = \varphi^{(d,+\infty)}(x), \quad x \in [0,d), \qquad [4.17]$$

and

$$\begin{aligned}\mu(1+\theta)(\mu+d)\big(\varphi^{(d,+\infty)}(x)\big)'_+ &= \mu\varphi^{(d,+\infty)}(x)\\ &- \mathrm{e}^{d/\mu}\int_d^x \varphi^{(d,+\infty)}(x-y)\,\mathrm{e}^{-y/\mu}\,\mathrm{d}y, \quad x \in [d,+\infty).\end{aligned} \qquad [4.18]$$

By theorem 2.1 and remark 2.1, $\varphi^{(d,+\infty)}(x)$ is continuously differentiable on $\mathbb{R}_+$ (see also section 1.1.3 and [ROL 99, pp. 162–163]). Let us introduce the functions $\varphi_{n+1}^{(d,+\infty)}(x)$, $n \geq 0$, in the following way: $\varphi_{n+1}^{(d,+\infty)}(x)$ is defined on $[nd,(n+1)d)$ and coincides with $\varphi^{(d,+\infty)}(x)$ on this interval.

Set $\gamma = (1+\theta)(\mu+d)$. Solving [4.17] gives [4.14], where the constant $C_{1,1}$ can be found from $\varphi_1^{(d,+\infty)}(0) = \theta/(1+\theta)$, which guarantees that the solution is the survival probability. Thus, $C_{1,1} = \theta/(1+\theta)$.

By [4.18], the function $\varphi^{(d,+\infty)}(x)$ is easily seen to have the second derivative, which is continuous. Moreover, integro-differential equation [4.18] can be reduced to the differential one

$$\begin{aligned}&\gamma\mu\big(\varphi^{(d,+\infty)}(x)\big)'' + (\gamma-\mu)\big(\varphi^{(d,+\infty)}(x)\big)' - \varphi^{(d,+\infty)}(x)\\ &= -\varphi^{(d,+\infty)}(x-d), \quad x \in [d,+\infty),\end{aligned} \qquad [4.19]$$

in a standard way (see section 1.1.6 and [ASM 10, BEA 84, BÜH 70, DIC 10, GER 79, GRA 91, ROL 99]).

If $x \in [d,2d)$, then $0 \leq x-d < d$ and the right-hand side of [4.19] has already been found. So, we only need to solve the linear differential equation

$$\begin{aligned}&\gamma\mu\big(\varphi_2^{(d,+\infty)}(x)\big)'' + (\gamma-\mu)\big(\varphi_2^{(d,+\infty)}(x)\big)' - \varphi_2^{(d,+\infty)}(x)\\ &= -\frac{\theta}{1+\theta}\,\mathrm{e}^{(x-d)/\gamma}\end{aligned} \qquad [4.20]$$

on $[d,2d)$ by standard techniques (see, for example [AGA 08, BIR 78, ROB 04]). Furthermore, since $\varphi^{(d,+\infty)}(x)$ and its first derivative are continuous at the point $x = d$, the following conditions must hold to guarantee that the solution is the survival probability:

$$\begin{cases} \lim\limits_{x\uparrow d} \varphi_1^{(d,\,+\infty)}(x) = \varphi_2^{(d,\,+\infty)}(d), \\ \lim\limits_{x\uparrow d} (\varphi_1^{(d,\,+\infty)}(x))' = (\varphi_2^{(d,\,+\infty)}(d))'. \end{cases} \qquad [4.21]$$

The solution to [4.20] can be found by writing

$$\varphi_2^{(d,\,+\infty)}(x) = \varphi_{2,\,\text{gen}}^{(d,\,+\infty)}(x) + \varphi_{2,\,\text{part}1}^{(d,\,+\infty)}(x),$$

where $\varphi_{2,\,\text{gen}}^{(d,\,+\infty)}(x)$ is a general solution to the linear homogeneous equation corresponding to [4.20] and $\varphi_{2,\,\text{part}1}^{(d,\,+\infty)}(x)$ is a particular solution to linear heterogeneous equation [4.20].

Note that $\varphi_{2,\,\text{part}1}^{(d,\,+\infty)}(x)$ can be written as

$$\varphi_{2,\,\text{part}1}^{(d,\,+\infty)}(x) = A_{2,\,0}\, x\, \mathrm{e}^{x/\gamma},$$

where $A_{2,\,0}$ is a constant. Substituting $\varphi_{2,\,\text{part}1}^{(d,\,+\infty)}(x)$ into [4.20] and applying the method of undetermined coefficients yield $A_{2,\,0}$. Moreover, we can write $\varphi_{2,\,\text{gen}}^{(d,\,+\infty)}(x)$ as

$$\varphi_{2,\,\text{gen}}^{(d,\,+\infty)}(x) = C_{2,\,1}\, \mathrm{e}^{x/\gamma} + C_{2,\,2}\, \mathrm{e}^{-x/\mu},$$

where the constants $C_{2,\,1}$ and $C_{2,\,2}$ are such that $\varphi_2^{(d,\,+\infty)}(x)$ satisfies the conditions [4.21].

Thus, $\varphi_2^{(d,\,+\infty)}(x)$ is defined by [4.15] and the constants $A_{2,\,0}$, $C_{2,\,1}$ and $C_{2,\,2}$ are given before the assertion of the theorem. It is easy to check that the second classical derivative of $\varphi^{(d,\,+\infty)}(x)$ does not exist at the point $x = d$.

If $x \in [2d, 3d)$, then $d \le x - d < 2d$ and the right-hand side of [4.19] has already been found. Therefore, $\varphi_3^{(d,\,+\infty)}(x)$ is a solution to

$$\begin{aligned} &\gamma\mu(\varphi_3^{(d,\,+\infty)}(x))'' + (\gamma-\mu)(\varphi_3^{(d,\,+\infty)}(x))' - \varphi_3^{(d,\,+\infty)}(x) \\ &= -\frac{\theta}{1+\theta}\left(1 + \frac{\gamma\mu + d(\gamma+\mu)}{(\gamma+\mu)^2}\,\mathrm{e}^{-d/\gamma} - \frac{1}{\gamma+\mu}\,\mathrm{e}^{-d/\gamma}\,(x-d)\right) \qquad [4.22] \\ &\quad \times \mathrm{e}^{(x-d)/\gamma} + \frac{\theta\mu(\mu+d)}{(\gamma+\mu)^2}\,\mathrm{e}^{2d/\mu}\,\mathrm{e}^{-x/\mu}. \end{aligned}$$

Furthermore, the continuity of $\varphi^{(d,+\infty)}(x)$ and its derivative at the point $x = 2d$ implies

$$\begin{cases} \lim\limits_{x \uparrow 2d} \varphi_2^{(d,+\infty)}(x) = \varphi_3^{(d,+\infty)}(2d), \\ \lim\limits_{x \uparrow 2d} \left(\varphi_2^{(d,+\infty)}(x)\right)' = \left(\varphi_3^{(d,+\infty)}(2d)\right)'. \end{cases} \qquad [4.23]$$

We can write $\varphi_3^{(d,+\infty)}(x)$ as

$$\varphi_3^{(d,+\infty)}(x) = \varphi_{3,\,\mathrm{gen}}^{(d,+\infty)}(x) + \varphi_{3,\,\mathrm{part1}}^{(d,+\infty)}(x) + \varphi_{3,\,\mathrm{part2}}^{(d,+\infty)}(x),$$

where $\varphi_{3,\,\mathrm{gen}}^{(d,+\infty)}(x)$ is a general solution to the linear homogeneous equation corresponding to [4.22], and $\varphi_{3,\,\mathrm{part1}}^{(d,+\infty)}(x)$ and $\varphi_{3,\,\mathrm{part2}}^{(d,+\infty)}(x)$ are particular solutions to the equations

$$\begin{aligned} &\gamma\mu\left(\varphi_3^{(d,+\infty)}(x)\right)'' + (\gamma-\mu)\left(\varphi_3^{(d,+\infty)}(x)\right)' - \varphi_3^{(d,+\infty)}(x) \\ &\quad = -\frac{\theta}{1+\theta}\left(1 + \frac{\gamma\mu + d(\gamma+\mu)}{(\gamma+\mu)^2}\,\mathrm{e}^{-d/\gamma} - \frac{1}{\gamma+\mu}\,\mathrm{e}^{-d/\gamma}\,(x-d)\right) \\ &\quad \times \mathrm{e}^{(x-d)/\gamma} \end{aligned} \qquad [4.24]$$

and

$$\begin{aligned} &\gamma\mu\left(\varphi_3^{(d,+\infty)}(x)\right)'' + (\gamma-\mu)\left(\varphi_3^{(d,+\infty)}(x)\right)' - \varphi_3^{(d,+\infty)}(x) \\ &\quad = \frac{\theta\mu(\mu+d)}{(\gamma+\mu)^2}\,\mathrm{e}^{2d/\mu}\,\mathrm{e}^{-x/\mu}, \end{aligned} \qquad [4.25]$$

respectively.

Next, $\varphi_{3,\,\mathrm{part1}}^{(d,+\infty)}(x)$ and $\varphi_{3,\,\mathrm{part2}}^{(d,+\infty)}(x)$ can be written as

$$\varphi_{3,\,\mathrm{part1}}^{(d,+\infty)}(x) = (A_{3,0} + A_{3,1}\,x)\,x\,\mathrm{e}^{x/\gamma}$$

and

$$\varphi_{3,\,\mathrm{part2}}^{(d,+\infty)}(x) = B_{3,0}\,x\,\mathrm{e}^{-x/\mu},$$

where $A_{3,0}$, $A_{3,1}$ and $B_{3,0}$ are constants. Substituting $\varphi_{3,\,\mathrm{part1}}^{(d,+\infty)}(x)$ and $\varphi_{3,\,\mathrm{part2}}^{(d,+\infty)}(x)$ into [4.24] and [4.25], respectively, and applying the method of undetermined coefficients yield $A_{3,0}$, $A_{3,1}$ and $B_{3,0}$. Moreover, we can write $\varphi_{3,\,\mathrm{gen}}^{(d,+\infty)}(x)$ as

$$\varphi_{3,\,\mathrm{gen}}^{(d,+\infty)}(x) = C_{3,1}\,\mathrm{e}^{x/\gamma} + C_{3,2}\,\mathrm{e}^{-x/\mu},$$

where the constants $C_{3,1}$ and $C_{3,2}$ are such that $\varphi_3^{(d,+\infty)}(x)$ satisfies the conditions [4.23], which guarantees that the solution is the survival probability.

All the constants are given before the assertion of the theorem. In particular, $A_{3,1}$, $A_{3,0}$ and $C_{3,1}$ are given by [4.6], [4.8] and [4.12], respectively, with $n = 2$.

In general, if we know $\varphi_n^{(d,+\infty)}(x)$, $n \geq 2$, we can find $\varphi_{n+1}^{(d,+\infty)}(x)$ by solving the equation

$$\begin{aligned}\gamma\mu\big(\varphi_{n+1}^{(d,+\infty)}(x)\big)'' + (\gamma-\mu)\big(\varphi_{n+1}^{(d,+\infty)}(x)\big)' - \varphi_{n+1}^{(d,+\infty)}(x)\\ = -\varphi_n^{(d,+\infty)}(x-d).\end{aligned} \qquad [4.26]$$

By the above, applying an induction argument yields

$$\varphi_n^{(d,+\infty)}(x) = \varphi_{n,\,\mathrm{gen}}^{(d,+\infty)}(x) + \varphi_{n,\,\mathrm{part1}}^{(d,+\infty)}(x) + \varphi_{n,\,\mathrm{part2}}^{(d,+\infty)}(x), \qquad [4.27]$$

where

$$\varphi_{n,\,\mathrm{gen}}^{(d,+\infty)}(x) = C_{n,1}\,\mathrm{e}^{x/\gamma} + C_{n,2}\,\mathrm{e}^{-x/\mu},$$

$$\varphi_{n,\,\mathrm{part1}}^{(d,+\infty)}(x) = \left(\sum_{i=0}^{n-2} A_{n,i}\,x^i\right) x\,\mathrm{e}^{x/\gamma},$$

$$\varphi_{n,\,\mathrm{part2}}^{(d,+\infty)}(x) = \left(\sum_{i=0}^{n-3} B_{n,i}\,x^i\right) x\,\mathrm{e}^{-x/\mu}.$$

Thus, [4.27] can be rewritten as

$$\begin{aligned}\varphi_n^{(d,+\infty)}(x) &= \left(C_{n,1} + \sum_{i=0}^{n-2} A_{n,i}\,x^{i+1}\right)\mathrm{e}^{x/\gamma}\\ &\quad + \left(C_{n,2} + \sum_{i=0}^{n-3} B_{n,i}\,x^{i+1}\right)\mathrm{e}^{-x/\mu}.\end{aligned} \qquad [4.28]$$

Therefore, we get

$$\varphi_{n+1}^{(d,+\infty)}(x) = \varphi_{n+1,\,\mathrm{gen}}^{(d,+\infty)}(x) + \varphi_{n+1,\,\mathrm{part1}}^{(d,+\infty)}(x) + \varphi_{n+1,\,\mathrm{part2}}^{(d,+\infty)}(x),$$

where

$$\varphi_{n+1,\,\mathrm{gen}}^{(d,+\infty)}(x) = C_{n+1,1}\,\mathrm{e}^{x/\gamma} + C_{n+1,2}\,\mathrm{e}^{-x/\mu},$$

$$\varphi_{n+1,\,\mathrm{part1}}^{(d,+\infty)}(x) = \left(\sum_{i=0}^{n-1} A_{n+1,i}\,x^i\right) x\,\mathrm{e}^{x/\gamma}, \qquad [4.29]$$

$$\varphi_{n+1,\,\mathrm{part2}}^{(d,+\infty)}(x) = \left(\sum_{i=0}^{n-2} B_{n+1,i}\,x^i\right) x\,\mathrm{e}^{-x/\mu}, \qquad [4.30]$$

which gives [4.16].

We now assume that we have already found $\varphi_n^{(d,+\infty)}(x)$, $n \geq 2$, which means that we know all the constants in [4.28]. We will now derive formulas to find the constants in [4.16].

If $x \in [nd, (n+1)d)$, then $(n-1)d \leq x - d < nd$ and

$$-\varphi_n^{(d,+\infty)}(x-d) = -\left(C_{n,1} + \sum_{i=0}^{n-2} A_{n,i}\,(x-d)^{i+1}\right)\mathrm{e}^{(x-d)/\gamma}$$

$$-\left(C_{n,2} + \sum_{i=0}^{n-3} B_{n,i}\,(x-d)^{i+1}\right)\mathrm{e}^{-(x-d)/\mu}. \qquad [4.31]$$

Applying the binomial theorem in [4.31] yields

$$-\varphi_n^{(d,+\infty)}(x-d)$$

$$= -\left[C_{n,1} + \sum_{i=0}^{n-2} A_{n,i}\left(\sum_{j=0}^{i+1}\binom{i+1}{j} x^j\,(-d)^{i-j+1}\right)\right]\mathrm{e}^{(x-d)/\gamma} \qquad [4.32]$$

$$-\left[C_{n,2} + \sum_{i=0}^{n-3} B_{n,i}\left(\sum_{j=0}^{i+1}\binom{i+1}{j} x^j\,(-d)^{i-j+1}\right)\right]\mathrm{e}^{-(x-d)/\mu}.$$

Interchanging the order of summation in [4.32] gives

$$\begin{aligned}
&-\varphi_n^{(d,+\infty)}(x-d)\\
&=-\Bigg[\bigg(C_{n,1}+\sum_{i=0}^{n-2}A_{n,i}\,(-d)^{i+1}\bigg)\\
&\quad+\sum_{j=1}^{n-1}\bigg(\sum_{i=j-1}^{n-2}A_{n,i}\,(-d)^{i-j+1}\binom{i+1}{j}\bigg)\,x^j\Bigg]\,\mathrm{e}^{(x-d)/\gamma}\\
&\quad-\Bigg[\bigg(C_{n,2}+\sum_{i=0}^{n-3}B_{n,i}\,(-d)^{i+1}\bigg)\\
&\quad+\sum_{j=1}^{n-2}\bigg(\sum_{i=j-1}^{n-3}B_{n,i}\,(-d)^{i-j+1}\binom{i+1}{j}\bigg)\,x^j\Bigg]\,\mathrm{e}^{-(x-d)/\mu}.
\end{aligned}\qquad [4.33]$$

To derive formulas for $A_{n+1,j}$, $0 \le j \le n-1$, we calculate $\big(\varphi_{n+1,\,\text{part1}}^{(d,+\infty)}(x)\big)'$ and $\big(\varphi_{n+1,\,\text{part1}}^{(d,+\infty)}(x)\big)''$ using [4.29] and substitute them, as well as [4.33], into [4.26]. Thus, we get

$$\begin{aligned}
&\mu\Bigg[(2A_{n+1,0}+2\gamma A_{n+1,1})\\
&+\sum_{j=1}^{n-2}\bigg(\frac{A_{n+1,j-1}}{\gamma}+2(j+1)A_{n+1,j}+(j+2)(j+1)\gamma A_{n+1,j+1}\bigg)\,x^j\\
&+\bigg(\frac{A_{n+1,n-2}}{\gamma}+2nA_{n+1,n-1}\bigg)\,x^{n-1}+\frac{A_{n+1,n-1}}{\gamma}\,x^n\Bigg]+(\gamma-\mu)\\
&\times\Bigg[A_{n+1,0}+\sum_{j=1}^{n-1}\bigg(\frac{A_{n+1,j-1}}{\gamma}+(j+1)A_{n+1,j}\bigg)\,x^j+\frac{A_{n+1,n-1}}{\gamma}\,x^n\Bigg]\\
&-\sum_{j=1}^{n}A_{n+1,j-1}\,x^j=-\Bigg[\bigg(C_{n,1}+\sum_{i=0}^{n-2}A_{n,i}\,(-d)^{i+1}\bigg)\\
&+\sum_{j=1}^{n-1}\bigg(\sum_{i=j-1}^{n-2}A_{n,i}\,(-d)^{i-j+1}\binom{i+1}{j}\bigg)\,x^j\Bigg]\,\mathrm{e}^{-d/\gamma}.
\end{aligned}\qquad [4.34]$$

To find $A_{n+1,j}$, $0 \le j \le n-1$, from [4.34], we apply the method of undetermined coefficients. To be more precise, we equate the expressions of x^{n-1}, x^j, $1 \le j \le n-2$, and x^0 in [4.34] and obtain [4.6], [4.7] and [4.8], respectively.

Thus, to find $A_{n+1,n-1}$, it suffices to know $A_{n,n-2}$. The constants $A_{n+1,j}$, $1 \le j \le n-2$, are expressed in a recurrent way in terms of $A_{n+1,j+1}$ and $A_{n,i}$, $j-1 \le i \le n-2$. Finally, $A_{n+1,0}$ can be found in terms of $A_{n+1,1}$, $C_{n,1}$ and $A_{n,i}$, $0 \le i \le n-2$.

Similarly, to derive formulas for $B_{n+1,j}$, $0 \le j \le n-2$, we calculate $\left(\varphi^{(d,+\infty)}_{n+1,\text{part2}}(x)\right)'$ and $\left(\varphi^{(d,+\infty)}_{n+1,\text{part2}}(x)\right)''$ using [4.30] and substitute them, as well as [4.34], into [4.26]. Thus, we obtain

$$
\begin{aligned}
&\gamma\Bigg[(-2B_{n+1,0} + 2\mu B_{n+1,1}) + \sum_{j=1}^{n-3}\Bigg(\frac{B_{n+1,j-1}}{\mu} \\
&\quad - 2(j+1)B_{n+1,j} + (j+2)(j+1)\mu B_{n+1,j+1}\Bigg)x^j \\
&\quad + \left(\frac{B_{n+1,n-3}}{\mu} - 2(n-1)B_{n+1,n-2}\right)x^{n-2} \\
&\quad + \frac{B_{n+1,n-2}}{\mu}x^{n-1}\Bigg] + (\gamma-\mu)\Bigg[B_{n+1,0} \\
&\quad + \sum_{j=1}^{n-2}\left(-\frac{B_{n+1,j-1}}{\mu} + (j+1)B_{n+1,j}\right)x^j - \frac{B_{n+1,n-2}}{\mu}x^{n-1}\Bigg] \\
&\quad - \sum_{j=1}^{n-1} B_{n+1,j-1}\,x^j = -\Bigg[\left(C_{n,2} + \sum_{i=0}^{n-3} B_{n,i}\,(-d)^{i+1}\right) \\
&\quad + \sum_{j=1}^{n-2}\left(\sum_{i=j-1}^{n-3} B_{n,i}\,(-d)^{i-j+1}\binom{i+1}{j}\right)x^j\Bigg]\mathrm{e}^{d/\mu}.
\end{aligned}
\qquad [4.35]
$$

Equating the expressions of x^{n-2}, x^j, $1 \le j \le n-3$, and x^0 in [4.35] yields [4.9], [4.10] and [4.11], respectively.

Since the continuity of $\varphi^{(d,\,+\infty)}(x)$ and its derivative at the point $x = nd$ implies

$$\begin{cases} \lim\limits_{x\uparrow nd} \varphi_n^{(d,\,+\infty)}(x) = \varphi_{n+1}^{(d,\,+\infty)}(nd), \\ \lim\limits_{x\uparrow nd} \left(\varphi_n^{(d,\,+\infty)}(x)\right)' = \left(\varphi_{n+1}^{(d,\,+\infty)}(nd)\right)', \end{cases}$$

the constants $C_{n+1,\,1}$ and $C_{n+1,\,2}$ can be found by solving the system of linear equations

$$\left(C_{n+1,\,1} + \sum_{i=0}^{n-1} A_{n+1,\,i}\,(nd)^{i+1}\right) \mathrm{e}^{nd/\gamma} + \left(C_{n+1,\,2} + \sum_{i=0}^{n-2} B_{n+1,\,i}\,(nd)^{i+1}\right) \mathrm{e}^{-nd/\mu}$$
$$= \left(C_{n,\,1} + \sum_{i=0}^{n-2} A_{n,\,i}\,(nd)^{i+1}\right) \mathrm{e}^{nd/\gamma} + \left(C_{n,\,2} + \sum_{i=0}^{n-3} B_{n,\,i}\,(nd)^{i+1}\right) \mathrm{e}^{-nd/\mu}$$

and

$$\left[\frac{C_{n+1,\,1}}{\gamma} + A_{n+1,\,0} + \sum_{i=0}^{n-2} \left(\frac{A_{n+1,\,i}}{\gamma} + (i+2)A_{n+1,\,i+1}\right)(nd)^{i+1}\right.$$
$$\left. + \frac{A_{n+1,\,n-1}}{\gamma}\,(nd)^{n}\right] \mathrm{e}^{nd/\gamma} + \left[-\frac{C_{n+1,\,2}}{\mu} + B_{n+1,\,0}\right.$$
$$+ \sum_{i=0}^{n-3} \left(-\frac{B_{n+1,\,i}}{\mu} + (i+2)B_{n+1,\,i+1}\right)(nd)^{i+1}$$
$$\left. - \frac{B_{n+1,\,n-2}}{\mu}\,(nd)^{n-1}\right] \mathrm{e}^{-nd/\mu}$$
$$= \left[\frac{C_{n,\,1}}{\gamma} + A_{n,\,0} + \sum_{i=0}^{n-3} \left(\frac{A_{n,\,i}}{\gamma} + (i+2)A_{n,\,i+1}\right)(nd)^{i+1}\right.$$
$$\left. + \frac{A_{n,\,n-2}}{\gamma}\,(nd)^{n-1}\right] \mathrm{e}^{nd/\gamma} + \left[-\frac{C_{n,\,2}}{\mu} + B_{n,\,0}\right.$$
$$\left. + \sum_{i=0}^{n-4} \left(-\frac{B_{n,\,i}}{\mu} + (i+2)B_{n,\,i+1}\right)(nd)^{i+1} - \frac{B_{n,\,n-3}}{\mu}\,(nd)^{n-2}\right] \mathrm{e}^{-nd/\mu},$$

which guarantees that the solution is the survival probability. Thus, $C_{n+1,1}$ and $C_{n+1,2}$ are given by [4.12] and [4.13], which completes the proof. □

REMARK 4.1.– Theorem 4.1 gives the analytic expression for $\varphi^{(d,+\infty)}(x)$ for all $x \geq 0$. However, the computations become too tedious for large initial surpluses. So, this theorem is useful when the initial surpluses are not too large, otherwise it is reasonable to use the Cramér–Lundberg approximation (see section 1.1.7 and [ASM 10, BEA 84, BÜH 70, DIC 10, GER 79, GRA 91, ROL 99, SCH 08b]).

EXAMPLE 4.1.– Let the surplus processes $\left(X_t(x)\right)_{t\geq 0}$ and $\left(X_t^{(2,+\infty)}(x)\right)_{t\geq 0}$ follow [4.1] and [4.4], respectively, under the above assumptions with $d = 2$ and $L = +\infty$. Furthermore, let $\varphi(x)$ and $\varphi^{(2,+\infty)}(x)$ be the corresponding survival probabilities, the claim sizes be exponentially distributed with mean μ and $\theta = 0.1$. To compute $\varphi(x)$ and $\varphi^{(2,+\infty)}(x)$, we apply formula [4.3] and theorem 4.1, respectively. Table 4.1 gives a comparison of $\varphi(x)$ and $\varphi^{(2,+\infty)}(x)$ for small initial surpluses and different μ. The computational results show that a franchise can considerably decrease the survival probability when d is too large for given μ ($d = \mu = 2$ in the example). If d is considerably less than μ, then a franchise decreases the survival probability very slightly. □

4.2.2. *Case of small and large enough initial surpluses*

We now investigate how a franchise changes the survival probability for small and large enough initial surpluses.

THEOREM 4.2.– Let the surplus processes $\left(X_t(x)\right)_{t\geq 0}$ and $\left(X_t^{(d,+\infty)}(x)\right)_{t\geq 0}$ follow [4.1] and [4.4], respectively, under the above assumptions with $0 < d < +\infty$ and $L = +\infty$. Moreover, let $\varphi(x)$ and $\varphi^{(d,+\infty)}(x)$ be the corresponding survival probabilities and the claim sizes be exponentially distributed with mean μ.

i) If

$$x \in \left[0,\ \min\left\{\frac{\mu(1+\theta)}{\theta}\ln\left(1+\frac{\theta d}{\mu(1+\theta)}\right),\ d\right\}\right], \qquad [4.36]$$

then $\varphi^{(d,+\infty)}(x) < \varphi(x)$ for any $0 < d < +\infty$.

ii) For

$$d \in \left(0,\ \frac{\mu(1+\theta)\,\ln(1+\theta)}{\theta}\right) \qquad [4.37]$$

and large enough initial surpluses, we have $\varphi^{(d,\,+\infty)}(x) > \varphi(x)$.

x	μ	$\varphi(x)$	$\varphi^{(2,\,+\infty)}(x)$	$\left(\frac{\varphi^{(2,\,+\infty)}(x)}{\varphi(x)} - 1\right) \times 100\%$
0.01	2	0.09132222	0.09111594	-0.2259%
0.20	2	0.09913610	0.09513668	-4.0343%
0.50	2	0.11133723	0.10184951	-8.5216%
1.00	2	0.13130633	0.11410656	-13.0990%
2.00	2	0.16990844	0.14322337	-15.7056%
4.00	2	0.24204280	0.20447048	-15.5230%
8.00	2	0.36805097	0.31354244	-14.8101%
0.01	10	0.09099173	0.09097799	-0.0151%
0.20	10	0.09256048	0.09229699	-0.2847%
0.50	10	0.09503195	0.09441867	-0.6453%
1.00	10	0.09913610	0.09806373	-1.0817%
2.00	10	0.10728866	0.10578145	-1.4048%
4.00	10	0.12337311	0.12165650	-1.3914%
8.00	10	0.15467783	0.15255999	-1.3692%
0.01	50	0.09092562	0.09092499	-0.0007%
0.20	50	0.09123961	0.09122751	-0.0133%
0.50	50	0.09173516	0.09170723	-0.0304%
1.00	50	0.09256048	0.09251238	-0.0520%
2.00	50	0.09420887	0.09414395	-0.0689%
4.00	50	0.09749668	0.09742971	-0.0687%
8.00	50	0.10403653	0.10396548	-0.0683%

Table 4.1. *Comparison of $\varphi(x)$ and $\varphi^{(2,\,+\infty)}(x)$ for different μ*

PROOF.– We now prove assertion i) of the theorem. Introduce the function

$$g_1(x) = \frac{\varphi(x)}{\varphi^{(d,\,+\infty)}(x)} \qquad [4.38]$$

for $x \in [0, d]$. Substituting [4.3] and [4.14] into [4.38] yields

$$g_1(x) = \frac{1+\theta-\exp\left(-\frac{\theta x}{\mu(1+\theta)}\right)}{\theta}\,\exp\left(-\frac{x}{(1+\theta)(\mu+d)}\right).$$

Taking the derivative gives

$$g_1'(x) = \frac{1}{\theta(\mu+d)} \exp\left(-\frac{x}{(1+\theta)(\mu+d)}\right) \times \left(\frac{\mu(1+\theta)+\theta d}{\mu(1+\theta)} \exp\left(-\frac{\theta x}{\mu(1+\theta)}\right) - 1\right).$$

Since

$$\frac{\mu(1+\theta)+\theta d}{\mu(1+\theta)} \exp\left(-\frac{\theta x}{\mu(1+\theta)}\right) - 1 > 0$$

for all x given by [4.36], we have $g_1'(x) > 0$ for these x. Furthermore, $g_1(0) = 1$, which gives $g_1(x) > 1$ for these x. Thus, assertion i) of the theorem follows and we now prove assertion ii).

By the Cramér–Lundberg approximation (see section 1.1.7 and [ASM 10, BEA 84, BÜH 70, DIC 10, GER 79, GRA 91, ROL 99, SCH 08b]), we have

$$\varphi^{(d,+\infty)}(x) \sim 1 - \frac{\theta}{(1+\theta)\, R^{(d,+\infty)}\, \mu^{(d,+\infty)}}\, \mathrm{e}^{-R^{(d,+\infty)}\, x}, \qquad [4.39]$$

where $R^{(d,+\infty)}$ is a unique positive solution (if it exists) to

$$\int_0^{+\infty} \mathrm{e}^{R^{(d,+\infty)}\, y} \left(1 - F^{(d,+\infty)}(y)\right) \mathrm{d}y = (1+\theta)(\mu+d)\, \mathrm{e}^{-d/\mu} \qquad [4.40]$$

and

$$\mu^{(d,+\infty)} = \frac{1}{(1+\theta)(\mu+d)\,\mathrm{e}^{-d/\mu}} \times \int_0^{+\infty} y\, \mathrm{e}^{R^{(d,+\infty)}\, y} \left(1 - F^{(d,+\infty)}(y)\right) \mathrm{d}y \qquad [4.41]$$

if the improper integral in the right-hand side of [4.41] is finite.

First, we show that there is a unique positive solution $R^{(d,+\infty)}$ to [4.40], such that the integral in the right-hand side of [4.41] is finite. If $R^{(d,+\infty)} < 1/\mu$, then substituting $F^{(d,+\infty)}(y)$ into [4.40] and doing elementary computations yield

$$\mathrm{e}^{dR^{(d,+\infty)}} = (1+\theta)(\mu+d)\, R^{(d,+\infty)} + 2 + \frac{1}{\mu R^{(d,+\infty)} - 1}. \qquad [4.42]$$

If $R^{(d,+\infty)} \geq 1/\mu$, then the integral in the left-hand side of [4.40] is infinite. So, in what follows, we consider the case $R^{(d,+\infty)} < 1/\mu$ only.

For $R \in [0, 1/\mu)$, introduce the functions

$$g_2(R) = \mathrm{e}^{dR}$$

and

$$g_3(R) = (1+\theta)(\mu+d)\,R + 2 + \frac{1}{\mu R - 1}.$$

Note that $g_2(0) = 1$ and $g_3(0) = 1$. Moreover, we have

$$g_2(R) \approx 1 + dR$$

and

$$g_3(R) \approx 1 + g_3'(0)\,R \approx 1 + (d + \theta d + \theta\mu)\,R$$

for small enough $R > 0$. Consequently, $g_3(R) > g_2(R)$ in some right semi-neighborhood of the point $R = 0$.

The function $g_2(R)$ is increasing on $[0, 1/\mu)$. The function $g_3(R)$ is increasing on $[0, R^*)$ and decreasing on $(R^*, 1/\mu)$, where

$$R^* = \frac{1}{\mu}\left(1 - \sqrt{\frac{\mu}{(1+\theta)(\mu+d)}}\right).$$

Thus, [4.42] has the unique solution $R^{(d,+\infty)}$ on $(0, 1/\mu)$. Note that $R^{(d,+\infty)}$ is the unique positive solution to [4.40]. It is evident that $\mu^{(d,+\infty)}$ is finite in this case.

Next, from [4.3] and [4.39] we conclude that $\varphi^{(d,+\infty)}(x) > \varphi(x)$ for large enough initial surpluses provided that

$$R^{(d,+\infty)} > \frac{\theta}{\mu(1+\theta)}.$$

Let $g_4(R) = g_2(R) - g_3(R)$ on $[0, 1/\mu)$. The function $g_4(R)$ is negative on $(0, R^{(d,+\infty)})$ and positive on $(R^{(d,+\infty)}, 1/\mu)$. Moreover, $g_4(0) = g_4(R^{(d,+\infty)}) = 0$ and

$$g_4\left(\frac{\theta}{\mu(1+\theta)}\right) = \exp\left(\frac{\theta d}{\mu(1+\theta)}\right) - \frac{\theta d}{\mu} - 1. \qquad [4.43]$$

Therefore, $\varphi^{(d,\,+\infty)}(x) > \varphi(x)$ for large enough initial surpluses provided that the expression in the right-hand side of [4.43] is negative.

Let

$$g_5(d) = \exp\left(\frac{\theta d}{\mu(1+\theta)}\right) - \frac{\theta d}{\mu} - 1.$$

Taking the derivative yields

$$g_5'(d) = \frac{\theta}{\mu}\left(\frac{1}{1+\theta}\exp\left(\frac{\theta d}{\mu(1+\theta)}\right) - 1\right).$$

Since $g_5(0) = 0$ and $g_5'(d) < 0$ for d given by [4.37], we prove assertion ii) of the theorem. □

4.3. Survival probability in the classical risk model with a liability limit

If the insurance company establishes a liability limit only and the claim sizes are exponentially distributed, then equation [4.2] for $\varphi^{(0,\,L)}(x)$ can be written as

$$c^{(0,\,L)}\big(\varphi^{(0,\,L)}(x)\big)'_+ = \lambda\varphi^{(0,\,L)}(x) - \lambda\int_0^x \varphi^{(0,\,L)}(x-y)\,\mathrm{d}F^{(0,\,L)}(y), \quad [4.44]$$

where

$$F^{(0,\,L)}(y) = \begin{cases} 0 & \text{if} \quad y < 0, \\ 1 - \mathrm{e}^{-y/\mu} & \text{if} \quad 0 \le y < L, \\ 1 & \text{if} \quad y \ge L, \end{cases}$$

and

$$\begin{aligned} c^{(0,\,L)} &= \lambda(1+\theta)\,\mathbb{E}\big[Y_i^{(0,\,L)}\big] \\ &= \lambda(1+\theta)\left(\int_0^L \frac{y\,\mathrm{e}^{-y/\mu}}{\mu}\,\mathrm{d}y + L\mathrm{e}^{-L/\mu}\right) = \lambda\mu(1+\theta)\big(1-\mathrm{e}^{-L/\mu}\big). \end{aligned}$$

In section 4.3.1, we derive analytic expressions for $\varphi^{(0,\,L)}(x)$ in the case of exponentially distributed claim sizes. In section 4.3.2, we investigate how a liability limit changes the survival probability for small enough and large enough initial surpluses.

4.3.1. *Analytic expression for the survival probability*

To formulate the next theorem, introduce the constants

$$\bar\gamma_1 = 1 - (1+\theta)\big(1 - \mathrm{e}^{-L/\mu}\big),$$

$$\bar\gamma_2 = \mu(1+\theta)\big(1 - \mathrm{e}^{-L/\mu}\big),$$

$$\bar C_{1,1} = -\frac{\theta\big(1 - \mathrm{e}^{-L/\mu}\big)}{\bar\gamma_1},$$

$$\bar C_{1,2} = \frac{\theta}{\bar\gamma_1(1+\theta)},$$

$$\bar A_{2,0} = -\frac{\theta}{\bar\gamma_1\bar\gamma_2(1+\theta)} \exp\left(-L\left(\frac{1}{\mu} + \frac{\bar\gamma_1}{\bar\gamma_2}\right)\right),$$

$$\bar C_{2,1} = \frac{\theta}{\bar\gamma_1(1+\theta)} \left(\left(1 - \frac{1}{\bar\gamma_1}\right)\mathrm{e}^{-L/\mu} - (1+\theta)\big(1 - \mathrm{e}^{-L/\mu}\big)\right),$$

$$\bar C_{2,2} = \frac{\theta}{\bar\gamma_1(1+\theta)} \left(1 + \left(\frac{1}{\bar\gamma_1} + \frac{L}{\bar\gamma_2} - 1\right) \exp\left(-L\left(\frac{1}{\mu} + \frac{\bar\gamma_1}{\bar\gamma_2}\right)\right)\right).$$

Moreover, let the constants $\bar A_{n+1,\,j}$, $0 \le j \le n-1$, be given in a recurrent way by formulas

$$\bar A_{n+1,\,n-1} = -\frac{\bar A_{n,\,n-2}}{n\bar\gamma_2} \exp\left(-L\left(\frac{1}{\mu} + \frac{\bar\gamma_1}{\bar\gamma_2}\right)\right), \quad n \ge 2, \qquad [4.45]$$

$$\begin{aligned} \bar A_{n+1,\,j} = {} & -\frac{(j+2)\bar\gamma_2 \bar A_{n+1,\,j+1}}{\bar\gamma_1} - \frac{1}{j+1} \exp\left(-L\left(\frac{1}{\mu} + \frac{\bar\gamma_1}{\bar\gamma_2}\right)\right) \\ & \times \left[\sum_{i=j-1}^{n-3} \left(\frac{\bar A_{n,\,i}}{\bar\gamma_2} + \frac{(i+2)\bar A_{n,\,i+1}}{\bar\gamma_1}\right) (-L)^{i-j+1} \binom{i+1}{j}\right. \\ & \left. + \frac{\bar A_{n,\,n-2}}{\bar\gamma_2} (-L)^{n-j+1} \binom{n-1}{j}\right], \\ & 1 \le j \le n-2, \quad n \ge 3, \end{aligned} \qquad [4.46]$$

$$\bar{A}_{n+1,0} = -\frac{2\bar{\gamma}_2\bar{A}_{n+1,1}}{\bar{\gamma}_1} - \exp\left(-L\left(\frac{1}{\mu}+\frac{\bar{\gamma}_1}{\bar{\gamma}_2}\right)\right)\left[\frac{\bar{C}_{n,2}}{\bar{\gamma}_2}+\frac{\bar{A}_{n,0}}{\bar{\gamma}_1}\right.$$
$$+\sum_{i=0}^{n-3}\left(\frac{\bar{A}_{n,i}}{\bar{\gamma}_2}+\frac{(i+2)\bar{A}_{n,i+1}}{\bar{\gamma}_1}\right)(-L)^{i+1} \qquad [4.47]$$
$$\left.+\frac{\bar{A}_{n,n-2}}{\bar{\gamma}_2}(-L)^{n-1}\right], \quad n \geq 2.$$

Finally, let the constants $\bar{C}_{n+1,1}$ and $\bar{C}_{n+1,2}$ be given by formulas

$$\bar{C}_{n+1,1} = \bar{C}_{n,1} + \frac{\bar{\gamma}_2}{\bar{\gamma}_1}\left(\sum_{i=0}^{n-2}(i+1)(\bar{A}_{n+1,i}-\bar{A}_{n,i})(nL)^i\right. \qquad [4.48]$$
$$\left.+n\bar{A}_{n+1,n-1}(nL)^{n-1}\right)e^{nL\bar{\gamma}_1/\bar{\gamma}_2}, \quad n \geq 2,$$

$$\bar{C}_{n+1,2} = \bar{C}_{n,2} + \frac{\bar{\gamma}_2(\bar{A}_{n,0}-\bar{A}_{n+1,0})}{\bar{\gamma}_1}$$
$$+\sum_{i=1}^{n-2}\left(\bar{A}_{n,i-1}-\bar{A}_{n+1,i-1}\right.$$
$$\left.+\frac{(i+1)\bar{\gamma}_2(\bar{A}_{n,i}-\bar{A}_{n+1,i})}{\bar{\gamma}_1}\right)(nL)^i \qquad [4.49]$$
$$+\left(\bar{A}_{n,n-2}-\bar{A}_{n+1,n-2}-\frac{n\bar{\gamma}_2\bar{A}_{n+1,n-1}}{\bar{\gamma}_1}\right)(nL)^{n-1}$$
$$-\bar{A}_{n+1,n-1}(nL)^n, \quad n \geq 2.$$

Note that to compute the constants by the formulas above for any $n \geq 2$, we have to know all the constants for $n-1$. Moreover, for any fixed $n \geq 2$, we start from the computation of $\bar{A}_{n+1,j}$ for j from $n-1$ to 0 and after that we can compute $\bar{C}_{n+1,1}$ and $\bar{C}_{n+1,2}$. We have introduced all the constants only to formulate the next theorem and will get them in the proof.

THEOREM 4.3.– Let the surplus process $\left(X_t^{(0,L)}(x)\right)_{t\geq 0}$ follow [4.4] under the above assumptions with $d=0$ and $0 < L < +\infty$, and the claim sizes be exponentially distributed with mean μ. Then,

$$\varphi^{(0,L)}(x) = \varphi_{n+1}^{(0,L)}(x) \quad \text{for all} \quad x \in [nL,(n+1)L), \quad n \geq 0,$$

where

$$\varphi_1^{(0,\,L)}(x) = \bar{C}_{1,\,1} + \bar{C}_{1,\,2}\,\mathrm{e}^{\bar{\gamma}_1 x/\bar{\gamma}_2}, \tag{4.50}$$

$$\varphi_2^{(0,\,L)}(x) = \bar{C}_{2,\,1} + \big(\bar{C}_{2,\,2} + \bar{A}_{2,\,0}\,x\big)\,\mathrm{e}^{\bar{\gamma}_1 x/\bar{\gamma}_2}, \tag{4.51}$$

$$\varphi_{n+1}^{(0,\,L)}(x) = \bar{C}_{n+1,\,1} + \left(\bar{C}_{n+1,\,2} + \sum_{j=0}^{n-1} \bar{A}_{n+1,\,j}\,x^{j+1}\right)\mathrm{e}^{\bar{\gamma}_1 x/\bar{\gamma}_2}, \quad n \geq 2. \tag{4.52}$$

PROOF.– By theorem 2.1 and remark 2.1, $\varphi^{(0,\,L)}(x)$ is continuous on $\mathbb{R}_+$ and continuously differentiable on $\mathbb{R}_+$, except at the point $x = L$, where there are only one-sided derivatives (see also section 1.1.3 and [ROL 99, pp. 162–163]).

Let us introduce the functions $\varphi_{n+1}^{(0,\,L)}(x)$, $n \geq 0$, in the following way: $\varphi_{n+1}^{(0,\,L)}(x)$ is defined on $[nL, (n+1)L)$ and coincides with $\varphi^{(0,\,L)}(x)$ on this interval.

If $x \in [0, L)$, then substituting $c^{(0,\,L)}$ and $F^{(0,\,L)}(y)$ into [4.44] yields

$$\begin{aligned} &\mu(1+\theta)\big(1-\mathrm{e}^{-L/\mu}\big)\big(\varphi_1^{(0,\,L)}(x)\big)' \\ &\quad = \varphi_1^{(0,\,L)}(x) - \frac{1}{\mu}\int_0^x \varphi_1^{(0,\,L)}(x-y)\,\mathrm{e}^{-y/\mu}\,\mathrm{d}y. \end{aligned} \tag{4.53}$$

This gives that $\varphi_1^{(0,\,L)}(x)$ has the continuous second derivative on $[0, L)$. So, integro-differential equation [4.53] can be reduced to the differential one

$$\begin{aligned} &\mu(1+\theta)\big(1-\mathrm{e}^{-L/\mu}\big)\big(\varphi_1^{(0,\,L)}(x)\big)'' \\ &\quad - \big(1-(1+\theta)\big(1-\mathrm{e}^{-L/\mu}\big)\big)\big(\varphi_1^{(0,\,L)}(x)\big)' = 0 \end{aligned} \tag{4.54}$$

in a standard way (see section 1.1.6 and [ASM 10, BEA 84, BÜH 70, DIC 10, GER 79, GRA 91, ROL 99]).

Let $\bar{\gamma}_1 = 1 - (1+\theta)\big(1-\mathrm{e}^{-L/\mu}\big)$ and $\bar{\gamma}_2 = \mu(1+\theta)\big(1-\mathrm{e}^{-L/\mu}\big)$. Solving [4.54] yields [4.50], where the constants $\bar{C}_{1,\,1}$ and $\bar{C}_{1,\,2}$ can be found from

$$\begin{cases} \varphi_1^{(0,\,L)}(0) = \dfrac{\theta}{1+\theta}, \\ \left(\varphi_1^{(0,\,L)}(0)\right)' = \dfrac{\varphi_1^{(0,\,L)}(0)}{\bar{\gamma}_2}. \end{cases} \qquad [4.55]$$

The first condition in [4.55] guarantees that the solution is the survival probability and the second one follows from [4.53]. Thus, the constants $\bar{C}_{1,\,1}$ and $\bar{C}_{1,\,2}$ are given before the assertion of the theorem.

If $x \in [L, +\infty)$, then substituting $c^{(0,\,L)}$ and $F^{(0,\,L)}(y)$ into [4.44] yields

$$\begin{aligned} &\mu(1+\theta)\left(1 - \mathrm{e}^{-L/\mu}\right)\left(\varphi^{(0,\,L)}(x)\right)' = \varphi^{(0,\,L)}(x) \\ &\quad - \frac{\mathrm{e}^{-x/\mu}}{\mu} \int_{x-L}^{x} \varphi^{(0,\,L)}(y)\, \mathrm{e}^{y/\mu}\, \mathrm{d}y - \mathrm{e}^{-L/\mu} \varphi^{(0,\,L)}(x-L). \end{aligned} \qquad [4.56]$$

Note that we imply the right derivative of $\varphi^{(0,\,L)}(x)$ at the point $x = L$. It is easily seen from [4.56] that the second classical derivative of $\varphi^{(0,\,L)}(x)$ exists on $[L, +\infty)$ except at the point $x = 2L$. So, integro-differential equation [4.56] can be reduced to the differential one

$$\begin{aligned} &\bar{\gamma}_2\left(\varphi^{(0,\,L)}(x)\right)'' - \bar{\gamma}_1\left(\varphi^{(0,\,L)}(x)\right)' \\ &\quad = -\mathrm{e}^{-L/\mu}\left(\varphi^{(0,\,L)}(x-L)\right)', \quad x \in [L, +\infty), \end{aligned} \qquad [4.57]$$

where we imply the right second derivative of $\varphi^{(0,\,L)}(x)$ at the point $x = 2L$.

If $x \in [L, 2L)$, then [4.57] can be rewritten as

$$\begin{aligned} &\bar{\gamma}_2\left(\varphi_2^{(0,\,L)}(x)\right)'' - \bar{\gamma}_1\left(\varphi_2^{(0,\,L)}(x)\right)' \\ &\quad = -\frac{\theta}{\bar{\gamma}_2(1+\theta)} \exp\left(-\frac{L}{\mu} + \frac{\bar{\gamma}_1(x-L)}{\bar{\gamma}_2}\right). \end{aligned} \qquad [4.58]$$

By theorem 2.1 and remark 2.1, the following conditions must hold to guarantee that the solution is the survival probability:

$$\begin{cases} \lim\limits_{x \uparrow L} \varphi_1^{(0,\,L)}(x) = \varphi_2^{(0,\,L)}(L), \\ \lim\limits_{x \uparrow L} \left(\varphi_1^{(0,\,L)}(x)\right)' = \left(\varphi_2^{(0,\,L)}(L)\right)' + \dfrac{\theta \mathrm{e}^{-L/\mu}}{\bar{\gamma}_2(1+\theta)}, \end{cases} \qquad [4.59]$$

(see also section 1.1.3 and [ROL 99, pp. 162–163]).

The solution to [4.58] can be found by writing

$$\varphi_2^{(0,\,L)}(x) = \varphi_{2,\,\text{gen}}^{(0,\,L)}(x) + \varphi_{2,\,\text{part}}^{(0,\,L)}(x),$$

where $\varphi_{2,\,\text{gen}}^{(0,\,L)}(x)$ is a general solution to the linear homogeneous equation corresponding to [4.58] and $\varphi_{2,\,\text{part}}^{(0,\,L)}(x)$ is a particular solution to linear heterogeneous equation [4.58].

Since $\varphi_{2,\,\text{part}}^{(0,\,L)}(x)$ can be written as

$$\varphi_{2,\,\text{part}}^{(0,\,L)}(x) = \bar{A}_{2,\,0}\, x\, \mathrm{e}^{\bar{\gamma}_1 x/\bar{\gamma}_2},$$

where $\bar{A}_{2,\,0}$ is a constant, substituting $\varphi_{2,\,\text{part}}^{(0,\,L)}(x)$ into [4.58] and applying the method of undetermined coefficients yield $\bar{A}_{2,\,0}$. Furthermore, we can write $\varphi_{2,\,\text{gen}}^{(0,\,L)}(x)$ as

$$\varphi_{2,\,\text{gen}}^{(0,\,L)}(x) = \bar{C}_{2,\,1} + \bar{C}_{2,\,2}\, \mathrm{e}^{\bar{\gamma}_1 x/\bar{\gamma}_2},$$

where the constants $\bar{C}_{2,\,1}$ and $\bar{C}_{2,\,2}$ are such that $\varphi_2^{(0,\,L)}(x)$ satisfies conditions [4.59].

Thus, $\varphi_2^{(0,\,L)}(x)$ is defined by [4.51] and the constants $\bar{A}_{2,\,0}$, $\bar{C}_{2,\,1}$ and $\bar{C}_{2,\,2}$ are given before the assertion of the theorem.

In general, if we know $\varphi_n^{(0,\,L)}(x)$, $n \geq 2$, we can find $\varphi_{n+1}^{(0,\,L)}(x)$ applying considerations similar to those in the proof of theorem 4.1. Since

$$\bar{\gamma}_2\big(\varphi_{n+1}^{(0,\,L)}(x)\big)'' - \bar{\gamma}_1\big(\varphi_{n+1}^{(0,\,L)}(x)\big)' = -\mathrm{e}^{-L/\mu}\big(\varphi_n^{(0,\,L)}(x-L)\big)',$$

we get [4.52] by induction on n.

Applying the method of undetermined coefficients to

$$\bar{\gamma}_2\Bigg[\bigg(\frac{2\bar{\gamma}_1\bar{A}_{n+1,\,0}}{\bar{\gamma}_2} + 2\bar{A}_{n+1,\,1}\bigg) + \sum_{j=1}^{n-2}\bigg(\frac{\bar{\gamma}_1^2\bar{A}_{n+1,\,j-1}}{\bar{\gamma}_2^2}$$

$$+ \frac{2(j+1)\bar{\gamma}_1\bar{A}_{n+1,\,j}}{\bar{\gamma}_2} + (j+2)(j+1)\bar{A}_{n+1,\,j+1}\bigg)\, x^j$$

$$+\left(\frac{\bar{\gamma}_1^2\bar{A}_{n+1,\,n-2}}{\bar{\gamma}_2^2}+\frac{2n\bar{\gamma}_1\bar{A}_{n+1,\,n-1}}{\bar{\gamma}_2}\right)x^{n-1}+\frac{\bar{\gamma}_1^2\bar{A}_{n+1,\,n-1}}{\bar{\gamma}_2^2}\,x^n\Bigg]$$

$$-\bar{\gamma}_1\Bigg[\bar{A}_{n+1,0}+\sum_{j=1}^{n-1}\left(\frac{\bar{\gamma}_1\bar{A}_{n+1,\,j-1}}{\bar{\gamma}_2}+(j+1)\bar{A}_{n+1,\,j}\right)x^j$$

$$+\frac{\bar{\gamma}_1\bar{A}_{n+1,\,n-1}}{\bar{\gamma}_2}\,x^n\Bigg]=-\Bigg[\Bigg(\frac{\bar{\gamma}_1\bar{C}_{n,\,2}}{\bar{\gamma}_2}+\bar{A}_{n,\,0}$$

$$+\sum_{i=0}^{n-3}\left(\frac{\bar{\gamma}_1\bar{A}_{n,\,i}}{\bar{\gamma}_2}+(i+2)\bar{A}_{n,\,i+1}\right)(-L)^{i+1}+\frac{\bar{\gamma}_1\bar{A}_{n,\,n-2}}{\bar{\gamma}_2}(-L)^{n-1}\Bigg)$$

$$+\sum_{j=1}^{n-1}\Bigg(\sum_{i=j-1}^{n-3}\left(\frac{\bar{\gamma}_1\bar{A}_{n,\,i}}{\bar{\gamma}_2}+(i+2)\bar{A}_{n,\,i+1}\right)(-L)^{i-j+1}\binom{i+1}{j}$$

$$+\frac{\bar{\gamma}_1\bar{A}_{n,\,n-2}}{\bar{\gamma}_2}(-L)^{n-j+1}\binom{n-1}{j}\Bigg)x^j\Bigg]\exp\left(-L\left(\frac{1}{\mu}+\frac{\bar{\gamma}_1}{\bar{\gamma}_2}\right)\right)$$

yields the constants $\bar{A}_{n+1,\,j}$, $0\le j\le n-1$, in [4.52]. Thus, they are given by [4.45], [4.46] and [4.47].

To find the constants $\bar{C}_{n+1,\,1}$ and $\bar{C}_{n+1,\,2}$ in [4.52], we use the continuity of $\varphi^{(0,\,L)}(x)$ and its derivative at the point $x=nL$, which implies

$$\begin{cases}\lim\limits_{x\uparrow nL}\varphi_n^{(0,\,L)}(x)=\varphi_{n+1}^{(0,\,L)}(nL),\\[2ex] \lim\limits_{x\uparrow nL}\left(\varphi_n^{(0,\,L)}(x)\right)'=\left(\varphi_{n+1}^{(0,\,L)}(nL)\right)',\end{cases}$$

and guarantees that the solution is the survival probability. Thus, $\bar{C}_{n+1,\,1}$ and $\bar{C}_{n+1,\,2}$ are given by [4.48] and [4.49], respectively. The proof is complete. □

EXAMPLE 4.2.– Let the surplus processes $\left(X_t(x)\right)_{t\ge0}$ and $\left(X_t^{(0,\,30)}(x)\right)_{t\ge0}$ follow [4.1] and [4.4], respectively, under the above assumptions with $d=0$ and $L=30$. Furthermore, let $\varphi(x)$ and $\varphi^{(0,\,30)}(x)$ be the corresponding survival probabilities, the claim sizes be exponentially distributed with mean μ and $\theta=0.1$. To compute $\varphi(x)$ and $\varphi^{(0,\,30)}(x)$, we apply formula [4.3] and theorem 4.3, respectively. Table 4.2 gives a comparison of $\varphi(x)$ and $\varphi^{(0,\,30)}(x)$ for different μ. The computational results show that a liability limit can considerably increase the survival probability when L is too small for given μ ($L = 2\mu = 30$ in the example). If L is considerably greater than μ, then a liability limit increases the survival probability very slightly. □

x	μ	$\varphi(x)$	$\varphi^{(0,\,30)}(x)$	$\left(\frac{\varphi^{(0,\,30)}(x)}{\varphi(x)} - 1\right) \times 100\%$
1	3	0.11804409	0.11804551	0.0012%
10	3	0.32856662	0.32859294	0.0080%
20	3	0.50409494	0.50416485	0.0139%
30	3	0.63373607	0.63385601	0.0189%
60	3	0.85243581	0.85257442	0.0163%
90	3	0.94054782	0.94064002	0.0098%
120	3	0.97604729	0.97609908	0.0053%
1	10	0.09913610	0.09958778	0.4556%
10	10	0.16990844	0.17602881	3.6022%
20	10	0.24204280	0.25754340	6.4041%
30	10	0.30790874	0.33560557	8.9951%
60	10	0.47311066	0.52122121	10.1690%
90	10	0.59887894	0.65518553	9.4020%
120	10	0.69462638	0.75054563	8.0503%
1	15	0.09640207	0.09729201	0.9232%
10	15	0.14436915	0.15573294	7.8713%
20	15	0.19468542	0.22281568	14.4491%
30	15	0.24204280	0.29223602	20.7373%
60	15	0.36805097	0.45896776	24.7022%
90	15	0.47311066	0.58698104	24.0684%
120	15	0.56070447	0.67711345	20.7612%

Table 4.2. *Comparison of $\varphi(x)$ and $\varphi^{(0,\,30)}(x)$ for different μ*

4.3.2. *Case of small enough and large enough initial surpluses*

We now investigate how a liability limit changes the survival probability for small enough and large enough initial surpluses.

THEOREM 4.4.– Let the surplus processes $\left(X_t(x)\right)_{t\geq 0}$ and $\left(X_t^{(0,\,L)}(x)\right)_{t\geq 0}$ follow [4.1] and [4.4], respectively, under the above assumptions with $d = 0$ and $0 < L < +\infty$. Moreover, let $\varphi(x)$ and $\varphi^{(0,\,L)}(x)$ be the corresponding survival probabilities and the claim sizes be exponentially distributed with mean μ. Then, $\varphi^{(0,\,L)}(x) > \varphi(x)$ for any $0 < L < +\infty$ and for small enough and large enough initial surpluses.

PROOF.– By [4.3] and [4.50], we have

$$\varphi(x) \approx \frac{\theta}{1+\theta} + \frac{\theta x}{\mu(1+\theta)^2}$$

and

$$\varphi_1^{(0,\,L)}(x) \approx \frac{\theta}{1+\theta} + \frac{\theta x}{\mu(1+\theta)^2\big(1-\mathrm{e}^{-L/\mu}\big)}$$

for small enough initial surpluses. This gives the assertion of the theorem for such initial surpluses.

By the Cramér–Lundberg approximation (see section 1.1.7 and [ASM 10, BEA 84, BÜH 70, DIC 10, GER 79, GRA 91, ROL 99, SCH 08b]), we have

$$\varphi^{(0,\,L)}(x) \sim 1 - \frac{\theta}{R^{(0,\,L)}\mu^{(0,\,L)}(1+\theta)}\,\mathrm{e}^{-R^{(0,\,L)}x} \qquad [4.60]$$

for large enough initial surpluses, where $R^{(0,\,L)}$ is a unique positive solution (if it exists) to

$$\int_0^{+\infty} \mathrm{e}^{R^{(0,\,L)}y}\big(1 - F^{(0,\,L)}(y)\big)\,\mathrm{d}y = \mu(1+\theta)\big(1-\mathrm{e}^{-L/\mu}\big) \qquad [4.61]$$

and

$$\mu^{(0,\,L)} = \frac{1}{\mu(1+\theta)\big(1-\mathrm{e}^{-L/\mu}\big)}\int_0^{+\infty} y\,\mathrm{e}^{R^{(0,\,L)}y}\big(1 - F^{(0,\,L)}(y)\big)\,\mathrm{d}y. \qquad [4.62]$$

It is obvious that the improper integral in the right-hand side of [4.62] is finite.

If $R^{(0,\,L)} = 1/\mu$ is a solution to [4.61], then

$$L = \mu(1+\theta)\big(1-\mathrm{e}^{-L/\mu}\big) \qquad [4.63]$$

must hold.

If $R^{(0,\,L)} \neq 1/\mu$, then substituting $F^{(0,\,L)}(y)$ into [4.61] yields

$$\mathrm{e}^{LR^{(0,\,L)}} = (1+\theta)\big(\mathrm{e}^{L/\mu}-1\big)\big(\mu R^{(0,\,L)}-1\big) + \mathrm{e}^{L/\mu}.$$

Consider the function

$$\bar{g}(R) = \mathrm{e}^{LR} - (1+\theta)\big(\mathrm{e}^{L/\mu} - 1\big)\big(\mu R - 1\big) - \mathrm{e}^{L/\mu}$$

on $\mathbb{R}_+$. It is decreasing on $[0, \bar{R}^*)$ and increasing on $(\bar{R}^*, +\infty)$, where

$$\bar{R}^* = \frac{1}{L}\ln\left(\frac{\mu(1+\theta)\big(\mathrm{e}^{L/\mu} - 1\big)}{L}\right).$$

Moreover, $\bar{g}(0) = \theta\big(\mathrm{e}^{L/\mu} - 1\big) > 0$, $\bar{g}(1/\mu) = 0$ and $\lim_{R\to+\infty} \bar{g}(R) = +\infty$. Note that $\bar{R}^* = 1/\mu$ if and only if [4.63] holds.

Thus, we conclude that $R = 1/\mu$ is a unique zero of $\bar{g}(R)$ if [4.63] is true. This means that [4.61] has the unique solution $R^{(0,\,L)} = 1/\mu$. Otherwise, if [4.63] is not true, then $R = R^{(0,\,L)}$ and $R = 1/\mu$ are positive zeros of $\bar{g}(R)$, which means that $R^{(0,\,L)}$ is a unique positive solution to [4.61], such that $R^{(0,\,L)} \neq 1/\mu$. Furthermore, in this case, $\bar{g}(R)$ is negative between its zeros $R = R^{(0,\,L)}$ and $R = 1/\mu$, otherwise it is positive.

Since

$$\bar{g}\left(\frac{\theta}{\mu(1+\theta)}\right) = \exp\left(\frac{\theta L}{\mu(1+\theta)}\right) - 1 > 0$$

and

$$\frac{\theta}{\mu(1+\theta)} < \frac{1}{\mu},$$

we get

$$\frac{\theta}{\mu(1+\theta)} < R^{(0,\,L)}.$$

Consequently, by [4.3] and [4.60], we have the assertion of the theorem for large enough initial surpluses. □

4.4. Survival probability in the classical risk model with both a franchise and a liability limit

If the insurance company establishes both a franchise and a liability limit and the claim sizes are exponentially distributed, then equation [4.2] for $\varphi^{(d,\,L)}(x)$ can be written as

$$c^{(d,\,L)}\big(\varphi^{(d,\,L)}(x)\big)'_+ = \lambda\varphi^{(d,\,L)}(x) - \lambda\int_0^x \varphi^{(d,\,L)}(x-y)\,\mathrm{d}F^{(d,\,L)}(y), \quad [4.64]$$

where

$$F^{(d,\,L)}(y) = \begin{cases} 0 & \text{if } \; y < 0, \\ 1 - \mathrm{e}^{-d/\mu} & \text{if } \; 0 \le y < d, \\ 1 - \mathrm{e}^{-y/\mu} & \text{if } \; d \le y < L, \\ 1 & \text{if } \; y \ge L, \end{cases}$$

and

$$\begin{aligned} c^{(d,\,L)} &= \lambda(1+\theta)\,\mathbb{E}\big[Y_i^{(d,\,L)}\big] \\ &= \lambda(1+\theta)\bigg(\int_d^L \frac{y\,\mathrm{e}^{-y/\mu}}{\mu}\,\mathrm{d}y + L\mathrm{e}^{-L/\mu}\bigg) \\ &= \lambda(1+\theta)\big((\mu+d)\mathrm{e}^{-d/\mu} - \mu\mathrm{e}^{-L/\mu}\big). \end{aligned}$$

In section 4.4.1, we derive analytic expressions for $\varphi^{(d,\,L)}(x)$ in the case of exponentially distributed claim sizes. In section 4.4.2, we investigate how a franchise and a liability limit change the survival probability for small initial surpluses.

4.4.1. *Analytic expression for the survival probability*

To formulate the next theorem, introduce the constants

$$\begin{aligned} \tilde\gamma &= (1+\theta)\big(\mu + d - \mu\mathrm{e}^{(d-L)/\mu}\big), \\ \tilde C_{1,1} &= \theta/(1+\theta), \\ \tilde A_{2,0} &= -\frac{\theta}{(1+\theta)(\tilde\gamma+\mu)}\,\mathrm{e}^{-d/\tilde\gamma}, \\ \tilde C_{2,1} &= \frac{\theta}{1+\theta}\bigg(1 + \frac{\tilde\gamma\mu + d(\tilde\gamma+\mu)}{(\tilde\gamma+\mu)^2}\,\mathrm{e}^{-d/\tilde\gamma}\bigg), \\ \tilde C_{2,2} &= -\frac{\theta\tilde\gamma\mu}{(1+\theta)(\tilde\gamma+\mu)^2}\,\mathrm{e}^{d/\mu}. \end{aligned}$$

Moreover, let the constants $\tilde A_{n+1,\,j}$, $0 \le j \le n-1$, be given in a recurrent way by formulas

$$\tilde A_{n+1,\,n-1} = -\frac{\tilde A_{n,\,n-2}}{n(\tilde\gamma+\mu)}\,\mathrm{e}^{-d/\tilde\gamma}, \quad n \ge 2, \qquad [4.65]$$

$$\tilde{A}_{n+1,\,j} = -\frac{1}{\tilde{\gamma}+\mu}\left[(j+2)\tilde{\gamma}\mu\tilde{A}_{n+1,\,j+1} + \frac{1}{j+1}\right.$$
$$\left.\times\left(\sum_{i=j-1}^{n-2}\tilde{A}_{n,\,i}\,(-d)^{i-j+1}\binom{i+1}{j}\right)\mathrm{e}^{-d/\tilde{\gamma}}\right], \qquad [4.66]$$
$$1 \le j \le n-2, \quad 3 \le n \le m,$$
$$\text{and } n-m+1 \le j \le n-2, \quad n \ge m+1,$$

$$\tilde{A}_{n+1,\,j} = -\frac{1}{\tilde{\gamma}+\mu}\left[(j+2)\tilde{\gamma}\mu\tilde{A}_{n+1,\,j+1} + \frac{1}{j+1}\right.$$
$$\times\left(\left(\sum_{i=j-1}^{n-2}\tilde{A}_{n,\,i}\,(-d)^{i-j+1}\binom{i+1}{j}\right)\mathrm{e}^{-d/\tilde{\gamma}}\right.$$
$$-\left(\sum_{i=j-1}^{n-m-1}\tilde{A}_{n-m+1,\,i}\,(-L)^{i-j+1}\binom{i+1}{j}\right) \qquad [4.67]$$
$$\left.\left.\times\exp\left(\frac{d-L}{\mu}-\frac{L}{\tilde{\gamma}}\right)\right)\right],$$
$$1 \le j \le n-m, \quad n \ge m+1,$$

$$\tilde{A}_{n+1,\,0} = -\frac{1}{\tilde{\gamma}+\mu}\left[2\tilde{\gamma}\mu\tilde{A}_{n+1,\,1}\right.$$
$$\left.+\left(\tilde{C}_{n,\,1}+\sum_{i=0}^{n-2}\tilde{A}_{n,\,i}\,(-d)^{i+1}\right)\mathrm{e}^{-d/\tilde{\gamma}}\right], \qquad [4.68]$$
$$2 \le n \le m-1,$$

$$\tilde{A}_{m+1,\,0} = -\frac{1}{\tilde{\gamma}+\mu}\left[2\tilde{\gamma}\mu\tilde{A}_{m+1,\,1}\right.$$
$$+\left(\tilde{C}_{m,\,1}+\sum_{i=0}^{m-2}\tilde{A}_{m,\,i}\,(-d)^{i+1}\right)\mathrm{e}^{-d/\tilde{\gamma}} \qquad [4.69]$$
$$\left.-\frac{\theta}{1+\theta}\exp\left(\frac{d-L}{\mu}-\frac{L}{\tilde{\gamma}}\right)\right],$$

$$\begin{aligned}\tilde{A}_{n+1,0} = -\frac{1}{\tilde{\gamma}+\mu}\Bigg[&2\tilde{\gamma}\mu\tilde{A}_{n+1,1}\\&+\left(\tilde{C}_{n,1}+\sum_{i=0}^{n-2}\tilde{A}_{n,i}(-d)^{i+1}\right)e^{-d/\tilde{\gamma}}\\&-\left(\tilde{C}_{n-m+1,1}+\sum_{i=0}^{n-m-1}\tilde{A}_{n-m+1,i}(-L)^{i+1}\right)\\&\times\exp\left(\frac{d-L}{\mu}-\frac{L}{\tilde{\gamma}}\right)\Bigg],\quad n\geq m+1.\end{aligned}\qquad [4.70]$$

Next, let the constants $\tilde{B}_{n+1,j}$, $0 \leq j \leq n-2$, be given in a recurrent way by formulas

$$\tilde{B}_{3,0} = \frac{\tilde{C}_{2,2}}{\tilde{\gamma}+\mu}\,e^{d/\mu},$$

$$\tilde{B}_{n+1,n-2} = \frac{\tilde{B}_{n,n-3}}{(n-1)(\tilde{\gamma}+\mu)}\,e^{d/\mu},\quad n\geq 3, \qquad [4.71]$$

$$\begin{aligned}\tilde{B}_{n+1,j} = \frac{1}{\tilde{\gamma}+\mu}\Bigg[&(j+2)\tilde{\gamma}\mu\tilde{B}_{n+1,j+1}+\frac{1}{j+1}\\&\times\left(\sum_{i=j-1}^{n-3}\tilde{B}_{n,i}(-d)^{i-j+1}\binom{i+1}{j}\right)e^{d/\mu}\Bigg],\end{aligned}\qquad [4.72]$$

$$1\leq j\leq n-3,\quad 4\leq n\leq m,$$

$$\text{and } n-m\leq j\leq n-3,\quad n\geq m+1,$$

$$\begin{aligned}\tilde{B}_{n+1,j} = \frac{1}{\tilde{\gamma}+\mu}\Bigg[&(j+2)\tilde{\gamma}\mu\tilde{B}_{n+1,j+1}+\frac{1}{j+1}\\&\times\left(\sum_{i=j-1}^{n-3}\tilde{B}_{n,i}(-d)^{i-j+1}\binom{i+1}{j}\right.\\&\left.-\sum_{i=j-1}^{n-m-2}\tilde{B}_{n-m+1,i}(-L)^{i-j+1}\binom{i+1}{j}\right)e^{d/\mu}\Bigg],\end{aligned}\qquad [4.73]$$

$$1\leq j\leq n-m-1,\quad n\geq m+1,$$

$$\tilde{B}_{n+1,0} = \frac{1}{\tilde{\gamma}+\mu}\left[2\tilde{\gamma}\mu\tilde{B}_{n+1,1} + \left(\tilde{C}_{n,2} + \sum_{i=0}^{n-3}\tilde{B}_{n,i}\,(-d)^{i+1}\right)\mathrm{e}^{d/\mu}\right], \quad [4.74]$$

$$3 \le n \le m,$$

$$\tilde{B}_{n+1,0} = \frac{1}{\tilde{\gamma}+\mu}\left[2\tilde{\gamma}\mu\tilde{B}_{n+1,1} + \left(\tilde{C}_{n,2} + \sum_{i=0}^{n-3}\tilde{B}_{n,i}\,(-d)^{i+1}\right.\right.$$
$$\left.\left. - \tilde{C}_{n-m+1,2} - \sum_{i=0}^{n-m-2}\tilde{B}_{n-m+1,i}\,(-L)^{i+1}\right)\mathrm{e}^{d/\mu}\right], \quad [4.75]$$

$$n \ge m+1.$$

Finally, let the constants $\tilde{C}_{n+1,1}$ and $\tilde{C}_{n+1,2}$ be given by formulas

$$\tilde{C}_{3,2} = \tilde{C}_{2,2} + \frac{\tilde{\gamma}\mu\tilde{B}_{3,0}}{\tilde{\gamma}+\mu} - 2d\tilde{B}_{3,0}$$
$$+ \frac{\tilde{\gamma}\mu(\tilde{A}_{3,0} - \tilde{A}_{2,0} + 4d\tilde{A}_{3,1})}{\tilde{\gamma}+\mu}\exp\left(2d\,\frac{\tilde{\gamma}+\mu}{\tilde{\gamma}\mu}\right),$$

$$\tilde{C}_{n+1,1} = \tilde{C}_{n,1} + \frac{\tilde{\gamma}\mu(\tilde{A}_{n,0} - \tilde{A}_{n+1,0})}{\tilde{\gamma}+\mu} + \sum_{i=0}^{n-3}\left(\tilde{A}_{n,i} - \tilde{A}_{n+1,i}\right.$$
$$\left. + \frac{(i+2)\tilde{\gamma}\mu(\tilde{A}_{n,i+1} - \tilde{A}_{n+1,i+1})}{\tilde{\gamma}+\mu}\right)(nd)^{i+1}$$
$$+ \left(\tilde{A}_{n,n-2} - \tilde{A}_{n+1,n-2} - \frac{n\tilde{\gamma}\mu\tilde{A}_{n+1,n-1}}{\tilde{\gamma}+\mu}\right)$$
$$\times (nd)^{n-1} - \tilde{A}_{n+1,n-1}\,(nd)^{n} \quad [4.76]$$
$$+ \frac{\tilde{\gamma}\mu}{\tilde{\gamma}+\mu}\left(\sum_{i=0}^{n-3}(i+1)(\tilde{B}_{n,i} - \tilde{B}_{n+1,i})(nd)^{i}\right.$$
$$\left. - (n-1)\tilde{B}_{n+1,n-2}\,(nd)^{n-2}\right)\exp\left(-nd\,\frac{\tilde{\gamma}+\mu}{\tilde{\gamma}\mu}\right),$$

$$n \ge 2, \quad n \ne m,$$

$$\begin{aligned}\tilde{C}_{n+1,2} = \tilde{C}_{n,2} &+ \frac{\tilde{\gamma}\mu(\tilde{B}_{n+1,0} - \tilde{B}_{n,0})}{\tilde{\gamma}+\mu} + \sum_{i=0}^{n-4}\Bigg(\tilde{B}_{n,i} - \tilde{B}_{n+1,i} \\ &+ \frac{(i+2)\tilde{\gamma}\mu(\tilde{B}_{n+1,i+1} - \tilde{B}_{n,i+1})}{\tilde{\gamma}+\mu}\Bigg)(nd)^{i+1} \\ &+ \left(\tilde{B}_{n,n-3} - \tilde{B}_{n+1,n-3} + \frac{(n-1)\tilde{\gamma}\mu\tilde{B}_{n+1,n-2}}{\tilde{\gamma}+\mu}\right) \\ &\times (nd)^{n-2} - \tilde{B}_{n+1,n-2}\,(nd)^{n-1} \\ &+ \frac{\tilde{\gamma}\mu}{\tilde{\gamma}+\mu}\Bigg(\sum_{i=0}^{n-2}(i+1)(\tilde{A}_{n+1,i} - \tilde{A}_{n,i})(nd)^i \\ &+ n\tilde{A}_{n+1,n-1}\,(nd)^{n-1}\Bigg)\exp\!\left(nd\,\frac{\tilde{\gamma}+\mu}{\tilde{\gamma}\mu}\right), \\ & n \geq 3, \quad n \neq m,\end{aligned} \quad [4.77]$$

$$\begin{aligned}\tilde{C}_{m+1,1} = \tilde{C}_{m,1} &+ \frac{\tilde{\gamma}\mu(\tilde{A}_{m,0} - \tilde{A}_{m+1,0})}{\tilde{\gamma}+\mu} + \sum_{i=0}^{m-3}\Bigg(\tilde{A}_{m,i} - \tilde{A}_{m+1,i} \\ &+ \frac{(i+2)\tilde{\gamma}\mu(\tilde{A}_{m,i+1} - \tilde{A}_{m+1,i+1})}{\tilde{\gamma}+\mu}\Bigg)(md)^{i+1} \\ &+ \left(\tilde{A}_{m,m-2} - \tilde{A}_{m+1,m-2} - \frac{m\tilde{\gamma}\mu\tilde{A}_{m+1,m-1}}{\tilde{\gamma}+\mu}\right) \\ &\times (md)^{m-1} - \tilde{A}_{m+1,m-1}\,(md)^{m} \\ &+ \frac{\tilde{\gamma}\mu}{\tilde{\gamma}+\mu}\Bigg(\sum_{i=0}^{m-3}(i+1)(\tilde{B}_{m,i} - \tilde{B}_{m+1,i})(md)^i \\ &- (m-1)\tilde{B}_{m+1,m-2}\,(md)^{m-2}\Bigg)\exp\!\left(-md\,\frac{\tilde{\gamma}+\mu}{\tilde{\gamma}\mu}\right) \\ &- \frac{\theta\mu}{(1+\theta)(\tilde{\gamma}+\mu)}\exp\!\left(\frac{d-L}{\mu} - \frac{L}{\tilde{\gamma}}\right),\end{aligned} \quad [4.78]$$

$$\begin{aligned}\tilde{C}_{m+1,\,2} &= \tilde{C}_{m,\,2} + \frac{\tilde{\gamma}\mu(\tilde{B}_{m+1,\,0} - \tilde{B}_{m,\,0})}{\tilde{\gamma}+\mu} + \sum_{i=0}^{m-4}\Bigg(\tilde{B}_{m,\,i} - \tilde{B}_{m+1,\,i} \\ &+ \frac{(i+2)\tilde{\gamma}\mu(\tilde{B}_{m+1,\,i+1} - \tilde{B}_{m,\,i+1})}{\tilde{\gamma}+\mu}\Bigg)(md)^{i+1} \\ &+ \Bigg(\tilde{B}_{m,\,m-3} - \tilde{B}_{m+1,\,m-3} \\ &+ \frac{(m-1)\tilde{\gamma}\mu\tilde{B}_{m+1,\,m-2}}{\tilde{\gamma}+\mu}\Bigg)(md)^{m-2} \\ &- \tilde{B}_{m+1,\,m-2}\,(md)^{m-1} + \frac{\tilde{\gamma}\mu}{\tilde{\gamma}+\mu} \\ &\times \Bigg(\sum_{i=0}^{m-2}(i+1)(\tilde{A}_{m+1,\,i} - \tilde{A}_{m,\,i})(md)^{i} \\ &+ m\tilde{A}_{m+1,\,m-1}\,(md)^{m-1}\Bigg)\exp\Bigg(md\,\frac{\tilde{\gamma}+\mu}{\tilde{\gamma}\mu}\Bigg) \\ &+ \frac{\theta\mu}{(1+\theta)(\tilde{\gamma}+\mu)}\,\mathrm{e}^{d/\mu}.\end{aligned} \qquad [4.79]$$

Note that in order to compute the constants by the formulas above for any $n \geq 2$, we have to know all the constants for $n-1$. Moreover, for any fixed $n \geq 2$, we start from the computation of $\tilde{A}_{n+1,\,j}$ for j from $n-1$ to 0, $\tilde{B}_{n+1,\,j}$ for j from $n-2$ to 0 and after that we can compute $\tilde{C}_{n+1,\,1}$ and $\tilde{C}_{n+1,\,2}$. We have introduced all the constants only to formulate the next theorem and we will get them in the proof.

THEOREM 4.5.– Let the surplus process $\left(X_t^{(d,\,L)}(x)\right)_{t\geq 0}$ follow [4.4] under the above assumptions with $0 < d < +\infty$ and $L = md$, where $m \geq 3$ is integer, and the claim sizes be exponentially distributed with mean μ. Then,

$$\varphi^{(d,\,L)}(x) = \varphi_{n+1}^{(d,\,L)}(x) \quad \text{for all} \quad x \in [nd, (n+1)d), \quad n \geq 0,$$

where

$$\varphi_1^{(d,\,L)}(x) = \tilde{C}_{1,\,1}\,\mathrm{e}^{x/\tilde{\gamma}}, \qquad [4.80]$$

$$\varphi_2^{(d,\,L)}(x) = \left(\tilde{C}_{2,\,1} + \tilde{A}_{2,\,0}\,x\right)\mathrm{e}^{x/\tilde{\gamma}} + \tilde{C}_{2,\,2}\,\mathrm{e}^{-x/\mu}, \qquad [4.81]$$

$$\varphi_{n+1}^{(d,\,L)}(x) = \left(\tilde{C}_{n+1,\,1} + \sum_{j=0}^{n-1} \tilde{A}_{n+1,\,j}\, x^{j+1}\right) \mathrm{e}^{x/\tilde{\gamma}} + \left(\tilde{C}_{n+1,\,2} + \sum_{j=0}^{n-2} \tilde{B}_{n+1,\,j}\, x^{j+1}\right) \mathrm{e}^{-x/\mu}, \quad n \geq 2. \qquad [4.82]$$

PROOF.– By theorem 2.1 and remark 2.1, $\varphi^{(d,\,L)}(x)$ is continuous on $\mathbb{R}_+$ and continuously differentiable on this interval, except at the point $x = L$, where there are only one-sided derivatives (see also section 1.1.3 and [ROL 99, pp. 162–163]).

Let us introduce the functions $\varphi_{n+1}^{(d,\,L)}(x)$, $n \geq 0$, in the following way: $\varphi_{n+1}^{(d,\,L)}(x)$ is defined on $[nd, (n+1)d)$ and coincides with $\varphi^{(d,\,L)}(x)$ on this interval.

If $x \in [0, d)$, then substituting $c^{(d,\,L)}$ and $F^{(d,\,L)}(y)$ into [4.64] yields

$$(1+\theta)\big(\mu + d - \mu \mathrm{e}^{(d-L)/\mu}\big)\big(\varphi_1^{(d,\,L)}(x)\big)' = \varphi_1^{(d,\,L)}(x). \qquad [4.83]$$

Let $\tilde{\gamma} = (1+\theta)\big(\mu + d - \mu \mathrm{e}^{(d-L)/\mu}\big)$. Solving [4.83] gives [4.80], where the constant $\tilde{C}_{1,\,1}$ can be found from $\varphi_1^{(d,\,L)}(0) = \theta/(1+\theta)$, which guarantees that the solution is the survival probability. Thus, $\tilde{C}_{1,\,1} = \theta/(1+\theta)$.

If $x \in [d, L)$, then [4.64] can be rewritten as

$$(1+\theta)\big(\mu + d - \mu \mathrm{e}^{(d-L)/\mu}\big)\big(\varphi^{(d,\,L)}(x)\big)' = \varphi^{(d,\,L)}(x) - \frac{\mathrm{e}^{(d-x)/\mu}}{\mu} \int_0^{x-d} \varphi^{(d,\,L)}(y)\, \mathrm{e}^{y/\mu}\, \mathrm{d}y, \qquad [4.84]$$

which shows that $\varphi^{(d,\,L)}(x)$ has the second derivative on this interval. So, integro-differential equation [4.84] can be reduced to the differential one

$$\tilde{\gamma}\mu\big(\varphi^{(d,\,L)}(x)\big)'' + (\tilde{\gamma} - \mu)\big(\varphi^{(d,\,L)}(x)\big)' - \varphi^{(d,\,L)}(x) = -\varphi^{(d,\,L)}(x-d). \qquad [4.85]$$

Comparison of [4.84] with [4.19] and [4.80] with [4.14] gives [4.81] for $\varphi_2^{(d,\,L)}(x)$ and [4.82] for $\varphi_{n+1}^{(d,\,L)}(x)$, $2 \leq n \leq m-1$, where the constants $\tilde{A}_{2,\,0}$,

$\tilde{C}_{2,1}$, $\tilde{C}_{2,2}$, $\tilde{B}_{3,0}$, $\tilde{C}_{3,2}$ are given before the assertion of the theorem and the constants $\tilde{A}_{n+1,n-1}$ for $2 \le n \le m-1$, $\tilde{A}_{n+1,j}$ for $1 \le j \le n-2$ and $3 \le n \le m-1$, $\tilde{A}_{n+1,0}$ for $2 \le n \le m-1$, $\tilde{B}_{n+1,n-2}$ for $3 \le n \le m-1$, $\tilde{B}_{n+1,j}$ for $1 \le j \le n-3$, $4 \le n \le m-1$, $\tilde{B}_{n+1,0}$ for $3 \le n \le m-1$ and $\tilde{C}_{n+1,1}$ for $2 \le n \le m-1$, $\tilde{C}_{n+1,2}$ for $3 \le n \le m-1$ are given by [4.65], [4.66], [4.68], [4.71], [4.72], [4.74], [4.76] and [4.77], respectively.

If $x \in [L, +\infty)$, then [4.64] can be rewritten as

$$\begin{aligned}&(1+\theta)\left(\mu + d - \mu e^{(d-L)/\mu}\right)\left(\varphi^{(d,L)}(x)\right)' \\ &\quad = \varphi^{(d,L)}(x) - \frac{e^{(d-x)/\mu}}{\mu}\int_{x-L}^{x-d} \varphi^{(d,L)}(y)\, e^{y/\mu}\, dy,\end{aligned} \qquad [4.86]$$

which shows that $\varphi^{(d,L)}(x)$ has the second derivative on this interval. Note that we imply the right second derivative of $\varphi^{(d,L)}(x)$ at the point $x = L$. Therefore, integro-differential equation [4.86] can be reduced to the differential one

$$\begin{aligned}&\tilde{\gamma}\mu\left(\varphi^{(d,L)}(x)\right)'' + (\tilde{\gamma} - \mu)\left(\varphi^{(d,L)}(x)\right)' - \varphi^{(d,L)}(x) \\ &\quad = e^{(d-L)/\mu}\varphi^{(d,L)}(x-L) - \varphi^{(d,L)}(x-d).\end{aligned} \qquad [4.87]$$

From [4.87], it follows that $\varphi_{n+1}^{(d,L)}(x)$, $n \ge m$, are also given by [4.82].

Next, if $x \in [md, (m+1)d)$, then $0 \le x - L < d$ and $(m-1)d \le x - d < md$. Hence, $\varphi^{(d,L)}(x-L) = \varphi_1^{(d,L)}(x)$ and $\varphi^{(d,L)}(x-d) = \varphi_m^{(d,L)}(x)$. So, to find the constants in [4.82] for $n = m$, we substitute $\varphi_1^{(d,L)}(x)$ and $\varphi_m^{(d,L)}(x)$, which have already been found, into [4.87]. Moreover, we take into account that

$$\begin{cases} \lim\limits_{x \uparrow L} \varphi_m^{(d,L)}(x) = \varphi_{m+1}^{(d,L)}(L), \\[2ex] \lim\limits_{x \uparrow L} \left(\varphi_m^{(d,L)}(x)\right)' = \left(\varphi_{m+1}^{(d,L)}(L)\right)' + \dfrac{\theta}{\tilde{\gamma}(1+\theta)} e^{(d-L)/\mu}, \end{cases}$$

must hold by theorem 2.1 and remark 2.1 to guarantee that the solution is the survival probability (see also section 1.1.3 and [ROL 99, pp. 162–163]). Arguments similar to those in the proofs of theorems 4.1 and 4.3 show that $\tilde{A}_{m+1,m-1}$, $\tilde{A}_{m+1,j}$ for $1 \le j \le m-2$, $\tilde{B}_{m+1,m-2}$, $\tilde{B}_{m+1,j}$ for $1 \le j \le m-3$ and $\tilde{B}_{m+1,0}$ are given by [4.65], [4.66], [4.71], [4.72] and

[4.74] with $n = m$; $\tilde{A}_{m+1,0}$, $\tilde{C}_{m+1,1}$ and $\tilde{C}_{m+1,2}$ are given by [4.69], [4.78], [4.79], respectively.

To find the constants in [4.82] for $n \geq m+1$, we also apply arguments similar to those in the proofs of theorems 4.1 and 4.3 and take into account that $\varphi^{(d,\,L)}(x)$ is continuously differentiable on $[L, +\infty)$. Thus, $\tilde{A}_{n+1,n-1}$, $\tilde{A}_{n+1,j}$ for $n-m+1 \leq j \leq n-2$, $\tilde{A}_{n+1,j}$ for $1 \leq j \leq n-m$, $\tilde{A}_{n+1,0}$, $\tilde{B}_{n+1,n-2}$, $\tilde{B}_{n+1,j}$ for $n-m \leq j \leq n-3$, $\tilde{B}_{n+1,j}$ for $1 \leq j \leq n-m-1$, $\tilde{B}_{n+1,0}$, $\tilde{C}_{n+1,1}$ and $\tilde{C}_{n+1,2}$ are given by [4.65], [4.66], [4.67], [4.70], [4.71], [4.72], [4.73], [4.75], [4.76] and [4.77], respectively. This proves the theorem. □

REMARK 4.2.– In the assertion of theorem 4.5 we assumed that $L = md$, where $m \geq 3$ is integer. Note that our considerations remain true if we remove the restriction that m is integer. But in this case the computations are much more tedious because we have to consider a finer partition of $\mathbb{R}_+$ into the intervals where $\varphi^{(d,\,L)}(x)$ is given by the same analytic expressions.

EXAMPLE 4.3.– Let the surplus processes $\left(X_t(x)\right)_{t\geq 0}$ and $\left(X_t^{(1,\,3)}(x)\right)_{t\geq 0}$ follow [4.1] and [4.4], respectively, under the above assumptions with $d = 1$ and $L = 3$. Furthermore, let $\varphi(x)$ and $\varphi^{(1,\,3)}(x)$ be the corresponding survival probabilities, the claim sizes be exponentially distributed with mean μ and $\theta = 0.1$. To compute $\varphi(x)$ and $\varphi^{(1,\,3)}(x)$, we apply formula [4.3] and theorem 4.5, respectively. The computational results in Table 4.3 show that a franchise and a liability limit can either decrease or increase the survival probability. □

4.4.2. *Case of small initial surpluses*

We now investigate how a franchise and a liability limit change the survival probability for small initial surpluses.

THEOREM 4.6.– Let the surplus processes $\left(X_t(x)\right)_{t\geq 0}$ and $\left(X_t^{(d,\,L)}(x)\right)_{t\geq 0}$ follow [4.1] and [4.4], respectively, under the above assumptions with $0 < d < +\infty$ and $L = md$, where $m \geq 3$ is integer. Moreover, let $\varphi(x)$ and $\varphi^{(d,\,L)}(x)$ be the corresponding survival probabilities and the claim sizes be exponentially distributed with mean μ.

i) If $d - \mu e^{(d-L)/\mu} > 0$, then $\varphi^{(d,\,L)}(x) < \varphi(x)$ for

$$x \in \left[0,\ \min\left\{\frac{\mu(1+\theta)}{\theta}\ \ln\left(1 + \frac{\theta\left(d - \mu e^{(d-L)/\mu}\right)}{\mu(1+\theta)}\right),\ d\right\}\right]. \qquad [4.88]$$

ii) If $d - \mu e^{(d-L)/\mu} < 0$, then $\varphi^{(d,\,L)}(x) > \varphi(x)$ for $x \in [0, d]$.

x	μ	$\varphi(x)$	$\varphi^{(1,\,3)}(x)$	$\left(\frac{\varphi^{(1,\,3)}(x)}{\varphi(x)} - 1\right) \times 100\%$
0.5	1	0.13130633	0.11600438	-11.6536%
1.0	1	0.16990844	0.14802717	-12.8783%
1.5	1	0.20679519	0.18329012	-11.3664%
2.0	1	0.24204280	0.21776307	-10.0312%
2.5	1	0.27572412	0.25190941	-8.6371%
3.0	1	0.30790874	0.28571710	-7.2072%
3.5	1	0.33866317	0.31724577	-6.3241%
4.0	1	0.36805097	0.35039170	-4.7981%
0.5	2	0.11133723	0.11111988	-0.1952%
1.0	2	0.13130633	0.13582390	3.4405%
1.5	2	0.15082671	0.16360737	8.4737%
2.0	2	0.16990844	0.19261398	13.3634%
2.5	2	0.18856139	0.22300123	18.2645%
3.0	2	0.20679519	0.25483819	23.2322%
3.5	2	0.22461926	0.28247501	25.7573%
4.0	2	0.24204280	0.31371490	29.6113%

Table 4.3. *Comparison of $\varphi(x)$ and $\varphi^{(1,\,3)}(x)$ for $\mu = 1$ and $\mu = 2$*

PROOF.– Introduce the function

$$\tilde{g}(x) = \frac{\varphi(x)}{\varphi^{(d,\,L)}(x)} \qquad [4.89]$$

for $x \in [0, d]$. Substituting [4.3] and [4.80] into [4.89] yields

$$\tilde{g}(x) = \frac{1}{\theta}\left(1 + \theta - \exp\left(-\frac{\theta x}{\mu(1+\theta)}\right)\right)$$
$$\times \exp\left(-\frac{x}{(1+\theta)\left(\mu + d - \mu e^{(d-L)/\mu}\right)}\right).$$

Taking the derivative gives

$$\tilde{g}'(x) = \frac{1}{\theta\left(\mu + d - \mu e^{(d-L)/\mu}\right)} \exp\left(-\frac{x}{(1+\theta)\left(\mu + d - \mu e^{(d-L)/\mu}\right)}\right)$$
$$\times \left(\frac{\mu(1+\theta) + \theta\left(d - \mu e^{(d-L)/\mu}\right)}{\mu(1+\theta)} \exp\left(-\frac{\theta x}{\mu(1+\theta)}\right) - 1\right).$$

Note that $\tilde{g}'(x) > 0$ for all x that are given by [4.88] if $d - \mu e^{(d-L)/\mu} > 0$ and $\tilde{g}'(x) < 0$ for $x \in [0, d]$ if $d - \mu e^{(d-L)/\mu} < 0$. Moreover, $\tilde{g}(0) = 1$. Therefore, $\tilde{g}(x) > 1$ for all x that are given by [4.88] if $d - \mu e^{(d-L)/\mu} > 0$ and $\tilde{g}(x) < 1$ for $x \in [0, d]$ if $d - \mu e^{(d-L)/\mu} < 0$, which proves the theorem. □

5

Optimal Control by the Franchise and Deductible Amounts in the Classical Risk Model

In this chapter, we consider the classical risk model where an insurance company is able to adjust a franchise amount continuously. The problem of optimal control by the franchise amount is solved from the viewpoint of survival probability maximization. We derive the Hamilton–Jacobi–Bellman equation for the optimal survival probability and prove the existence of the solution to this equation with certain properties. The verification theorem gives the connection between this solution and the optimal survival probability, which differ in a constant multiplier. Then, we concentrate on the case of exponentially distributed claim sizes. Finally, we extend these results to the problem of optimal control by a deductible amount.

5.1. Introduction

We deal with the problem of survival probability maximization in the *classical risk model*, where an insurance company is able to choose franchise and deductible amounts continuously. A *franchise*, as mentioned before, is a provision in an insurance policy whereby an insurer does not pay unless damage exceeds the franchise amount, whereas, a *deductible* is a provision whereby an insurer pays any amounts of damage that exceed the deductible amount. As a rule, these provisions are applied when an insured's losses are relatively small to prevent a large number of trivial claims. Moreover, a deductible encourages an insured to take more care of the insured property.

Normally, a franchise and a deductible also imply reduction of insurance premiums. Thus, changes of claim and premium sizes have an influence on the survival probability of the insurance company. Our problems are maximizing the survival probability adjusting the franchise or deductible amounts. To solve these problems, we apply stochastic control theory.

We assume that the *net profit condition* holds, i.e. $c > \lambda\mu$, in the classical risk model described in section 1.1.1. Moreover, let the insurance company use the *expected value principle* for premium calculation, which means that $c = \lambda\mu(1+\theta)$, where $\theta > 0$ is a *safety loading*.

In addition, we suppose that the insurance company is able to choose either a franchise amount or a deductible amount at every time $t \geq 0$. The problems of optimal control by the franchise and deductible amounts are considered in sections 5.2 and 5.3, respectively. Section 5.4 contains some bibliographical notes regarding optimal control problems in insurance.

5.2. Optimal control by the franchise amount

In this section, we solve the problem of optimal control by the franchise amount from viewpoint of survival probability maximization.

5.2.1. *Problem statement*

We assume that the insurance company adjusts the franchise amount d_t at every time $t \geq 0$ on the basis of the information available up to time t, i.e. every *admissible strategy* $(d_t)_{t\geq 0}$ of the franchise amount choice is a predictable process w.r.t. the natural filtration generated by $(N_t)_{t\geq 0}$ and $(Y_i)_{i\geq 1}$. Let $0 \leq d_t \leq d_{\max}$, where $d_{\max}$ is the maximum allowed franchise amount, such that $0 < F(d_{\max}) < 1$. In particular, if $d_t = 0$, then a franchise is not used at time t. In what follows, we will denote any admissible strategy $(d_t)_{t\geq 0}$ briefly by (d_t). We suppose that the safety loading $\theta > 0$ is constant, but insurance compensation is variable and depends on the franchise amount. Then, in contrast to the classical risk model, in this case the premium intensity at time t also depends on the franchise amount at this time and is given by

$$c(d_t) = \lambda(1+\theta) \int_{d_t}^{+\infty} y \, \mathrm{d}F(y)$$

provided that the insurance company applies the expected value principle for premium calculation.

Let $X_t^{(d_t)}(x)$ be the surplus of the insurance company at time t provided that its initial surplus is x and the strategy (d_t) is used. Then, the *surplus process* $\big(X_t^{(d_t)}(x)\big)_{t\geq 0}$ follows

$$X_t^{(d_t)}(x) = x + \int_0^t c(d_s) \, \mathrm{d}s - \sum_{i=1}^{N_t} Y_i \, \mathbb{1}_{\{Y_i > d_{T_i}\}}, \quad t \geq 0. \qquad [5.1]$$

The *ruin time* under the admissible strategy (d_t) is defined by

$$\tau^{(d_t)}(x) = \inf\{t \geq 0 \colon X_t^{(d_t)}(x) < 0\}$$

and the corresponding *infinite-horizon survival probability* is given by

$$\varphi^{(d_t)}(x) = \mathbb{P}\big[\tau^{(d_t)}(x) = \infty\big].$$

Our aim is to maximize the survival probability over all admissible strategies (d_t), i.e. to find

$$\varphi^*(x) = \sup_{(d_t)} \varphi^{(d_t)}(x)$$

and show that there is an *optimal strategy* (d_t^*), such that $\varphi^*(x) = \varphi^{(d_t^*)}(x)$ for all $x \geq 0$. We will later show that the optimal strategy is a function of an initial surplus only.

To solve this problem, we apply approaches of [AND 08, HIP 00a, HIP 03a, HIP 03b, SCH 01, SCH 02b], which were used to solve optimal control problems by investments and/or reinsurance. In section 5.2.2, we derive the Hamilton–Jacobi–Bellman equation for $\varphi^*(x)$ under the additional assumption that this function is differentiable. In section 5.2.3, we prove the existence of a solution to this equation with certain properties. In section 5.2.4, we establish a connection between this solution and $\varphi^*(x)$, which allows us to find the optimal survival probability. Section 5.2.5 deals with the case of exponentially distributed claim sizes.

5.2.2. *Hamilton–Jacobi–Bellman equation*

For any $\delta > 0$ and $\varepsilon > 0$, consider the strategy (d_t), such that the franchise amount at time t equals

$$d_t = \begin{cases} d & \text{if } \ 0 \le t \le \delta \wedge \tau_1, \\ \tilde{d}_{t-(\delta\wedge\tau_1)}\big(X^{(d)}_{\delta\wedge\tau_1}(x)\big) & \text{if } \ t > \delta\wedge\tau_1 \ \text{ and } \ \delta\wedge\tau_1 < \tau^{(d)}(x), \end{cases}$$

where (d) is any admissible constant strategy, $\big(\tilde{d}_t(x)\big)$ ($(\tilde{d}_t)$ for short) is an admissible strategy, such that $\varphi^{(\tilde{d}_t)}(x) > \varphi^*(x) - \varepsilon$. Note that we do not need to determine the strategy (d_t) for $t > \delta\wedge\tau_1$ if $\delta\wedge\tau_1 = \tau^{(d)}(x)$, i.e. ruin occurs at the time τ_1, and the case $\delta\wedge\tau_1 > \tau^{(d)}(x)$ is impossible.

For that strategy (d_t), by the law of total probability, we have

$$\begin{aligned} \varphi^*(x) \ge \varphi^{(d_t)}(x) &= \mathrm{e}^{-\lambda\delta}\,\varphi^{(d_t)}\left(x + c(d)\delta\right) \\ &+ \int_0^\delta \lambda\mathrm{e}^{-\lambda s}\bigg(F(d)\,\varphi^{(d_t)}\left(x + c(d)s\right) \\ &+ \int_d^{d\vee(x+c(d)s)} \varphi^{(d_t)}\left(x + c(d)s - y\right)\mathrm{d}F(y)\bigg)\,\mathrm{d}s \\ &\ge \mathrm{e}^{-\lambda\delta}\,\varphi^*(x + c(d)\delta) + \int_0^\delta \lambda\mathrm{e}^{-\lambda s}\bigg(F(d)\,\varphi^*(x + c(d)s) \\ &+ \int_d^{d\vee(x+c(d)s)} \varphi^*(x + c(d)s - y)\,\mathrm{d}F(y)\bigg)\,\mathrm{d}s - \varepsilon. \end{aligned} \qquad [5.2]$$

Letting $\varepsilon \to 0$ and doing elementary calculations in [5.2], we get

$$\begin{aligned} &\frac{\varphi^*(x + c(d)\delta) - \varphi^*(x)}{\delta}\,\mathrm{e}^{-\lambda\delta} - \frac{1 - \mathrm{e}^{-\lambda\delta}}{\delta}\,\varphi^*(x) \\ &+ \frac{1}{\delta}\int_0^\delta \lambda\mathrm{e}^{-\lambda s}\bigg(F(d)\,\varphi^*(x + c(d)s) \\ &+ \int_d^{d\vee(x+c(d)s)} \varphi^*(x + c(d)s - y)\,\mathrm{d}F(y)\bigg)\,\mathrm{d}s \le 0. \end{aligned} \qquad [5.3]$$

Assuming that $\varphi^*(x)$ is differentiable on $\mathbb{R}_+$ and letting $\delta \to 0$ in [5.3], we obtain

$$\begin{aligned}&(1+\theta)\int_d^{+\infty} y\,\mathrm{d}F(y)\,\left(\varphi^*(x)\right)' + \left(F(d)-1\right)\varphi^*(x)\\&+\int_d^{d\vee x}\varphi^*(x-y)\,\mathrm{d}F(y) \le 0.\end{aligned} \qquad [5.4]$$

Inequality [5.4] is true for all $d \in [0, d_{\max}]$, and equality in [5.4] is attained when d is the value of some optimal strategy at the time $t = 0$. This yields the *Hamilton–Jacobi–Bellman equation* for $\varphi^*(x)$:

$$\begin{aligned}&\sup_{d\in[0,\,d_{\max}]}\Bigg((1+\theta)\int_d^{+\infty} y\,\mathrm{d}F(y)\,\left(\varphi^*(x)\right)'\\&+\left(F(d)-1\right)\varphi^*(x)+\int_d^{d\vee x}\varphi^*(x-y)\,\mathrm{d}F(y)\Bigg)=0.\end{aligned} \qquad [5.5]$$

Note that [5.5] can be rewritten as

$$\sup_{d\in[0,\,d_{\max}]}\left(A^d\varphi^*(x)\right)=0,$$

where A^d is the infinitesimal generator of $\left(X_t^{(d_t)}(x)\right)_{t\ge 0}$ as $d_t = d$ for all $t \ge 0$.

Rewrite [5.5] in a more convenient form. Since $0 < F(d_{\max}) < 1$, expressing $\left(\varphi^*(x)\right)'$ from [5.4] gives

$$\left(\varphi^*(x)\right)' = \inf_{d\in[0,\,d_{\max}]}\left(\frac{\left(1-F(d)\right)\varphi^*(x)-\int_d^{d\vee x}\varphi^*(x-y)\,\mathrm{d}F(y)}{(1+\theta)\int_d^{+\infty} y\,\mathrm{d}F(y)}\right). \qquad [5.6]$$

PROPOSITION 5.1.– Let the surplus process $\left(X_t^{(d_t)}(x)\right)_{t\ge 0}$ follow [5.1] under the above assumptions. If $\varphi^*(x)$ is differentiable on $\mathbb{R}_+$, then it satisfies the Hamilton–Jacobi–Bellman equation [5.5], which is equivalent to [5.6].

REMARK 5.1.– Note that if there is one solution to [5.5] or [5.6], then there are infinitely many solutions to these equations, which differ with a multiplicative constant.

5.2.3. *Existence theorem*

We now prove the existence of a solution to [5.6] with certain properties.

THEOREM 5.1.– If the r.v.'s Y_i, $i \geq 1$, have a p.d.f. $f(y)$, then there is a solution $G(x)$ to [5.6] with $G(0) = \theta/(1+\theta)$, which is non-decreasing and continuously differentiable on $\mathbb{R}_+$ and $\theta/(1+\theta) \leq \lim_{x\to+\infty} G(x) \leq 1$.

PROOF.– Define a sequence of functions $\big(G_n(x)\big)_{n\geq 0}$ on $\mathbb{R}_+$ in the following way. Let $G_0(x) = \varphi^{(0)}(x)$, where $\varphi^{(0)}(x)$ is the survival probability provided that $d_t = 0$ for all $t \geq 0$, and

$$G_n'(x) = \inf_{d\in[0,\, d_{\max}]} \left(\frac{\big(1 - F(d)\big)G_{n-1}(x) - \int_d^{d\vee x} G_{n-1}(x-y)\,\mathrm{d}F(y)}{(1+\theta)\int_d^{+\infty} y\,\mathrm{d}F(y)} \right),$$
$$G_n(0) = \theta/(1+\theta), \quad n \geq 1. \qquad [5.7]$$

This gives

$$G_n(x) = \frac{\theta}{1+\theta} + \int_0^x G_n'(u)\,\mathrm{d}u, \quad n \geq 1. \qquad [5.8]$$

Since Y_i, $i \geq 1$, have a p.d.f. $f(y)$ and $\mathbb{E}[Y_i] < \infty$, the function $G_0(x)$ is continuously differentiable on $\mathbb{R}_+$ and satisfies the equation

$$G_0'(x) = \frac{G_0(x) - \int_0^x G_0(x-y)f(y)\,\mathrm{d}y}{(1+\theta)\int_0^{+\infty} yf(y)\,\mathrm{d}y}, \quad G_0(0) = \frac{\theta}{1+\theta}, \qquad [5.9]$$

by theorem 2.1 and remark 2.1 (see also section 1.1.3 and [ROL 99, pp. 162–163]).

Since $\theta > 0$, there is a unique solution to [5.9], which is a non-decreasing function with $\lim_{x\to+\infty} G_0(x) = 1$ (see section 1.1.5 and [ASM 10, BEA 84, BÜH 70, DIC 10, GER 79, GRA 91, ROL 99]).

Equation [5.9] and equality [5.7] with $n = 1$ yield $G_1'(x) \leq G_0'(x)$ for all $x \geq 0$. Therefore, from [5.8] we get $G_1(x) \leq G_0(x)$. Moreover, by equality [5.7] with $n = 1$ and the properties of $G_0(x)$, we conclude that $G_1'(x)$ is non-negative and continuous on $\mathbb{R}_+$.

Let $G_n'(x) \leq G_{n-1}'(x)$ for all $x \geq 0$, which implies $G_n(x) \leq G_{n-1}(x)$, and $G_n'(x)$ be non-negative and continuous on $\mathbb{R}_+$. Then, for all $d \in [0, d_{\max}]$ and $x \geq 0$, we have

$$\begin{aligned}
&\big(1 - F(d)\big)G_{n-1}(x) - \int_d^{d\vee x} G_{n-1}(x-y) f(y)\,\mathrm{d}y \\
&= \int_{d\vee x}^{+\infty} G_{n-1}(x) f(y)\,\mathrm{d}y + \int_d^{d\vee x} \left(\int_{x-y}^{x} G_{n-1}'(u)\,\mathrm{d}u\right) f(y)\,\mathrm{d}y \\
&\geq \int_{d\vee x}^{+\infty} G_n(x) f(y)\,\mathrm{d}y + \int_d^{d\vee x} \left(\int_{x-y}^{x} G_n'(u)\,\mathrm{d}u\right) f(y)\,\mathrm{d}y \\
&= \big(1 - F(d)\big)G_n(x) - \int_d^{d\vee x} G_n(x-y) f(y)\,\mathrm{d}y.
\end{aligned}$$

This together with [5.7] yields $G_{n+1}'(x) \leq G_n'(x)$ for all $x \geq 0$ (therefore, $G_{n+1}(x) \leq G_n(x)$) and $G_{n+1}'(x)$ is non-negative and continuous on $\mathbb{R}_+$.

Thus, by induction, $\big(G_n'(x)\big)_{n\geq 0}$ is a non-increasing sequence of functions, which are non-negative and continuous on $\mathbb{R}_+$. Hence, there is a pointwise limit of these functions. We denote it by $g(x)$.

From [5.8], we deduce that $\big(G_n(x)\big)_{n\geq 0}$ is a non-increasing sequence of non-decreasing functions. Furthermore, it is bounded from below by the value of these functions at $x = 0$, i.e. $\theta/(1+\theta)$. Hence, there is a pointwise limit of these functions. By the Cauchy criterion, it follows that $\big(G_n(\tilde{x})\big)_{n\geq 0}$ is a Cauchy sequence for any fixed $\tilde{x} \geq 0$.

We now use the Cauchy criterion to show that $\big(G_n'(x)\big)_{n\geq 0}$ converges uniformly on any $[0, \tilde{x}]$, where $0 < \tilde{x} < +\infty$. For all $k \geq 1$ and $n \geq 1$, we have

$$\sup_{x\in[0,\,\tilde{x}]} \left|G_k'(x) - G_n'(x)\right| \leq \sup_{\substack{x\in[0,\,\tilde{x}]\\ d\in[0,\,d_{\max}]}} \left| \frac{\big(1 - F(d)\big)\big(G_{k-1}(x) - G_{n-1}(x)\big)}{(1+\theta)\int_d^{+\infty} y f(y)\,\mathrm{d}y}\right.$$

$$
\left. - \frac{\int_d^{d\vee x} (G_{k-1}(x-y) - G_{n-1}(x-y)) f(y)\,\mathrm{d}y}{(1+\theta)\int_d^{+\infty} y f(y)\,\mathrm{d}y} \right|
$$

$$
\leq \sup_{d\in[0,\,d_{\max}]} \left(\frac{1-F(d)}{(1+\theta)\int_d^{+\infty} y f(y)\,\mathrm{d}y} \right) \sup_{x\in[0,\,\tilde{x}]} \left| G_{k-1}(x) - G_{n-1}(x) \right| \quad [5.10]
$$

$$
= \frac{\left| G_{k-1}(\tilde{x}) - G_{n-1}(\tilde{x}) \right|}{\mu(1+\theta)}.
$$

Here we used the facts that

$$
\sup_{d\in[0,\,d_{\max}]} \left(\frac{1-F(d)}{(1+\theta)\int_d^{+\infty} y f(y)\,\mathrm{d}y} \right)
$$

is attained at $d = 0$ and

$$
\sup_{x\in[0,\,\tilde{x}]} \left| G_{k-1}(x) - G_{n-1}(x) \right| = \sup_{x\in[0,\,\tilde{x}]} \left| \int_0^x (G'_{k-1}(u) - G'_{n-1}(u))\,\mathrm{d}u \right|
$$

$$
= \left| \int_0^{\tilde{x}} (G'_{k-1}(u) - G'_{n-1}(u))\,\mathrm{d}u \right| = \left| G_{k-1}(\tilde{x}) - G_{n-1}(\tilde{x}) \right|.
$$

By [5.10], the sequence $\left(G'_n(x)\right)_{n\geq 0}$ converges uniformly on any interval $[0, \tilde{x}]$. Thus, $g(x)$ is continuous on any interval $[0, \tilde{x}]$. Hence, it is continuous on $\mathbb{R}_+$. Let

$$
G(x) = \frac{\theta}{1+\theta} + \int_0^x g(u)\,\mathrm{d}u. \quad [5.11]
$$

From [5.11], we conclude that $G(x)$ is continuously differentiable on $\mathbb{R}_+$ and $G'(x) = g(x)$. Since $0 \leq g(x) \leq G'_0(x)$, from [5.11] we get $\theta/(1+\theta) \leq \lim_{x\to+\infty} G(x) \leq 1$.

Furthermore,

$$\begin{aligned}
&\Bigg| \inf_{d\in[0,\,d_{\max}]} \left(\frac{\big(1-F(d)\big)G_{n-1}(x) - \int_d^{d\vee x} G_{n-1}(x-y)\,\mathrm{d}F(y)}{(1+\theta)\int_d^{+\infty} y\,\mathrm{d}F(y)} \right) \\
&\quad - \inf_{d\in[0,\,d_{\max}]} \left(\frac{\big(1-F(d)\big)G(x) - \int_d^{d\vee x} G(x-y)\,\mathrm{d}F(y)}{(1+\theta)\int_d^{+\infty} y\,\mathrm{d}F(y)} \right) \Bigg| \\
&\quad \le \sup_{d\in[0,\,d_{\max}]} \Bigg| \frac{\big(1-F(d)\big)\big(G_{n-1}(x)-G(x)\big)}{(1+\theta)\int_d^{+\infty} y\,\mathrm{d}F(y)} \\
&\quad - \frac{\int_d^{d\vee x} \big(G_{n-1}(x-y)-G(x-y)\big)\,\mathrm{d}F(y)}{(1+\theta)\int_d^{+\infty} y\,\mathrm{d}F(y)} \Bigg| \\
&\quad \le \sup_{d\in[0,\,d_{\max}]} \left| \frac{\big(1-F(d)\big)\big(G_{n-1}(x)-G(x)\big)}{(1+\theta)\int_d^{+\infty} y\,\mathrm{d}F(y)} \right| \\
&\quad \le \frac{1}{\mu(1+\theta)\int_{d_{\max}}^{+\infty} y\,\mathrm{d}F(y)} \int_0^x \left| G'_{n-1}(u) - G'(u) \right| \mathrm{d}u \\
&\quad \le \frac{x}{\mu(1+\theta)\int_{d_{\max}}^{+\infty} y\,\mathrm{d}F(y)} \sup_{u\in[0,\,x]} \left| G'_{n-1}(u) - G'(u) \right|
\end{aligned}$$

and

$$\lim_{n\to\infty} \sup_{u\in[0,\,x]} \left| G'_{n-1}(u) - G'(u) \right| = 0.$$

Letting $n \to \infty$ in [5.7] shows that $G(x)$ is a solution to [5.6], which completes the proof. □

5.2.4. *Verification theorem*

We now prove that $\varphi^*(x)$ coincides with $\frac{G(x)}{\lim_{x\to+\infty} G(x)}$.

THEOREM 5.2.– Let the surplus process $\left(X_t^{(d_t)}(x)\right)_{t\ge 0}$ follow [5.1] and let $G(x)$ be the solution to [5.6] that satisfies the conditions of theorem 5.1. Then, for any $x \ge 0$ and arbitrary admissible strategy (d_t), we have

$$\varphi^{(d_t)}(x) \le \frac{G(x)}{\lim_{x\to+\infty} G(x)}, \qquad [5.12]$$

and equality in [5.12] is attained under the strategy $(d_t^*) = \left(d_t^*(X_{t-}^{(d_t^*)}(x))\right)$, where $\left(d_t^*(x)\right)$ minimizes the right-hand side of [5.6], i.e.

$$\varphi^*(x) = \varphi^{(d_t^*)}(x) = \frac{G(x)}{\lim_{x\to+\infty} G(x)}.$$

PROOF.– We denote by $G(x)$ the solution to [5.6] that satisfies the conditions of theorem 5.1. Let $(d_t^*) = \left(d_t^*(x)\right)$ be the corresponding strategy and (d_t) be any admissible strategy. Moreover, we suppose that $G(x) = 0$ for all $x < 0$ and $G'(x)$ at $x = 0$ means the right derivative of $G(x)$ at this point. Applying Dynkin's formula (see, for example [ØKS 03, p. 127] and [ØKS 07, pp. 11–12]) to the processes $\left(G\left(X_{t\wedge\tau^{(d_t)}(x)}^{(d_t)}(x)\right)\right)_{t\geq 0}$ and $\left(G\left(X_{t\wedge\tau^{(d_t^*)}(x)}^{(d_t^*)}(x)\right)\right)_{t\geq 0}$ yields

$$\begin{aligned}
&\mathbb{E}\left[G\left(X_{t\wedge\tau^{(d_t)}(x)}^{(d_t)}(x)\right)\right] \\
&= G(x) + \mathbb{E}\left[\int\limits_0^{t\wedge\tau^{(d_t)}(x)} \left((1+\theta)\int_{d_s}^{+\infty} y\,\mathrm{d}F(y)\; G'\left(X_s^{(d_t)}(x)\right)\right)\mathrm{d}s\right. \\
&\quad + \int\limits_0^{t\wedge\tau^{(d_t)}(x)} \left(\left(F(d_s)-1\right) G(X_s^{(d_t)}(x))\right. \\
&\quad \left.\left. + \int_{d_s}^{d_s\vee X_s^{(d_t)}(x)} G\left(X_s^{(d_t)}(x)-y\right)\mathrm{d}F(y)\right)\mathrm{d}s\right]
\end{aligned}$$

and

$$\begin{aligned}
&\mathbb{E}\left[G\left(X_{t\wedge\tau^{(d_t^*)}(x)}^{(d_t^*)}(x)\right)\right] \\
&= G(x) + \mathbb{E}\left[\int\limits_0^{t\wedge\tau^{(d_t^*)}(x)} \left((1+\theta)\int_{d_s^*}^{+\infty} y\,\mathrm{d}F(y)\; G'\left(X_s^{(d_t^*)}(x)\right)\right)\mathrm{d}s\right. \\
&\quad + \int\limits_0^{t\wedge\tau^{(d_t^*)}(x)} \left(\left(F(d_s^*)-1\right) G(X_s^{(d_t^*)}(x))\right. \\
&\quad \left.\left. + \int_{d_s^*}^{d_s^*\vee X_s^{(d_t^*)}(x)} G\left(X_s^{(d_t^*)}(x)-y\right)\mathrm{d}F(y)\right)\mathrm{d}s\right].
\end{aligned}$$

Since $G(x)$ is a solution to [5.6], it is also a solution to [5.5]. Therefore, it follows from the above that for all $x \geq 0$ and $t \geq 0$,

$$\mathbb{E}\Big[G\Big(X^{(d_t)}_{t\wedge\tau^{(d_t)}(x)}(x)\Big)\Big] \leq G(x) = \mathbb{E}\Big[G\Big(X^{(d^*_t)}_{t\wedge\tau^{(d^*_t)}(x)}(x)\Big)\Big]. \qquad [5.13]$$

We now show that if ruin does not occur, then for any admissible strategy (d_t), the process $\big(X^{(d_t)}_t(x)\big)_{t\geq 0}$ is unbounded from above a.s., i.e. for all $\tilde{x} > 0$,

$$\mathbb{P}\big[X^{(d_t)}_t(x) \leq \tilde{x} \quad \text{for all} \quad t \geq 0,\ \tau^{(d_t)}(x) = \infty\big] = 0. \qquad [5.14]$$

Let c be the premium intensity when the franchise amount equals 0. The probability of more than $(\tilde{x} + c)/d_{\max}$ claims of size larger than $d_{\max}$ on any unit time interval is positive. In addition, the claim process has stationary independent increments. Hence, by the Borel–Cantelli lemma, there is a random integer value $\tilde{t} > 0$, such that there are more than $(\tilde{x} + c)/d_{\max}$ claims of this kind on the time interval $[\tilde{t}, \tilde{t} + 1]$. If in addition $X^{(d_t)}_t(x) \leq \tilde{x}$ for all $t \in [0, \tilde{t}]$, then

$$X^{(d_t)}_{\tilde{t}}(x) < \tilde{x} + c - d_{\max}\big((\tilde{x} + c)/d_{\max}\big) = 0,$$

i.e. $\tau^{(d_t)}(x) < \infty$. This yields [5.14].

We now fix any admissible strategy (d_t) and $\varepsilon > 0$, which is small enough. Recall that $\varphi^{(0)}(x)$, which was introduced at the beginning of the proof of theorem 5.1, is the survival probability provided that $d_t = 0$ for all $t \geq 0$. Let $x_\varepsilon > x$ be a large enough number, such that $1 - \varphi^{(0)}(x_\varepsilon) < \varepsilon$. Set

$$\hat{\tau}_{x_\varepsilon} = \inf\big\{t > 0\colon X^{(d_t)}_t(x) = x_\varepsilon\big\}.$$

By [5.14], $\hat{\tau}_{x_\varepsilon} < \infty$ a.s. if $\tau^{(d_t)}(x) = \infty$ a.s. Define the strategy (d^ε_t) in the following way:

$$d^\varepsilon_t = \begin{cases} d_t & \text{if} \quad t \leq \hat{\tau}_{x_\varepsilon}, \\ 0 & \text{if} \quad t > \hat{\tau}_{x_\varepsilon}. \end{cases}$$

Then,

$$\mathbb{P}\big[\tau^{(d_t)}(x) = \infty,\ \tau^{(d^\varepsilon_t)}(x) < \infty\big] \leq 1 - \varphi^{(0)}(x_\varepsilon) < \varepsilon. \qquad [5.15]$$

Furthermore, since $\mathbb{P}[\hat{\tau}_{x_\varepsilon} < \infty] = 1$ and $\mathbb{P}\left[\lim_{t\to+\infty} X_t^{(0)}(x) = +\infty,\ \tau^{(0)}(x) = \infty\right] = 1$, we get $\mathbb{P}\left[\lim_{t\to+\infty} X_t^{(d_t^\varepsilon)}(x) = +\infty,\ \tau^{(d_t^\varepsilon)}(x) = \infty\right] = 1$.

Applying [5.13] to the strategy $\left(d_t^{\hat{\tau}_{x_\varepsilon}}\right)$ gives

$$\mathbb{E}\left[G\left(X^{(d_t^\varepsilon)}_{t\wedge\tau^{(d_t^\varepsilon)}(x)}(x)\right)\right] \leq G(x) = \mathbb{E}\left[G\left(X^{(d_t^*)}_{t\wedge\tau^{(d_t^*)}(x)}(x)\right)\right]. \qquad [5.16]$$

Letting $t \to \infty$ in [5.16] yields

$$\begin{aligned}\mathbb{P}\left[\tau^{(d_t)}(x) = \infty\right] - \varepsilon &\leq \mathbb{P}\left[\tau^{(d_t)}(x) = \infty,\ \tau^{(d_t^\varepsilon)}(x) = \infty\right] \\ &\leq \frac{G(x)}{\lim_{x\to+\infty} G(x)} \leq \mathbb{P}\left[\tau^{(d_t^*)}(x) = \infty\right],\end{aligned} \qquad [5.17]$$

where $\theta/(1+\theta) \leq \lim_{x\to+\infty} G(x) \leq 1$. Here we used [5.15] and the facts that $\left[G\left(X^{(d_t^\varepsilon)}_{\tau^{(d_t^\varepsilon)}(x)}(x)\right)\right] = 0$ and $\left[G\left(X^{(d_t^*)}_{\tau^{(d_t^*)}(x)}(x)\right)\right] = 0$.

Since $\varepsilon > 0$ is arbitrary in [5.17], the proof is complete. □

REMARK 5.2.– In the proof of theorem 5.2, we used any solution to [5.6] that satisfies the conditions of theorem 5.1. However, theorem 5.2 also implies uniqueness of such a solution. The corresponding strategy (d_t^*) may not be unique in general.

5.2.5. *Exponentially distributed claim sizes*

We now deal with the case of exponentially distributed claim sizes.

THEOREM 5.3.– Let the surplus process $\left(X_t^{(d_t)}(x)\right)_{t\geq 0}$ follow [5.1], the claim sizes be exponentially distributed with mean μ and $d_{\max} = \mu$. Then, the strategy (d_t) with $d_t = 0$ for all $t \geq 0$ is not optimal.

PROOF.– If the strategy (d_t) with $d_t = 0$ for all $t \geq 0$ is optimal, then $\varphi^{(0)}(x)$ is the solution to [5.6] that satisfies the conditions of theorem 5.1, i.e. $G_1'(x) = G_0'(x)$ for all $x \geq 0$. We now show that this is not true.

Indeed, note that

$$G_0(x) = \varphi^{(0)}(x) = 1 - \frac{1}{1+\theta}\exp\left(-\frac{\theta x}{(1+\theta)\mu}\right),$$

$$G_1'(x) = \inf_{d\in[0,\,\mu]} \left(\frac{\exp\left(-\frac{d}{\mu}\right)\left(1 - \frac{1}{1+\theta}\exp\left(-\frac{\theta x}{(1+\theta)\mu}\right)\right)}{(1+\theta)(d+\mu)\exp\left(-\frac{d}{\mu}\right)} \right.$$
$$\left. - \frac{\int_d^{d\vee x}\left(1 - \frac{1}{1+\theta}\exp\left(-\frac{\theta(x-y)}{(1+\theta)\mu}\right)\right)\frac{1}{\mu}\exp\left(-\frac{y}{\mu}\right)dy}{(1+\theta)(d+\mu)\exp\left(-\frac{d}{\mu}\right)} \right).$$

Let $x \geq 0$ be any fixed number. Consider two cases.

1) If $d \geq x$ (in addition, here $x \in [0,\ \mu]$), then

$$\inf_{d\in[x,\mu]} \left(\frac{1+\theta-\exp\left(-\frac{\theta x}{(1+\theta)\mu}\right)}{(1+\theta)^2(d+\mu)} \right) = \frac{1+\theta-\exp\left(-\frac{\theta x}{(1+\theta)\mu}\right)}{2\mu(1+\theta)^2}$$

as $d = \mu$.

2) If $d < x$, then

$$\inf_{d\in[0,\,\mu\wedge x]} \left(\frac{\exp\left(-\frac{\theta x}{(1+\theta)\mu}\right)\left(\exp\left(\frac{\theta d}{(1+\theta)\mu}\right) - \frac{1}{1+\theta}\right)}{(1+\theta)(d+\mu)} \right)$$
$$= \frac{\theta}{\mu(1+\theta)^2}\exp\left(-\frac{\theta x}{(1+\theta)\mu}\right)$$

as $d = 0$ because for all $d > 0$,

$$\left(\frac{\exp\left(\frac{\theta d}{(1+\theta)\mu}\right) - \frac{1}{1+\theta}}{d+\mu} \right)_d' = \frac{\left(\frac{\theta d}{\mu} - 1\right)\exp\left(\frac{\theta d}{(1+\theta)\mu}\right) + 1}{((1+\theta)(d+\mu)^2)} > 0.$$

Notice that

$$\frac{1+\theta-\exp\left(-\frac{\theta x}{(1+\theta)\mu}\right)}{2\mu(1+\theta)^2} \leq \frac{\theta}{\mu(1+\theta)^2}\exp\left(-\frac{\theta x}{(1+\theta)\mu}\right)$$

for $x \leq \hat{x}$, where $\hat{x} = \frac{(1+\theta)\mu}{\theta}\ln\left(1 + \frac{\theta}{1+\theta}\right)$. Moreover,

$$\hat{x} = \frac{(1+\theta)\mu}{\theta}\left(\frac{\theta}{1+\theta} + \sum_{i=1}^{\infty}\frac{(-1)^i}{i}\left(\frac{\theta}{1+\theta}\right)^i\right) < \mu.$$

Here we used the fact that $\sum_{i=1}^{\infty} \frac{(-1)^i}{i} \left(\frac{\theta}{1+\theta}\right)^i$ is a Leibniz series; hence, its sum is negative.

Therefore, we get

$$G_1'(x) = \begin{cases} \dfrac{1+\theta-\exp\left(-\frac{\theta x}{(1+\theta)\mu}\right)}{2\mu(1+\theta)^2} & \text{if} \quad x \in [0, \hat{x}], \\ \dfrac{\theta}{\mu(1+\theta)^2} \exp\left(-\dfrac{\theta x}{(1+\theta)\mu}\right) & \text{if} \quad x \in (\hat{x}, +\infty). \end{cases}$$

This shows that the assertion $G_1'(x) = G_0'(x)$ for all $x \in [0, \hat{x}]$ is not true, which completes the proof. □

REMARK 5.3.– Theorem 5.3 implies that we can always increase the survival probability adjusting the franchise amount if the claim sizes are exponentially distributed.

EXAMPLE 5.1.– Let the claim sizes be exponentially distributed with mean $\mu = 10$, $d_{\max} = \mu$ and $\theta = 0.1$. By theorem 1.5, we have

$$\varphi^{(0)}(x) \approx 1 - 0.90909090\,\mathrm{e}^{-x/110}, \quad x \geq 0.$$

Applying theorems 5.1 and 5.2 yields

$$\varphi^*(x) \approx \begin{cases} 0.11104877\,\mathrm{e}^{x/22} & \text{if} \quad x \leq 8.93258, \\ 1 - 0.90382792\,\mathrm{e}^{-x/110} & \text{if} \quad x > 8.93258, \end{cases}$$

$$d_t^*(x) = \begin{cases} 10 & \text{if} \quad x \leq 8.93258, \\ 0 & \text{if} \quad x > 8.93258. \end{cases}$$

Note that the optimal strategy $\left(d_t^*(x)\right)$ is simple enough in this example. Some computational results are given in Table 5.1. They show that we can considerably increase the survival probability for small initial surpluses applying the optimal strategy of the franchise amount choice. On the contrary, the difference between $\varphi^{(0)}(x)$ and $\varphi^*(x)$ is very slight for large initial surpluses. □

x	$\varphi^{(0)}(x)$	$\varphi^{*}(x)$	$\left(\frac{\varphi^{*}(x)}{\varphi^{(0)}(x)} - 1\right) \times 100\%$
0	0.09090909	0.11104877	22.1536%
1	0.09913610	0.11621292	17.2256%
2	0.10728866	0.12161722	13.3551%
3	0.11536744	0.12727283	10.3195%
4	0.12337311	0.13319146	7.9583%
5	0.13130633	0.13938532	6.1528%
10	0.16990844	0.17471407	2.8284%
20	0.24204280	0.24643083	1.8129%
30	0.30790874	0.31191545	1.3013%
40	0.36805097	0.37170951	0.9940%
50	0.42296689	0.42630750	0.7898%
60	0.47311066	0.47616097	0.6447%
70	0.51889696	0.52168220	0.5368%
80	0.56070447	0.56324768	0.4536%
90	0.59887894	0.60120114	0.3878%
100	0.63373607	0.63585648	0.3346%
200	0.85243581	0.85329010	0.1002%
300	0.94054782	0.94089200	0.0366%
400	0.97604729	0.97618596	0.0142%
500	0.99034969	0.99040555	0.0056%

Table 5.1. *Comparison of $\varphi^{(0)}(x)$ and $\varphi^{*}(x)$ for different initial surpluses*

5.3. Optimal control by the deductible amount

In this section, we solve the problem of optimal control by the deductible amount from the viewpoint of survival probability maximization.

5.3.1. *Problem statement*

We now assume that the insurance company adjusts the deductible amount $\bar{d}_t$ at every time $t \geq 0$ on the basis of the information available up to time t, i.e. every admissible strategy $(\bar{d}_t)_{t\geq 0}$ of the deductible amount choice is a predictable process w.r.t. the natural filtration generated by $(N_t)_{t\geq 0}$ and $(Y_i)_{i\geq 1}$. Let $0 \leq \bar{d}_t \leq \bar{d}_{\max}$, where $\bar{d}_{\max}$ is the maximum allowed deductible amount, such that $0 < F(\bar{d}_{\max}) < 1$. In particular, if $\bar{d}_t = 0$, then a deductible is not used at time t. In what follows, we will denote any admissible strategy $(\bar{d}_t)_{t\geq 0}$ briefly by $(\bar{d}_t)$. We suppose that the safety loading $\theta > 0$ is constant, but insurance compensation is variable and depends on the deductible amount. Then, the premium intensity at time t also depends on the deductible amount at this time and is given by

$$c(\bar{d}_t) = \lambda(1+\theta)\int_{\bar{d}_t}^{+\infty}(y-\bar{d}_t)\,\mathrm{d}F(y).$$

Let $X_t^{(\bar{d}_t)}(x)$ be the surplus of the insurance company at time t if its initial surplus is x and the strategy $(\bar{d}_t)$ is used. Then, the *surplus process* $\big(X_t^{(\bar{d}_t)}(x)\big)_{t\geq 0}$ follows

$$X_t^{(\bar{d}_t)}(x) = x + \int_0^t c(\bar{d}_s)\,\mathrm{d}s - \sum_{i=1}^{N_t}(Y_i - \bar{d}_{T_i})^+. \qquad [5.18]$$

The *ruin time* under the admissible strategy $(\bar{d}_t)$ is defined by

$$\tau^{(\bar{d}_t)}(x) = \inf\{t \geq 0\colon X_t^{(\bar{d}_t)}(x) < 0\}$$

and the corresponding *infinite-horizon survival probability* is given by

$$\varphi^{(\bar{d}_t)}(x) = \mathbb{P}\big[\tau^{(\bar{d}_t)}(x) = \infty\big].$$

Our aim is to maximize the survival probability over all admissible strategies $(\bar{d}_t)$, i.e. to find

$$\varphi^*(x) = \sup_{(\bar{d}_t)}\varphi^{(\bar{d}_t)}(x)$$

and show that there is an *optimal strategy* $(\bar{d}_t^*)$, such that $\varphi^*(x) = \varphi^{(\bar{d}_t^*)}(x)$ for all $x \geq 0$.

To solve this problem, we apply techniques similar to those in section 5.2. In section 5.3.2, we derive the Hamilton–Jacobi–Bellman equation for $\varphi^*(x)$ under the additional assumption that this function is differentiable. In section 5.3.3, we formulate the existence and verification theorems. Section 5.3.4 is devoted to the case of exponentially distributed claim sizes.

5.3.2. *Hamilton–Jacobi–Bellman equation*

For any $\delta > 0$ consider the strategy $(\bar{d}_t)$, such that the deductible amount at time t equals

$$\bar{d}_t = \begin{cases} \bar{d} & \text{if} \quad 0 \le t \le \delta \wedge \tau_1, \\ \tilde{\bar{d}}_{t-(\delta\wedge\tau_1)}\big(X^{(\bar{d})}_{\delta\wedge\tau_1}(x)\big) & \text{if} \quad t > \delta \wedge \tau_1 \ \text{ and } \ \delta \wedge \tau_1 < \tau^{(\bar{d})}(x), \end{cases}$$

where $(\bar{d})$ is an arbitrary admissible constant strategy, $\big(\tilde{\bar{d}}_t(x)\big)$ ($(\tilde{\bar{d}}_t)$ for brevity) is an admissible strategy, such that $\varphi^{(\tilde{\bar{d}}_t)}(x) > \varphi^*(x) - \varepsilon$. Here $\varepsilon > 0$ is any fixed number.

Applying the law of total probability gives

$$\begin{aligned} \varphi^*(x) \ge \varphi^{(\bar{d}_t)}(x) &= \mathrm{e}^{-\lambda\delta}\, \varphi^{(\bar{d}_t)}\left(x + c(\bar{d})\delta\right) \\ &\quad + \int_0^\delta \lambda \mathrm{e}^{-\lambda s}\Bigg(F(\bar{d})\, \varphi^{(\bar{d}_t)}\left(x + c(\bar{d})s\right) \\ &\quad + \int_{\bar{d}}^{x+c(\bar{d})s+\bar{d}} \varphi^{(\bar{d}_t)}\left(x + c(\bar{d})s + \bar{d} - y\right) \mathrm{d}F(y)\Bigg)\, \mathrm{d}s \\ &\ge \mathrm{e}^{-\lambda\delta}\, \varphi^*\left(x + c(\bar{d})\delta\right) + \int_0^\delta \lambda \mathrm{e}^{-\lambda s}\Bigg(F(\bar{d})\, \varphi^*\left(x + c(\bar{d})s\right) \\ &\quad + \int_{\bar{d}}^{x+c(\bar{d})s+\bar{d}} \varphi^*\left(x + c(\bar{d})s + \bar{d} - y\right) \mathrm{d}F(y)\Bigg)\, \mathrm{d}s - \varepsilon. \end{aligned} \qquad [5.19]$$

Letting $\varepsilon \to 0$ and doing elementary calculations in [5.19] yield

$$\begin{aligned} &\frac{\varphi^*\left(x + c(\bar{d})\delta\right) - \varphi^*(x)}{\delta}\, \mathrm{e}^{-\lambda\delta} - \frac{1 - \mathrm{e}^{-\lambda\delta}}{\delta}\, \varphi^*(x) \\ &+ \frac{1}{\delta}\int_0^\delta \lambda \mathrm{e}^{-\lambda s}\Bigg(F(\bar{d})\, \varphi^*\left(x + c(\bar{d})s\right) \\ &+ \int_{\bar{d}}^{x+c(\bar{d})s+\bar{d}} \varphi^*\left(x + c(\bar{d})s + \bar{d} - y\right) \mathrm{d}F(y)\Bigg)\, \mathrm{d}s \le 0. \end{aligned} \qquad [5.20]$$

Assuming that $\varphi^*(x)$ is differentiable on $\mathbb{R}_+$ and letting $\delta \to 0$ in [5.20], we get

$$\begin{aligned}(1+\theta)\int_{\bar d}^{+\infty}(y-\bar d)\,\mathrm{d}F(y)\,\bigl(\varphi^*(x)\bigr)' + \bigl(F(\bar d)-1\bigr)\,\varphi^*(x)\\ + \int_{\bar d}^{x+\bar d}\varphi^*(x+\bar d-y)\,\mathrm{d}F(y) \le 0.\end{aligned} \qquad [5.21]$$

Inequality [5.21] is true for all $\bar d \in [0, \bar d_{\max}]$ and equality in [5.21] is attained when $\bar d$ is the value of some optimal strategy at the time $t = 0$. Thus, we have the *Hamilton–Jacobi–Bellman equation*

$$\begin{aligned}\sup_{\bar d\in[0,\,\bar d_{\max}]}\Bigl((1+\theta)\int_{\bar d}^{+\infty}(y-\bar d)\,\mathrm{d}F(y)\,\bigl(\varphi^*(x)\bigr)'\\ + \bigl(F(\bar d)-1\bigr)\,\varphi^*(x) + \int_{\bar d}^{x+\bar d}\varphi^*(x+\bar d-y)\,\mathrm{d}F(y)\Bigr) = 0.\end{aligned} \qquad [5.22]$$

To rewrite [5.22] in a more convenient form, we express $\bigl(\varphi^*(x)\bigr)'$ from [5.21] and get

$$\bigl(\varphi^*(x)\bigr)' = \inf_{\bar d\in[0,\,\bar d_{\max}]}\left(\frac{\bigl(1-F(\bar d)\bigr)\varphi^*(x) - \int_{\bar d}^{x+\bar d}\varphi^*(x+\bar d-y)\,\mathrm{d}F(y)}{(1+\theta)\int_{\bar d}^{+\infty}(y-\bar d)\,\mathrm{d}F(y)}\right). \qquad [5.23]$$

PROPOSITION 5.2.– Let the surplus process $\bigl(X_t^{(\bar d_t)}(x)\bigr)_{t\ge 0}$ follow [5.18] under the above assumptions. If $\varphi^*(x)$ is differentiable on $\mathbb{R}_+$, then it satisfies the Hamilton–Jacobi–Bellman equation [5.22], which is equivalent to [5.23].

5.3.3. *Existence and verification theorems*

We now formulate the existence and verification theorems.

THEOREM 5.4.– If the r.v.'s Y_i, $i \ge 1$, have a p.d.f. $f(y)$, then there is a solution $G(x)$ to [5.23] with $G(0) = \theta/(1+\theta)$, which is non-decreasing and continuously differentiable on $\mathbb{R}_+$ and $\theta/(1+\theta) \le \lim_{x\to+\infty} G(x) \le 1$.

THEOREM 5.5.– Let the surplus process $\left(X_t^{(\bar{d}_t)}(x)\right)_{t\geq 0}$ follow [5.18] and let $G(x)$ be the solution to [5.23] that satisfies the conditions of theorem 5.4. Then, for any $x \geq 0$ and arbitrary admissible strategy $(\bar{d}_t)$, we have

$$\varphi^{(\bar{d}_t)}(x) \leq \frac{G(x)}{\lim_{x\to+\infty} G(x)}, \qquad [5.24]$$

and equality in [5.24] is attained under the strategy $(\bar{d}_t^*) = \left(\bar{d}_t^*\left(X_{t-}^{(\bar{d}_t^*)}(x)\right)\right)$, where $\left(\bar{d}_t^*(x)\right)$ minimizes the right-hand side of [5.23], i.e.

$$\varphi^*(x) = \varphi^{(\bar{d}_t^*)}(x) = \frac{G(x)}{\lim_{x\to+\infty} G(x)}.$$

The proofs of these theorems are similar to the corresponding ones in section 5.2. We only note that the solution $G(x)$ to [5.23] that satisfies the conditions of theorem 5.4 can be found as the limit of the sequence $\left(G_n(x)\right)_{n\geq 0}$ on $\mathbb{R}_+$, where $G_0(x) = \varphi^{(0)}(x)$ is the survival probability provided that $\bar{d}_t = 0$ for all $t \geq 0$ and

$$G_n'(x) = \inf_{\bar{d}\in[0,\,\bar{d}_{\max}]} \left(\frac{(1-F(\bar{d}))G_{n-1}(x) - \int_{\bar{d}}^{x+\bar{d}} G_{n-1}(x+\bar{d}-y)\,\mathrm{d}F(y)}{(1+\theta)\int_{\bar{d}}^{+\infty}(y-\bar{d})\,\mathrm{d}F(y)} \right),$$

$$G_n(0) = \theta/(1+\theta), \quad n \geq 1.$$

5.3.4. *Exponentially distributed claim sizes*

THEOREM 5.6.– Let the surplus process $\left(X_t^{(\bar{d}_t)}(x)\right)_{t\geq 0}$ follow [5.18] and the claim sizes be exponentially distributed. Then $\varphi^*(x) = \varphi^{(\bar{d}_t)}(x)$ for every admissible strategy $(\bar{d}_t)$, i.e. every admissible strategy is optimal.

PROOF.– If the claim sizes are exponentially distributed with mean μ, then $G_0(x)$ is given in the proof of theorem 5.3 and

$$G_1'(x) = \inf_{\bar{d}\in[0,\,\bar{d}_{\max}]} \left(\frac{\exp\left(-\frac{\bar{d}}{\mu}\right)\left(1-\frac{1}{1+\theta}\exp\left(-\frac{\theta x}{(1+\theta)\mu}\right)\right)}{(1+\theta)\int_{\bar{d}}^{+\infty}(y-\bar{d})\frac{1}{\mu}\exp\left(-\frac{y}{\mu}\right)\mathrm{d}y} - \frac{\int_{\bar{d}}^{x+\bar{d}}\left(1-\frac{1}{1+\theta}\exp\left(-\frac{\theta(x+\bar{d}-y)}{(1+\theta)\mu}\right)\right)\frac{1}{\mu}\exp\left(-\frac{y}{\mu}\right)\mathrm{d}y}{(1+\theta)\int_{\bar{d}}^{+\infty}(y-\bar{d})\frac{1}{\mu}\exp\left(-\frac{y}{\mu}\right)\mathrm{d}y} \right)$$

$$= \inf_{\bar{d}\in[0,\,\bar{d}_{\max}]} \left(\frac{\exp\left(-\frac{\bar{d}}{\mu}\right)\left(1-\frac{1}{1+\theta}\exp\left(-\frac{\theta x}{(1+\theta)\mu}\right)\right)}{(1+\theta)\left((\bar{d}+\mu)\exp\left(-\frac{\bar{d}}{\mu}\right)-\bar{d}\exp\left(-\frac{\bar{d}}{\mu}\right)\right)} - \frac{\exp\left(-\frac{\bar{d}}{\mu}\right)\left(1-\frac{1}{1+\theta}\exp\left(-\frac{\theta x}{(1+\theta)\mu}\right)\right)}{(1+\theta)\left((\bar{d}+\mu)\exp\left(-\frac{\bar{d}}{\mu}\right)-\bar{d}\exp\left(-\frac{\bar{d}}{\mu}\right)\right)} \right)$$

$$= \inf_{\bar{d}\in[0,\,\bar{d}_{\max}]} \left(\frac{\frac{\theta}{1+\theta}\exp\left(-\frac{\bar{d}}{\mu}\right)\exp\left(-\frac{\theta x}{(1+\theta)\mu}\right)}{(1+\theta)\mu\exp\left(-\frac{\bar{d}}{\mu}\right)} \right)$$

$$= \frac{\theta}{(1+\theta)^2\mu}\exp\left(-\frac{\theta x}{(1+\theta)\mu}\right) = G_0'(x),$$

where the infimum is attained at any $\bar{d} \in [0,\, \bar{d}_{\max}]$. This yields $G_1(x) = G_0(x)$. Hence, $G(x) = \varphi^{(0)}(x)$, which is our assertion. □

REMARK 5.4.– Theorem 5.6 implies that we cannot increase the survival probability adjusting the deductible amount for exponentially distributed claim sizes.

5.4. Bibliographical notes

The problems of minimization of the ruin probability, which is equivalent to maximization of the survival probability, were mainly considered when investment and/or reinsurance are available. In the problems of optimal control by investments the price of a risky asset usually follows a geometric Brownian motion.

The problem of optimal control by investments was studied by Browne [BRO 95] in the diffusion setting. He focused on obtaining investment strategies which are optimal in the sense of minimizing the ruin probability. The problem of optimal control by investments in the classical risk model was solved by Hipp and Plum [HIP 00a] without borrowing constraints and by Azcue and Muler [AZC 09] under borrowing constraints. Their aim was also to find the investment strategies which minimize the ruin probability. Liu and Yang [LIU 04] generalized results of [HIP 00a] by including a risk-free asset and also dealt with the cases where borrowing constraints or reinsurance are present.

In the classical risk model, Hipp [HIP 04a] obtained the asymptotics of the ruin probability under the optimal investment strategy in the case of light-tailed distribution of the claim sizes. For sub-exponentially distributed claim sizes, the asymptotics of the ruin probability under the optimal investment strategy was obtained by Schmidli [SCH 05]. Some results concerning the asymptotics of the ruin probabilities in optimal investment problems were obtained by Gaier and Grandits [GAI 02], Gaier, Grandits and Schachermayer [GAI 03] and Grandits [GRA 04a]; see also [GAI 04, GRA 05] for the case where a risk-free asset is available.

Optimal investment problems in the risk models where the premium intensity depends on the current surplus were solved in [AND 08, HIP 03a]. Optimal investment strategies for an insurer with state-dependent constraints were studied by Edalati and Hipp [EDA 13].

The problem of optimal proportional reinsurance was solved by Schmidli [SCH 01] in a diffusion setup and the classical risk model in the sense of minimizing the ruin probability. The problem of optimal control by excess of loss reinsurance in the classical risk model was investigated by Hipp and Vogt [HIP 03b]. Asymptotics of the ruin probabilities under optimal reinsurance policies were found by Schmidli [SCH 02a, SCH 04b, SCH 08a]. Hald and Schmidli [HAL 04] and Waters [WAT 79, WAT 83] considered the problem of minimizing the ruin probability in an asymptotic sense and maximized the adjustment coefficient.

The problem of minimizing the ruin probability when both investment and reinsurance are available was considered in [HIP 10, LUO 08, PRO 05, SCH 02b, SCH 04a, TAK 03]. Luo and Taksar [LUO 11] minimized the probability of absolute ruin, i.e. the probability that the inferior limit of the surplus process is $-\infty$, by investment and reinsurance in a diffusion setting. Bai and Guo [BAI 08] solved the problem of minimizing the ruin probability when the basic claim process follows a Brownian motion with drift and the insurer is allowed to invest in a risk-free asset and several risky assets and to purchase proportional reinsurance. Hipp and Taksar [HIP 00b] considered the possibility of investing in new business to minimize the probability of technical ruin.

A systematic description of optimal control methods applied to insurance problems was provided by Schmidli [SCH 08b]; see also [HIP 04b] for a short survey in this direction. Azcue and Muler [AZC 14] applied the viscosity approach to control problems in insurance. Note that optimality does not necessarily imply minimizing the ruin probability. One of the most widely distributed approaches is the maximization of dividends (see for example [AZC 14, SCH 08b] and references therein). For general methods and approaches in continuous-time stochastic control theory, we refer the reader, for instance, to [DAV 93, FLE 06, ØKS 07].

For more information on various benefits that can be provided by insurance policies, see, for example [PIT 14] and references therein.

6

Risk Models with Investments in Risk-Free and Risky Assets

In this chapter, we deal with generalizations of the classical risk model and the risk model with stochastic premiums where an insurance company invests all surplus in risk-free and risky assets proportionally. The price of the risky asset follows a jump process. We get upper and lower bounds for the infinite-horizon survival probability and investigate the continuity and differentiability of the infinite- and finite-horizon survival probabilities in the generalization of the classical risk model. Moreover, we extend these results to the generalization of the risk model with stochastic premiums. Finally, we obtain relations connecting accuracy and reliability of uniform approximations of the survival probabilities by their statistical estimates.

6.1. Description of the models

We consider generalizations of the *classical risk model* and the *risk model with stochastic premiums* described in sections 1.1.1 and 1.2.1, respectively: an insurance company invests all surplus in risk-free and risky assets proportionally. The price of the *risk-free asset* equals $B_t = B_0 \, \mathrm{e}^{rt}$ at time t, where B_0 is the price of the asset at the time $t = 0$ and $r > 0$ is a risk-free interest rate. The price of the *risky asset* equals $S_t = S_0 \exp\left(\tilde{r}t + \sum_{i=1}^{\tilde{N}_t} \tilde{Y}_i\right)$ at time t, where S_0 is the price of the asset at the time $t = 0$, $\tilde{r} > 0$ is a constant, $(\tilde{Y}_i)_{i \geq 1}$ is a sequence of i.i.d. r.v.'s with c.d.f. $\tilde{F}(y) = \mathbb{P}[\tilde{Y}_i \leq y]$, such that $\mathbb{P}[\tilde{Y}_i > 0] > 0$ and $\mathbb{P}[\tilde{Y}_i < 0] > 0$, $(\tilde{N}_t)_{t \geq 0}$ is a Poisson process with constant intensity $\tilde{\lambda} > 0$. We denote by $\tilde{\tau}_i$ the time of the ith jump of $(\tilde{N}_t)_{t \geq 0}$

and set $\sum_{i=1}^{0} \tilde{Y}_i = 0$ if $\tilde{N}_t = 0$. All the r.v.'s and processes in the models considered are mutually independent.

We suppose that all surplus is invested in the assets proportionally at any time $t \geq 0$: a part α is invested in the risky asset and a part $1 - \alpha$ is invested in the risk-free asset, where $0 < \alpha \leq 1$. The case $\alpha = 0$ is considered in Chapters 2 and 3. Set $\hat{r} = \alpha\tilde{r} + (1 - \alpha)r$.

Then, the *surplus process* $\left(X_t(x)\right)_{t\geq 0}$ follows either the equation

$$X_t(x) = x + \int_0^t \left(\hat{r}X_s(x) + c\right) \mathrm{d}s + \alpha \sum_{i=1}^{\tilde{N}_t} X_{\tilde{\tau}_i-}(x) \left(\mathrm{e}^{\tilde{Y}_i} - 1\right) - \sum_{i=1}^{N_t} Y_i, \quad t \geq 0, \qquad [6.1]$$

if we deal with the classical risk model, or the equation

$$X_t(x) = x + \hat{r} \int_0^t X_s(x) \, \mathrm{d}s + \alpha \sum_{i=1}^{\tilde{N}_t} X_{\tilde{\tau}_i-}(x) \left(\mathrm{e}^{\tilde{Y}_i} - 1\right) + \sum_{i=1}^{\bar{N}_t} \bar{Y}_i - \sum_{i=1}^{N_t} Y_i, \quad t \geq 0, \qquad [6.2]$$

if we deal with the risk model with stochastic premiums.

The rest of the chapter is organized as follows. In section 6.2, we obtain upper and lower power bounds for the infinite-horizon survival probability and investigate the continuity and differentiability of the infinite- and finite-horizon survival probabilities in the generalization of the classical risk model described above. In section 6.3, we extend these results to the generalization of the risk model with stochastic premiums. Section 6.4 is devoted to derivation of relations connecting accuracy and reliability of uniform approximations of the survival probabilities by their statistical estimates. In section 6.5, we give some bibliographical notes regarding different risk models with investments in a risky asset.

6.2. Classical risk model with investments in risk-free and risky assets

In section 6.2.1, we obtain upper and lower bounds for the infinite-horizon survival probability. In sections 6.2.2 and 6.2.3, we investigate the continuity and differentiability of the infinite- and finite-horizon survival probabilities, respectively.

6.2.1. *Upper and lower bounds for the infinite-horizon survival probability*

We now obtain upper and lower power bounds for the infinite-horizon survival probability using some techniques of [FRO 02, KAL 02]. To this end, we introduce the discrete-time process $X_n^x = X_{\tau_n}(x)$, $n \geq 0$, where $X_0^x = x$. Since ruin can occur only when a claim arrives, we conclude that

$$\varphi(x) = \mathbb{P}\big[X_n^x \geq 0,\ n \geq 1\big].$$

For $n \geq 1$, we define

$$\tilde{\tau}_{1,n} = \begin{cases} \min\limits_{i \geq 1}\{\tilde{\tau}_i \colon \tilde{\tau}_i > \tau_{n-1}\} & \text{if} \quad \min\limits_{i \geq 1}\{\tilde{\tau}_i \colon \tilde{\tau}_i > \tau_{n-1}\} < \tau_n, \\ \tau_n & \text{if} \quad \min\limits_{i \geq 1}\{\tilde{\tau}_i \colon \tilde{\tau}_i > \tau_{n-1}\} \geq \tau_n, \end{cases}$$

$$\tilde{\tau}_{2,n} = \begin{cases} \max\limits_{i \geq 1}\{\tilde{\tau}_i \colon \tilde{\tau}_i \leq \tau_n\} & \text{if} \quad \max\limits_{i \geq 1}\{\tilde{\tau}_i \colon \tilde{\tau}_i \leq \tau_n\} > \tau_{n-1}, \\ \tau_n & \text{if} \quad \max\limits_{i \geq 1}\{\tilde{\tau}_i \colon \tilde{\tau}_i \leq \tau_n\} \leq \tau_{n-1}. \end{cases}$$

From [6.1], it follows that

$$X_n^x = P_{1,n} X_{n-1}^x + P_{2,n} - Y_n, \quad n \geq 1, \qquad [6.3]$$

where

$$P_{1,n} = \mathrm{e}^{\hat{r}(\tau_n - \tau_{n-1})} \prod_{i \colon \tau_{n-1} < \tilde{\tau}_i \leq \tau_n} \big(1 - \alpha + \alpha \mathrm{e}^{\tilde{Y}_i}\big), \qquad [6.4]$$

$$\begin{aligned} P_{2,\,n} = {} & \left(\frac{c}{\hat{r}}\,\mathrm{e}^{\hat{r}(\tilde{\tau}_{1,\,n}-\tau_{n-1})} - \frac{c}{\hat{r}}\right)\mathrm{e}^{\hat{r}(\tau_n-\tilde{\tau}_{1,\,n})} \\ & \times \prod_{i:\,\tilde{\tau}_{1,\,n}\le\tilde{\tau}_i\le\tau_n} \left(1-\alpha+\alpha\mathrm{e}^{\tilde{Y}_i}\right) \\ & + \sum_{i:\,\tau_{n-1}<\tilde{\tau}_{i-1}<\tilde{\tau}_i\le\tau_n} \left(\left(\frac{c}{\hat{r}}\,\mathrm{e}^{\hat{r}(\tilde{\tau}_i-\tilde{\tau}_{i-1})} - \frac{c}{\hat{r}}\right)\mathrm{e}^{\hat{r}(\tau_n-\tilde{\tau}_i)}\right. \\ & \left. \times \prod_{j:\,\tilde{\tau}_i\le\tilde{\tau}_j\le\tau_n} \left(1-\alpha+\alpha\mathrm{e}^{\tilde{Y}_j}\right)\right) + \left(\frac{c}{\hat{r}}\,\mathrm{e}^{\hat{r}(\tau_n-\tilde{\tau}_{2,\,n})} - \frac{c}{\hat{r}}\right). \end{aligned} \quad [6.5]$$

Here $P_{1,n}$ describes changes in the surplus that the company has at time τ_{n-1} on $[\tau_{n-1}, \tau_n)$ and $P_{2,n}$ stands for total premiums on $[\tau_{n-1}, \tau_n)$ and investment income due to these premiums on this time interval. Note that we have the second summand in [6.5] if and only if there are at least two jumps of $(\tilde{N}_t)_{t\ge 0}$ on $(\tau_{n-1}, \tau_n]$.

It is easily seen that $P_{1,n}$, $n \ge 1$, are non-negative i.i.d. r.v.'s. Moreover, $P_{2,n}$, $n \ge 1$, are also non-negative i.i.d. r.v.'s.

Let $P_n^* = \prod_{i=1}^n P_{1,\,i}$. From [6.3] by induction on n, we get

$$X_n^x = P_n^* x + P_n^* \sum_{i=1}^n \frac{P_{2,\,i} - Y_i}{P_i^*}, \quad n \ge 1. \quad [6.6]$$

THEOREM 6.1.– Let the surplus process $\left(X_t(x)\right)_{t\ge 0}$ follow [6.1] under the above assumptions. Then there are constants $\beta_1 > 0$ and $\beta_2 > 0$, such that

$$\varphi(x) \le 1 - \beta_1 x^{-\beta_2} \quad [6.7]$$

for large enough x.

PROOF.– Since $\mathbb{P}[\tilde{Y}_i < 0] > 0$ and $\mathbb{P}[\tilde{Y}_i > 0] > 0$, $i \ge 1$, from [6.4] we deduce that

$$0 < \mathbb{P}[P_{1,n} \le \beta_3] < 1, \quad n \ge 1,$$

for any $\beta_3 \in (0,1)$. Consequently, there are constants $\beta_3 \in (0,1)$, $\beta_4 > 0$ and $\tilde{p}_1 \in (0,1)$, such that

$$\mathbb{P}[P_{1,n} \le \beta_3,\ P_{2,n} - Y_n \le \beta_4] = \tilde{p}_1, \quad n \ge 1. \quad [6.8]$$

Furthermore, there are constants $\beta_5 > 0$ and $\tilde{p}_2 \in (0,1)$, such that

$$\mathbb{P}[P_{1,n} \le \beta_5,\ \ P_{2,n} - Y_n \le -2\beta_4\beta_5/(1-\beta_3)] = \tilde{p}_2, \quad n \ge 1. \qquad [6.9]$$

Here we used the facts that $P_{1,n}$ and $P_{2,n}$, $n \ge 1$, can be small enough positive and we can choose small enough positive β_5 in [6.9].

For $x \ge \beta_4(1-\beta_3)$, we define

$$\kappa(x) = 1 + \left\lfloor \frac{\ln\big(\beta_4(1-\beta_3)\big) - \ln x}{\ln \beta_3} \right\rfloor.$$

Next, we introduce the following events:

$$E_1 = \bigcap_{n=1}^{\kappa(x)} \{P_{1,n} \le \beta_3,\ \ P_{2,n} - Y_n \le \beta_4\} \qquad [6.10]$$

and

$$E_2 = \big\{P_{1,\kappa(x)+1} \le \beta_5,\ \ P_{2,\kappa(x)+1} - Y_{\kappa(x)+1} \le -2\beta_4\beta_5/(1-\beta_3)\big\}. \qquad [6.11]$$

Note that the events $\big\{P_{1,n} \le \beta_3,\ \ P_{2,n} - Y_n \le \beta_4\big\}$, $1 \le n \le \kappa(x)$, and E_2 are mutually independent.

Next, if E_1 occurs, then by [6.6] and [6.10], we have

$$X^x_{\kappa(x)} \le \beta_3^{\kappa(x)}\, x + \beta_4 \sum_{n=1}^{\kappa(x)} \beta_3^{\kappa(x)-n} < \beta_3^{\kappa(x)}\, x + \beta_4/(1-\beta_3) \quad \text{a.s.} \qquad [6.12]$$

A simple computation shows that

$$\beta_3^{\kappa(x)}\, x \le x\, \beta_3^{\frac{\ln(\beta_4(1-\beta_3)) - \ln x}{\ln \beta_3}} = \exp\!\left(\ln \beta_3 \left(\frac{\ln\big(\beta_4(1-\beta_3)\big) - \ln x}{\ln \beta_3}\right)\right)$$
$$= \beta_4(1-\beta_3) < \beta_4/(1-\beta_3).$$

[6.13]

Moreover, if $E_1 \cap E_2$ occurs, then by [6.3] and [6.11]–[6.13], we get

$$\begin{aligned} X^x_{\kappa(x)+1} &\le \beta_5 X^x_{\kappa(x)} - 2\beta_4\beta_5/(1-\beta_3) \\ &< \beta_5\big(\beta_3^{\kappa(x)}\, x + \beta_4/(1-\beta_3) - 2\beta_4/(1-\beta_3)\big) \\ &= \beta_5\big(\beta_3^{\kappa(x)}\, x - \beta_4/(1-\beta_3)\big) < 0 \quad \text{a.s.,} \end{aligned} \qquad [6.14]$$

which implies ruin.

Hence, from [6.8]–[6.11] and [6.14], we obtain

$$\begin{aligned} \varphi(x) &\le 1 - \mathbb{P}[E_1 \cap E_2] = 1 - \tilde{p}_2\tilde{p}_1^{\kappa(x)} \\ &\le 1 - \tilde{p}_2\tilde{p}_1^{1+\frac{\ln(\beta_4(1-\beta_3))-\ln x}{\ln \beta_3}} = 1 - \beta_1 x^{-\beta_2}, \end{aligned}$$

where $\beta_1 = \tilde{p}_2\tilde{p}_1^{1+\frac{\ln(\beta_4(1-\beta_3))}{\ln \beta_3}}$ and $\beta_2 = \ln \tilde{p}_2/\ln \beta_3$. The proof is complete. □

REMARK 6.1.– Bound [6.7] implies that if the current surplus is invested in the risk-free and risky assets proportionally, then the survival probability grows no faster than a power function when initial surpluses are large enough. Hence, such investments may decrease the survival probability in comparison to the case without any investments. In particular, exponential asymptotics for the corresponding ruin probability, which holds in the classical risk model without any investments, is not true in this case.

To formulate and prove the next theorem, we define the sets Z and Z^* as follows:

$$\begin{aligned} Z = \Big\{ z \colon &\mathbb{E}\big[(1-\alpha+\alpha e^{\tilde{Y}_1})^{-z}\big] < \infty, \\ &\lambda + \hat{r}z - \tilde{\lambda}\big(\mathbb{E}\big[(1-\alpha+\alpha e^{\tilde{Y}_1})^{-z}\big] - 1\big) > 0 \Big\} \end{aligned}$$

and

$$Z^* = \Big\{ z \colon \mathbb{E}\big[(1-\alpha+\alpha e^{\tilde{Y}_1})^{-z}\big] < \infty,\ 0 < z \le 1 \Big\}.$$

Note that Z is always non-empty since it contains 0.

THEOREM 6.2.– Let the surplus process $\big(X_t(x)\big)_{t\geq 0}$ follow [6.1] under the above assumptions. If there is $z_0 \in Z^*$, such that

$$\tilde{\lambda}\big(\mathbb{E}\big[(1-\alpha+\alpha \mathrm{e}^{\tilde{Y}_1})^{-z_0}\big]-1\big)-\hat{r}z_0<0, \qquad [6.15]$$

then there is a constant $\beta_0>0$, such that

$$\varphi(x)\geq 1-\beta_0 x^{-z_0} \qquad [6.16]$$

for all $x>0$.

PROOF.– Let $z\in Z^*$. A simple computation shows that

$$\begin{aligned}
\mathbb{E}\big[P_{1,1}^{-z}\big] &= \mathbb{E}\Bigg[\Bigg(\mathrm{e}^{\hat{r}\tau_1}\prod_{i:\,\tilde{\tau}_i\leq\tau_1}\big(1-\alpha+\alpha\mathrm{e}^{\tilde{Y}_i}\big)\Bigg)^{-z}\Bigg]\\
&= \mathbb{E}\Bigg[\int_0^{+\infty}\lambda\mathrm{e}^{-(\lambda+\hat{r}z)s}\prod_{i:\,\tilde{\tau}_i\leq s}\big(1-\alpha+\alpha\mathrm{e}^{\tilde{Y}_i}\big)^{-z}\,\mathrm{d}s\Bigg]\\
&= \int_0^{+\infty}\lambda\mathrm{e}^{-(\lambda+\hat{r}z)s}\sum_{i=0}^{\infty}\Bigg(\frac{(\tilde{\lambda}s)^i\,\mathrm{e}^{-\tilde{\lambda}s}}{i!}\Big(\mathbb{E}\big[(1-\alpha+\alpha\mathrm{e}^{\tilde{Y}_1})^{-z}\big]\Big)^i\Bigg)\,\mathrm{d}s\\
&= \lambda\int_0^{+\infty}\exp\Big(\Big(-\lambda-\hat{r}z-\tilde{\lambda}+\tilde{\lambda}\mathbb{E}\big[(1-\alpha+\alpha\mathrm{e}^{\tilde{Y}_1})^{-z}\big]\Big)\,s\Big)\,\mathrm{d}s\\
&= \frac{\lambda}{\lambda+\hat{r}z-\tilde{\lambda}\big(\mathbb{E}\big[(1-\alpha+\alpha\mathrm{e}^{\tilde{Y}_1})^{-z}\big]-1\big)}.
\end{aligned}$$

Obviously, $\mathbb{E}\big[P_{1,1}^{-z}\big]$ is well-defined on Z. Since $z_0\in Z^*$ and [6.15] holds, from the above we conclude that $z_0\in Z$ and $0<\mathbb{E}\big[P_{1,1}^{-z_0}\big]<1$. Furthermore, since $\mathbb{E}\big[Y_1\big]<\infty$ and $0<z_0\leq 1$, we see that $\mathbb{E}\big[Y_1^{z_0}\big]<\infty$.

By [6.6] and the fact that $P_{2,i}$, $i\geq 1$, are non-negative, we have

$$X_n^x \geq P_n^* x - P_n^*\sum_{i=1}^{\infty}\frac{Y_i}{P_i^*}\quad \text{a.s.,}\quad n\geq 1.$$

Therefore, we get

$$\varphi(x)\geq 1-\mathbb{P}\Bigg[\sum_{i=1}^{\infty}\frac{Y_i}{P_i^*}>x\Bigg]. \qquad [6.17]$$

Since $P_{1,i}$, $i \geq 1$, are i.i.d. r.v.'s, Y_i, $i \geq 1$, are also i.i.d. r.v.'s, which are independent of $P_{1,i}$, $i \geq 1$, and $0 < z_0 \leq 1$, applying Chebyshev's inequality yields

$$\mathbb{P}\left[\sum_{i=1}^{\infty} \frac{Y_i}{P_i^*} > x\right] \leq \frac{\mathbb{E}\left[\left(\sum_{i=1}^{\infty} \frac{Y_i}{P_i^*}\right)^{z_0}\right]}{x^{z_0}} \leq \frac{\sum_{i=1}^{\infty} \mathbb{E}\left[\left(\frac{Y_i}{P_i^*}\right)^{z_0}\right]}{x^{z_0}}$$
$$= \frac{\mathbb{E}\left[Y_1^{z_0}\right]}{x^{z_0}} \sum_{i=1}^{\infty} \left(\mathbb{E}\left[P_{1,1}^{-z_0}\right]\right)^i = \frac{\mathbb{E}\left[P_{1,1}^{-z_0}\right] \mathbb{E}\left[Y_1^{z_0}\right]}{x^{z_0}\left(1 - \mathbb{E}\left[P_{1,1}^{-z_0}\right]\right)} = \beta_0 x^{-z_0}, \qquad [6.18]$$

where

$$\beta_0 = \frac{\mathbb{E}\left[P_{1,1}^{-z_0}\right] \mathbb{E}\left[Y_1^{z_0}\right]}{1 - \mathbb{E}\left[P_{1,1}^{-z_0}\right]}.$$

Finally, [6.17] and [6.18] give [6.16], which is the desired conclusion. □

REMARK 6.2.– Theorem 6.2 gives a sufficient condition for $\lim_{x \to +\infty} \varphi(x) = 1$. We will use this condition in theorem 6.3 below.

We now consider two examples, which show that the condition of theorem 6.2 can be fulfilled, i.e. there is $z_0 \in Z^*$, such that [6.15] holds.

EXAMPLE 6.1.– Let $\alpha = 1$ and $\tilde{Y}_i$, $i \geq 1$, be normally distributed with parameters 0 and $\tilde{\sigma}^2$. For all $z \in \mathbb{R}$, we have

$$\mathbb{E}\left[\left(1 - \alpha + \alpha e^{\tilde{Y}_1}\right)^{-z}\right] = e^{\tilde{\sigma}^2 z^2/2} < \infty,$$

which gives $Z^* = \{z : 0 < z \leq 1\}$.

Next, for all $z \in \mathbb{R}$, we get

$$\tilde{\lambda}\left(\mathbb{E}\left[\left(1 - \alpha + \alpha e^{\tilde{Y}_1}\right)^{-z}\right] - 1\right) - \hat{r} z = \tilde{\lambda}\left(e^{\tilde{\sigma}^2 z^2/2} - 1\right) - \tilde{r} z.$$

Moreover, for small enough $z > 0$, we have

$$\tilde{\lambda}\left(e^{\tilde{\sigma}^2 z^2/2} - 1\right) - \tilde{r} z \sim \frac{\tilde{\lambda}\tilde{\sigma}^2 z^2}{2} - \tilde{r} z,$$

which is negative. Thus, there is $z_0 \in Z^*$, such that [6.15] holds. □

EXAMPLE 6.2.– Let $\alpha = 1$ and $\tilde{Y}_i$, $i \geq 1$, have the Laplace distribution with p.d.f. $\tilde{f}(y) = \frac{\mathrm{e}^{-|y|/\tilde{\mu}}}{2\tilde{\mu}}$, where $y \in \mathbb{R}$ and $\tilde{\mu} > 0$. For all $|z| < 1/\tilde{\mu}$, we have

$$\mathbb{E}\big[\big(1-\alpha+\alpha\mathrm{e}^{\tilde{Y}_1}\big)^{-z}\big] = \frac{1}{1-\tilde{\mu}^2 z^2} < \infty,$$

which gives $Z^* = \big\{z\colon -1/\tilde{\mu} < z < 1/\tilde{\mu},\ 0 < z \leq 1\big\}$.

For all $|z| < 1/\tilde{\mu}$, we obtain

$$\tilde{\lambda}\big(\mathbb{E}\big[\big(1-\alpha+\alpha\mathrm{e}^{\tilde{Y}_1}\big)^{-z}\big]-1\big) - \hat{r}z = \frac{\tilde{\lambda}\tilde{\mu}^2 z^2}{1-\tilde{\mu}^2 z^2} - \tilde{r}z.$$

It is easy to check that

$$\tilde{\lambda}\tilde{\mu}^2 z^2 - \tilde{r}z\big(1-\tilde{\mu}^2 z^2\big) < 0$$

if either

$$z < \frac{-\tilde{\lambda}\tilde{\mu} - \sqrt{\tilde{\lambda}^2\tilde{\mu}^2 + 4\tilde{r}^2}}{2\tilde{r}\tilde{\mu}}$$

or

$$0 < z < \frac{\tilde{\lambda}\tilde{\mu} + \sqrt{\tilde{\lambda}^2\tilde{\mu}^2 + 4\tilde{r}^2}}{2\tilde{r}\tilde{\mu}}.$$

Moreover, note that

$$\frac{-\tilde{\lambda}\tilde{\mu} + \sqrt{\tilde{\lambda}^2\tilde{\mu}^2 + 4\tilde{r}^2}}{2\tilde{r}\tilde{\mu}} < \frac{1}{\tilde{\mu}}.$$

Therefore, the condition of theorem 6.2 is fulfilled for all

$$z_0 \in \left(0,\ \min\left\{1,\ \frac{-\tilde{\lambda}\tilde{\mu} + \sqrt{\tilde{\lambda}^2\tilde{\mu}^2 + 4\tilde{r}^2}}{2\tilde{r}\tilde{\mu}}\right\}\right).$$

6.2.2. *Continuity and differentiability of the infinite-horizon survival probability*

THEOREM 6.3.–

1) Let the surplus process $\left(X_t(x)\right)_{t\geq 0}$ follow [6.1] under the above assumptions. Then, the function $\varphi(x)$ is continuous on $\mathbb{R}_+$.

2) Moreover, let there be $z_0 \in Z^*$, such that [6.15] holds. Then, the following assertions hold:

i) the function $\varphi(x)$ is continuously differentiable on $\mathbb{R}_+$, except at positive points of discontinuity of $F(y)$;

ii) if $x > 0$ is a point of discontinuity of $F(y)$ and $F(x) - F(x_-) = p$, then $\varphi(x)$ has the left and right derivatives $\varphi'_-(x)$ and $\varphi'_+(x)$, respectively, and

$$\varphi'_-(x) - \varphi'_+(x) = \frac{\lambda p \varphi(0)}{\hat{r}x + c} > 0; \qquad [6.19]$$

iii) the function $\varphi(x)$ satisfies the integro-differential equation

$$\begin{aligned}(\hat{r}x + c)\varphi'_+(x) = (\lambda + \tilde{\lambda})\varphi(x) - \lambda \int_0^x \varphi(x-y)\,\mathrm{d}F(y) \\ - \tilde{\lambda} \int_{-\infty}^{+\infty} \varphi\left(\left((1-\alpha) + \alpha \mathrm{e}^y\right) x\right) \mathrm{d}\tilde{F}(y)\end{aligned} \qquad [6.20]$$

on $\mathbb{R}_+$ with the boundary condition $\lim_{x\to+\infty} \varphi(x) = 1$;

iv) for all $x_* > 0$, we have

$$\sup_{x\in[x_*,+\infty)} \left|\varphi'_+(x)\right| \leq \frac{\lambda + \tilde{\lambda}}{\hat{r}x_* + c}. \qquad [6.21]$$

PROOF.– It follows immediately from [6.1] that $X_t(x) = (x + c/\hat{r})\,\mathrm{e}^{\hat{r}t} - c/\hat{r}$ up to the time $\tau_1 \wedge \tilde{\tau}_1$. Let $\tau_1 \wedge \tilde{\tau}_1 = s$. Moreover, let $Y_1 = y$ if $\tau_1 \wedge \tilde{\tau}_1 = \tau_1$ and $\tilde{Y}_1 = y$ if $\tau_1 \wedge \tilde{\tau}_1 = \tilde{\tau}_1$. It is clear that ruin does not occur up to the time $\tau_1 \wedge \tilde{\tau}_1$ and it will not take place after that time if and only if one of the following two conditions holds:

– $\tau_1 \wedge \tilde{\tau}_1 = \tau_1$, $y \leq (x + c/\hat{r})\,\mathrm{e}^{\hat{r}t} - c/\hat{r}$ and ruin does not occur on the time interval $(s, +\infty)$ with the initial surplus $(x + c/\hat{r})\,\mathrm{e}^{\hat{r}t} - c/\hat{r} - y$;

– $\tau_1 \wedge \tilde{\tau}_1 = \tilde{\tau}_1$ and ruin does not occur on the time interval $(s, +\infty)$ with the initial surplus $\left((1-\alpha) + \alpha\mathrm{e}^y\right)\left((x + c/\hat{r})\,\mathrm{e}^{\hat{r}t} - c/\hat{r}\right)$.

Since $\mathbb{P}\big[\tau_1 \wedge \tilde{\tau}_1 = \tau_1\big] = \lambda/(\lambda + \tilde{\lambda})$, $\mathbb{P}\big[\tau_1 \wedge \tilde{\tau}_1 = \tilde{\tau}_1\big] = \tilde{\lambda}/(\lambda + \tilde{\lambda})$ and $\tau_1 \wedge \tilde{\tau}_1$ is exponentially distributed with mean $1/(\lambda + \tilde{\lambda})$, by the law of total probability, for all $x \geq 0$, we have

$$\begin{aligned}
\varphi(x) = \int_0^{+\infty} & \mathrm{e}^{-(\lambda+\tilde{\lambda})s} \\
& \times \left(\lambda \int_0^{(x+c/\hat{r})\,\mathrm{e}^{\hat{r}s}-c/\hat{r}} \varphi\big((x + c/\hat{r})\,\mathrm{e}^{\hat{r}s} - c/\hat{r} - y\big)\,\mathrm{d}F(y)\right. \\
& + \tilde{\lambda} \int_{-\infty}^{+\infty} \varphi\Big(\big((1-\alpha) + \alpha\mathrm{e}^{y}\big) \\
& \left.\times \big((x + c/\hat{r})\,\mathrm{e}^{\hat{r}s} - c/\hat{r}\big)\Big)\,\mathrm{d}\tilde{F}(y)\right)\mathrm{d}s.
\end{aligned} \qquad [6.22]$$

Changing the variable $(x + c/\hat{r})\,\mathrm{e}^{\hat{r}s} - c/\hat{r} = u$ in the outer integral in the right-hand side of [6.22], we get

$$\begin{aligned}
\varphi(x) = {} & (\hat{r}x + c)^{(\lambda+\tilde{\lambda})/\hat{r}} \\
& \times \int_x^{+\infty} \left(\frac{1}{\hat{r}u + c}\right)^{(\lambda+\tilde{\lambda})/\hat{r}+1} \left(\lambda \int_0^u \varphi(u - y)\,\mathrm{d}F(y)\right. \\
& \left. + \tilde{\lambda} \int_{-\infty}^{+\infty} \varphi\big(\big((1-\alpha) + \alpha\mathrm{e}^{y}\big)\,u\big)\,\mathrm{d}\tilde{F}(y)\right)\mathrm{d}u.
\end{aligned} \qquad [6.23]$$

Next, the inner integrals in the right-hand side of [6.23] are non-decreasing and bounded functions of u. Hence, the integrand in the outer integral is integrable on any $[x, +\infty)$, where $x \geq 0$. Thus, $\varphi(x)$ is continuous on $\mathbb{R}_+$ and the first assertion of the theorem is proved.

By Lebesgue's decomposition theorem, we have

$$F(y) = F^{\mathrm{c}}(y) + F^{\mathrm{d}}(y),$$

where $F^{\mathrm{c}}(y)$ and $F^{\mathrm{d}}(y)$ are continuous and discrete components of $F(y)$, respectively.

Denote by $(y_k)_{k\geq 1}$ a sequence of all points of discontinuity of $F(y)$, if any. Set $p_k = F(y_k) - \lim_{y\uparrow y_k} F(y)$. Then, we can rewrite [6.23] as

$$\varphi(x) = \varphi_1(x) + \varphi_2(x), \qquad [6.24]$$

where

$$\varphi_1(x) = (\hat{r}x + c)^{(\lambda+\tilde{\lambda})/\hat{r}} \times \int_x^{+\infty} \left(\frac{1}{\hat{r}u + c}\right)^{(\lambda+\tilde{\lambda})/\hat{r}+1} (\lambda I(u) + \tilde{\lambda}\tilde{I}(u))\,\mathrm{d}u, \qquad [6.25]$$

$$\varphi_2(x) = \lambda(\hat{r}x + c)^{(\lambda+\tilde{\lambda})/\hat{r}} \times \int_x^{+\infty} \left(\frac{1}{\hat{r}u + c}\right)^{(\lambda+\tilde{\lambda})/\hat{r}+1} \Bigg(\sum_{k:\, y_k \le u} p_k\, \varphi(u - y_k)\Bigg)\,\mathrm{d}u, \qquad [6.26]$$

$$I(u) = \int_0^u \varphi(u - y)\,\mathrm{d}F^{\mathrm{c}}(y),$$

$$\tilde{I}(u) = \int_{-\infty}^{+\infty} \varphi\big(\big((1-\alpha) + \alpha \mathrm{e}^y\big)\, u\big)\,\mathrm{d}\tilde{F}(y).$$

We showed the continuity of $I(u)$ in the proof of theorem 2.1. Next, for all $u \ge 0$, $u_0 \ge 0$ and $M_1 < M_2$, we have

$$\begin{aligned}
&\big|\tilde{I}(u) - \tilde{I}(u_0)\big| \\
&\le \int_{-\infty}^{+\infty} \Big|\varphi\big(\big((1-\alpha) + \alpha \mathrm{e}^y\big)\, u\big) - \varphi\big(\big((1-\alpha) + \alpha \mathrm{e}^y\big)\, u_0\big)\Big|\,\mathrm{d}\tilde{F}(y) \\
&\le \tilde{F}(M_1) + \big(1 - \tilde{F}(M_2)\big) \\
&+ \sup_{y \in [M_1,\, M_2]} \Big|\varphi\big(\big((1-\alpha) + \alpha \mathrm{e}^y\big)\, u\big) - \varphi\big(\big((1-\alpha) + \alpha \mathrm{e}^y\big)\, u_0\big)\Big|. \qquad [6.27]
\end{aligned}$$

Choosing $M_1 < 0$ and $M_2 > 0$ with large enough absolute values, we can first make $\tilde{F}(M_1)$ and $1 - \tilde{F}(M_2)$ small enough, and then we can make the third summand in [6.27] small enough when u is close to u_0. Hence, the continuity of $\tilde{I}(u)$ follows on $\mathbb{R}_+$.

Thus, the integrand in the right-hand side of [6.25] is continuous on $\mathbb{R}_+$ w.r.t. u. By [6.25], theorem A.1 and remark A.1, $\varphi_1(x)$ is differentiable on $\mathbb{R}_+$ and

$$\varphi_1'(x) = \frac{\lambda + \tilde{\lambda}}{\hat{r}x + c}\, \varphi_1(x) - \frac{\lambda I(x) + \tilde{\lambda}\tilde{I}(x)}{\hat{r}x + c}. \qquad [6.28]$$

Note that the integrand in the right-hand side of [6.26] is right-continuous on $\mathbb{R}_+$ w.r.t. u. Consequently, there is a right derivative of $\varphi_2(x)$ on $\mathbb{R}_+$ (we denote it by $\big(\varphi_2(x)\big)'_+$) and

$$\big(\varphi_2(x)\big)'_+ = \frac{\lambda+\tilde{\lambda}}{\hat{r}x+c}\,\varphi_2(x) - \frac{\lambda}{\hat{r}x+c}\sum_{k:\, y_k\le x} p_k\,\varphi(x-y_k). \qquad [6.29]$$

Since the Riemann integral does not depend on values of the integrand at a countable number of points, we can rewrite [6.26] as

$$\varphi_2(x) = \lambda(\hat{r}x+c)^{(\lambda+\tilde{\lambda})/\hat{r}} \times \int_x^{+\infty}\left(\frac{1}{\hat{r}u+c}\right)^{(\lambda+\tilde{\lambda})/\hat{r}+1}\Bigg(\sum_{k:\, y_k<u} p_k\,\varphi(u-y_k)\Bigg)\,\mathrm{d}u. \qquad [6.30]$$

The integrand in the right-hand side of [6.30] is left-continuous on $(0,+\infty)$ w.r.t. u. Therefore, there is a left derivative of $\varphi_2(x)$ on $(0,+\infty)$ (we denote it by $\big(\varphi_2(x)\big)'_-$) and

$$\big(\varphi_2(x)\big)'_- = \frac{\lambda+\tilde{\lambda}}{\hat{r}x+c}\,\varphi_2(x) - \frac{\lambda}{\hat{r}x+c}\sum_{k:\, y_k<x} p_k\,\varphi(x-y_k). \qquad [6.31]$$

From [6.29] and [6.31], we get

$$\big(\varphi_2(x)\big)'_- - \big(\varphi_2(x)\big)'_+ = \frac{\lambda p_k\,\varphi(0)\,\mathbb{1}_{\{x=y_k\}}}{\hat{r}x+c} \qquad [6.32]$$

for $x>0$. If there is $z_0\in Z^*$, such that [6.15] holds, then by theorem 6.2 and remark 6.2, we have $\lim_{x\to+\infty}\varphi(x)=1$. From this and [6.22], we deduce that $\varphi(0)>0$.

Thus, from [6.32], we conclude that $\varphi_2(x)$ is differentiable on $\mathbb{R}_+$, except at positive points y_k, $k\ge 1$. Therefore, by [6.24], $\varphi(x)$ is also differentiable on $\mathbb{R}_+$, except at positive points of discontinuity of $F(y)$, and [6.19] holds. Moreover, from [6.24], [6.28] and [6.29], we obtain

$$\begin{aligned}\varphi'_+(x) &= \frac{\lambda+\tilde{\lambda}}{\hat{r}x+c}\left(\varphi_1(x)+\varphi_2(x)\right)\\ &\quad - \frac{\lambda}{\hat{r}x+c}\left(I(x)+\sum_{k:\, y_k\le x} p_k\,\varphi(x-y_k)\right) - \frac{\tilde{\lambda}}{\hat{r}x+c}\tilde{I}(x)\\ &= \frac{\lambda+\tilde{\lambda}}{\hat{r}x+c}\varphi(x) - \frac{\lambda}{\hat{r}x+c}\int_0^x \varphi(x-y)\,\mathrm{d}F(y)\\ &\quad - \frac{\tilde{\lambda}}{\hat{r}x+c}\int_{-\infty}^{+\infty}\varphi\left(\left((1-\alpha)+\alpha \mathrm{e}^{y}\right)x\right)\mathrm{d}\tilde{F}(y),\end{aligned}$$

which yields [6.20]. Since the right-hand side of [6.20] is continuous on $\mathbb{R}_+$, except at positive points of discontinuity of $F(y)$, we conclude that $\varphi'_+(x)$ is also continuous on $\mathbb{R}_+$, except at these points. Bound [6.21] follows immediately from [6.20], which completes the proof. □

6.2.3. *Continuity of the finite-horizon survival probability and existence of its partial derivatives*

THEOREM 6.4.– Let the surplus process $\left(X_t(x)\right)_{t\ge 0}$ follow [6.1] under the above assumptions.

1) If Y_i, $i\ge 1$, have a p.d.f. $f(y)$, which is continuous on $\mathbb{R}_+$, and $\tilde{Y}_i$, $i\ge 1$, have a p.d.f. $\tilde{f}(y)$, which is continuous on $\mathbb{R}$, then $\varphi(x,t)$ is continuous on $(0,+\infty)\times[0,+\infty)$ as a function of two variables.

2) If the p.d.f. $f(y)$ of Y_i, $i\ge 1$, has the derivative $f'(y)$ on $\mathbb{R}_+$, such that $|f'(y)|$ is integrable and bounded on $\mathbb{R}_+$, and the p.d.f. $\tilde{f}(y)$ of $\tilde{Y}_i$, $i\ge 1$, have the derivative $\tilde{f}'(y)$ on $\mathbb{R}$, such that $|\tilde{f}'(y)|$, $\tilde{f}(y)\,\mathrm{e}^{-y}$ and $|\tilde{f}'(y)|\,\mathrm{e}^{-y}$ are integrable and bounded on $\mathbb{R}$, then:

i) $\varphi(x,t)$ has partial derivatives w.r.t. x and t on $(0,+\infty)\times[0,+\infty)$, which are continuous as functions of two variables;

ii) $\varphi(x,t)$ satisfies the partial integro-differential equation

$$\begin{aligned}&\frac{\partial\varphi(x,t)}{\partial t} - (\hat{r}x + c)\,\frac{\partial\varphi(x,t)}{\partial x}\\ &\quad + (\lambda+\tilde{\lambda})\,\varphi(x,t) - \lambda\int_0^x \varphi(x-y,\,t)\,\mathrm{d}F(y)\\ &\quad - \tilde{\lambda}\int_{-\infty}^{+\infty} \varphi\big(((1-\alpha)+\alpha\mathrm{e}^y)\,x,\,t\big)\,\mathrm{d}\tilde{F}(y) = 0\end{aligned} \qquad [6.33]$$

on $(0,+\infty)\times[0,+\infty)$ with the boundary conditions $\varphi(x,0) = 1$ and $\lim_{x\to+\infty}\varphi(x,t) = 1$;

iii) for all $x_* > 0$ and $T > 0$, we have

$$\sup_{\substack{x\in[x_*,+\infty),\\ t\in[0,T]}} \left|\frac{\partial\varphi(x,t)}{\partial x}\right| \le C_1\left(C_2 + \frac{C_3}{x_*}\right), \qquad [6.34]$$

where

$$C_1 = \begin{cases} \dfrac{1-\mathrm{e}^{(\hat{r}-\lambda-\tilde{\lambda})T}}{\lambda+\tilde{\lambda}-\hat{r}} & \text{if } \lambda+\tilde{\lambda}\ne\hat{r},\\ T & \text{if } \lambda+\tilde{\lambda}=\hat{r},\end{cases}$$

$$C_2 = \lambda\left(f(0) + \int_0^{+\infty} |f'(y)|\,\mathrm{d}y\right),$$

$$C_3 = \frac{\tilde{\lambda}}{\alpha}\left((1-\alpha)\int_{-\infty}^{+\infty}\tilde{f}(y)\,\mathrm{e}^{-y}\,\mathrm{d}y + \int_{-\infty}^{+\infty}|\tilde{f}'(y)|\,((1-\alpha)\,\mathrm{e}^{-y}+\alpha)\,\mathrm{d}y\right).$$

PROOF.– Arguments similar to those in the proofs of theorems 2.2, 3.2 and 6.3 show that

$$\begin{aligned}\varphi(x,t) = \int_0^t \mathrm{e}^{-(\lambda+\tilde{\lambda})s}\Bigg(\lambda\int_0^{(x+c/\hat{r})\,\mathrm{e}^{\hat{r}s}-c/\hat{r}} &\varphi\big((x+c/\hat{r})\,\mathrm{e}^{\hat{r}s}\\ - c/\hat{r} - y,\, t-s\big)\,\mathrm{d}F(y) + \tilde{\lambda}\int_{-\infty}^{+\infty}&\varphi\Big(((1-\alpha)+\alpha\mathrm{e}^y)\\ \times\big((x+c/\hat{r})\,\mathrm{e}^{\hat{r}s} - c/\hat{r}\big),\, t-s\Big)\,\mathrm{d}\tilde{F}(y)\Bigg)&\,\mathrm{d}s + \mathrm{e}^{-(\lambda+\tilde{\lambda})t}\end{aligned} \qquad [6.35]$$

for all $x\ge 0$ and $t\ge 0$.

Changing the variable $t-s=v$ in the outer integral on the right-hand side of [6.35], we get

$$\varphi(x,t)=\mathrm{e}^{-(\lambda+\tilde{\lambda})t}\int_0^t \mathrm{e}^{(\lambda+\tilde{\lambda})v}\Bigg(\lambda\int_0^{(x+c/\hat{r})\,\mathrm{e}^{\hat{r}(t-v)}-c/\hat{r}}\varphi\big((x+c/\hat{r})$$
$$\times \mathrm{e}^{\hat{r}(t-v)}-c/\hat{r}-y,\,v\big)\,\mathrm{d}F(y)+\tilde{\lambda}\int_{-\infty}^{+\infty}\varphi\Big(\big((1-\alpha)+\alpha\mathrm{e}^{y}\big) \qquad [6.36]$$
$$\times\big((x+c/\hat{r})\,\mathrm{e}^{\hat{r}(t-v)}-c/\hat{r}\big),\,v\Big)\,\mathrm{d}\tilde{F}(y)\Bigg)\,\mathrm{d}v+\mathrm{e}^{-(\lambda+\tilde{\lambda})t}.$$

We now use that Y_i and $\tilde{Y}_i$, $i\geq 1$, have p.d.f.'s $f(y)$ and $\tilde{f}(y)$, respectively. For $x>0$, changing the variables $(x+c/\hat{r})\,\mathrm{e}^{\hat{r}(t-v)}-c/\hat{r}-y=u$ and $\big((1-\alpha)+\alpha\mathrm{e}^{y}\big)\big((x+c/\hat{r})\,\mathrm{e}^{\hat{r}(t-v)}-c/\hat{r}\big)=u$ in the first and second inner integrals, respectively, on the right-hand side of [6.36], we obtain

$$\varphi(x,t)=\mathrm{e}^{-(\lambda+\tilde{\lambda})t}\int_0^t \mathrm{e}^{(\lambda+\tilde{\lambda})v}\Bigg(\lambda\int_0^{(x+c/\hat{r})\,\mathrm{e}^{\hat{r}(t-v)}-c/\hat{r}}\varphi(u,v)$$
$$\times f\big((x+c/\hat{r})\,\mathrm{e}^{\hat{r}(t-v)}-c/\hat{r}-u\big)\,\mathrm{d}u$$
$$+\tilde{\lambda}\int_{(1-\alpha)\big((x+c/\hat{r})\,\mathrm{e}^{\hat{r}(t-v)}-c/\hat{r}\big)}^{+\infty}\varphi(u,v)$$
$$\times\frac{1}{u+(1-\alpha)\big((x+c/\hat{r})\,\mathrm{e}^{\hat{r}(t-v)}-c/\hat{r}\big)} \qquad [6.37]$$
$$\times\tilde{f}\Bigg(\ln\Bigg(\frac{u}{\alpha\big((x+c/\hat{r})\,\mathrm{e}^{\hat{r}(t-v)}-c/\hat{r}\big)}$$
$$+\frac{\alpha-1}{\alpha}\Bigg)\Bigg)\,\mathrm{d}u\Bigg)\,\mathrm{d}v+\mathrm{e}^{-(\lambda+\tilde{\lambda})t}.$$

Note that we excluded the case $x=0$ to avoid division by zero in [6.37].

Let

$$I(x,t)=\int_0^t \mathrm{e}^{(\lambda+\tilde{\lambda})v}$$
$$\times\Bigg(\lambda\int_0^{(x+c/\hat{r})\,\mathrm{e}^{\hat{r}(t-v)}-c/\hat{r}}\varphi(u,v)f\big((x+c/\hat{r})\,\mathrm{e}^{\hat{r}(t-v)}-c/\hat{r}-u\big)\,\mathrm{d}u$$

$$+\tilde{\lambda}\int_{(1-\alpha)\left((x+c/\hat{r})\,\mathrm{e}^{\hat{r}(t-v)}-c/\hat{r}\right)}^{+\infty}\frac{\varphi(u,v)}{u+(1-\alpha)\left((x+c/\hat{r})\,\mathrm{e}^{\hat{r}(t-v)}-c/\hat{r}\right)}$$

$$\times\tilde{f}\left(\ln\left(\frac{u}{\alpha\left((x+c/\hat{r})\,\mathrm{e}^{\hat{r}(t-v)}-c/\hat{r}\right)}+\frac{\alpha-1}{\alpha}\right)\right)\mathrm{d}u\Big)\,\mathrm{d}v.$$

If $f(y)$ is continuous on $\mathbb{R}_+$ and $\tilde{f}(y)$ is continuous on $\mathbb{R}$, then applying arguments similar to those in the proof of theorem 2.2 establishes the continuity of $\varphi(x,t)$ as a function of two variables. Indeed, for any fixed $(x_0,t_0)\in(0,+\infty)\times(0,+\infty)$ and all $0<x\le x_0$ and $0\le t\le t_0$, we have

$$\left|I(x,t)-I(x_0,t_0)\right|$$

$$\le\int_0^t \mathrm{e}^{(\lambda+\tilde{\lambda})v}\Big(\lambda\int_0^{(x+c/\hat{r})\,\mathrm{e}^{\hat{r}(t-v)}-c/\hat{r}}\Big|f\left((x_0+c/\hat{r})\,\mathrm{e}^{\hat{r}(t_0-v)}-c/\hat{r}-u\right)$$

$$-f\left((x+c/\hat{r})\,\mathrm{e}^{\hat{r}(t-v)}-c/\hat{r}-u\right)\Big|\,\mathrm{d}u$$

$$+\lambda\int_{(x+c/\hat{r})\,\mathrm{e}^{\hat{r}(t-v)}-c/\hat{r}}^{(x_0+c/\hat{r})\,\mathrm{e}^{\hat{r}(t_0-v)}-c/\hat{r}}f\left((x_0+c/\hat{r})\,\mathrm{e}^{\hat{r}(t_0-v)}-c/\hat{r}-u\right)\mathrm{d}u$$

$$+\tilde{\lambda}\int_{(1-\alpha)\left((x_0+c/\hat{r})\,\mathrm{e}^{\hat{r}(t_0-v)}-c/\hat{r}\right)}^{+\infty}\Bigg|\frac{1}{u+(1-\alpha)\left((x_0+c/\hat{r})\,\mathrm{e}^{\hat{r}(t_0-v)}-c/\hat{r}\right)}$$

$$\times\tilde{f}\left(\ln\left(\frac{u}{\alpha\left((x_0+c/\hat{r})\,\mathrm{e}^{\hat{r}(t_0-v)}-c/\hat{r}\right)}+\frac{\alpha-1}{\alpha}\right)\right)$$

$$-\frac{1}{u+(1-\alpha)\left((x+c/\hat{r})\,\mathrm{e}^{\hat{r}(t-v)}-c/\hat{r}\right)}$$

$$\times\tilde{f}\left(\ln\left(\frac{u}{\alpha\left((x+c/\hat{r})\,\mathrm{e}^{\hat{r}(t-v)}-c/\hat{r}\right)}+\frac{\alpha-1}{\alpha}\right)\right)\Bigg|\,\mathrm{d}u$$

$$+\tilde{\lambda}\int_{(1-\alpha)\left((x+c/\hat{r})\,\mathrm{e}^{\hat{r}(t-v)}-c/\hat{r}\right)}^{(1-\alpha)\left((x_0+c/\hat{r})\,\mathrm{e}^{\hat{r}(t_0-v)}-c/\hat{r}\right)}\frac{1}{u+(1-\alpha)\left((x+c/\hat{r})\,\mathrm{e}^{\hat{r}(t-v)}-c/\hat{r}\right)}$$

$$\times\tilde{f}\left(\ln\left(\frac{u}{\alpha\left((x+c/\hat{r})\,\mathrm{e}^{\hat{r}(t-v)}-c/\hat{r}\right)}+\frac{\alpha-1}{\alpha}\right)\right)\mathrm{d}u\Big)\mathrm{d}v$$

$$+(\lambda+\tilde{\lambda})\int_t^{t_0}\mathrm{e}^{(\lambda+\tilde{\lambda})v}\,\mathrm{d}v.$$

From this we get

$$
\begin{aligned}
&\left|I(x,t)-I(x_0,t_0)\right| \\
&\le \int_0^t e^{(\lambda+\tilde{\lambda})v}\Bigg(\lambda\int_0^{(x+c/\hat{r})\,e^{\hat{r}(t-v)}-c/\hat{r}} \Big|f\big((x_0+c/\hat{r})\,e^{\hat{r}(t_0-v)} \\
&-c/\hat{r}-u\big)-f\big((x+c/\hat{r})\,e^{\hat{r}(t-v)}-c/\hat{r}-u\big)\Big|\,du \\
&+\lambda F\big((x_0+c/\hat{r})\,e^{\hat{r}(t_0-v)}-(x+c/\hat{r})\,e^{\hat{r}(t-v)}\big) \\
&+\tilde{\lambda}\int_{M_1(v)}^{M_2}\Bigg|\frac{1}{u+(1-\alpha)\big((x_0+c/\hat{r})\,e^{\hat{r}(t_0-v)}-c/\hat{r}\big)} \\
&\times\tilde{f}\Bigg(\ln\Bigg(\frac{u}{\alpha\big((x_0+c/\hat{r})\,e^{\hat{r}(t_0-v)}-c/\hat{r}\big)}+\frac{\alpha-1}{\alpha}\Bigg)\Bigg) \\
&-\frac{1}{u+(1-\alpha)\big((x+c/\hat{r})\,e^{\hat{r}(t-v)}-c/\hat{r}\big)} \\
&\times\tilde{f}\Bigg(\ln\Bigg(\frac{u}{\alpha\big((x+c/\hat{r})\,e^{\hat{r}(t-v)}-c/\hat{r}\big)}+\frac{\alpha-1}{\alpha}\Bigg)\Bigg)\Bigg|\,du \\
&+\tilde{\lambda}\tilde{F}\Bigg(\ln\Bigg(\frac{(1-\alpha)\big((x_0+c/\hat{r})\,e^{\hat{r}(t_0-v)}-c/\hat{r}\big)}{\alpha\big((x+c/\hat{r})\,e^{\hat{r}(t-v)}-c/\hat{r}\big)}+\frac{\alpha-1}{\alpha}\Bigg)\Bigg) \\
&+2\tilde{\lambda}\Bigg(1-\tilde{F}\Bigg(\ln\Bigg(\frac{M_2}{\alpha\big((x_0+c/\hat{r})\,e^{\hat{r}(t_0-v)}-c/\hat{r}\big)}+\frac{\alpha-1}{\alpha}\Bigg)\Bigg) \\
&+\tilde{F}\Bigg(\ln\Bigg(\frac{M_1(v)}{\alpha\big((x_0+c/\hat{r})\,e^{\hat{r}(t_0-v)}-c/\hat{r}\big)}+\frac{\alpha-1}{\alpha}\Bigg)\Bigg)\Bigg)\Bigg)\,dv \\
&+e^{(\lambda+\tilde{\lambda})t_0}-e^{(\lambda+\tilde{\lambda})t},
\end{aligned}
\tag{6.38}
$$

where the function $M_1(v)$ and the number M_2 are such that

$$
(1-\alpha)\big((x_0+c/\hat{r})\,e^{\hat{r}(t_0-v)}-c/\hat{r}\big)<M_1(v)<M_2<\infty
$$

for all $v\in[0,t_0]$. Therefore, choosing $M_1(v)$, such that

$$
\sup_{v\in[0,t_0]}\left|(1-\alpha)\big((x_0+c/\hat{r})\,e^{\hat{r}(t_0-v)}-c/\hat{r}\big)-M_1(v)\right|
$$

is small enough, and large enough M_2 we can first make

$$1-\tilde{F}\left(\ln\left(\frac{M_2}{\alpha\big((x_0+c/\hat{r})\,\mathrm{e}^{\hat{r}(t_0-v)}-c/\hat{r}\big)}+\frac{\alpha-1}{\alpha}\right)\right)$$
$$+\tilde{F}\left(\ln\left(\frac{M_1(v)}{\alpha\big((x_0+c/\hat{r})\,\mathrm{e}^{\hat{r}(t_0-v)}-c/\hat{r}\big)}+\frac{\alpha-1}{\alpha}\right)\right)$$

small enough and, then, by the continuity of $f(y)$ and $\tilde{f}(y)$, we can make the right-hand side of [6.38] small enough when x is close to x_0 and t is close to t_0. Hence, $\varphi(x,t)$ is continuous at (x_0,t_0) for $x \le x_0$ and $t \le t_0$. Similarly, we can consider other cases. Thus, we proved the continuity of $\varphi(x,t)$ on $(0,+\infty)\times[0,+\infty)$ as a function of two variables.

If $f(y)$ and $\tilde{f}(y)$ have the derivatives $f'(y)$ and $\tilde{f}'(y)$, which satisfy conditions of the theorem, then we can show the existence of partial derivatives of $\varphi(x,t)$ as in the proof of theorem 2.2. Furthermore, from [6.37] we get

$$\begin{aligned}\frac{\partial\varphi(x,t)}{\partial x} &= \mathrm{e}^{(r-\lambda-\tilde{\lambda})t}\int_0^t \mathrm{e}^{(\lambda+\tilde{\lambda}-r)v}\Bigg(\lambda f(0)\,\varphi\big((x+c/\hat{r})\,\mathrm{e}^{\hat{r}(t-v)} \\ &\quad -c/\hat{r},\,v\big)+\lambda\int_0^{(x+c/\hat{r})\,\mathrm{e}^{\hat{r}(t-v)}-c/\hat{r}}\varphi(u,v) \\ &\quad \times f'\big((x+c/\hat{r})\,\mathrm{e}^{\hat{r}(t-v)}-c/\hat{r}-u\big)\,\mathrm{d}u \\ &\quad -\tilde{\lambda}\int_{(1-\alpha)\left((x+c/\hat{r})\,\mathrm{e}^{\hat{r}(t-v)}-c/\hat{r}\right)}^{+\infty}\varphi(u,v) \\ &\quad \times\frac{1}{\big(u+(1-\alpha)\big((x+c/\hat{r})\,\mathrm{e}^{\hat{r}(t-v)}-c/\hat{r}\big)\big)^2} \\ &\quad \times\Bigg((\alpha-1)\,\tilde{f}\Bigg(\ln\Bigg(\frac{u}{\alpha\big((x+c/\hat{r})\,\mathrm{e}^{\hat{r}(t-v)}-c/\hat{r}\big)} \\ &\quad +\frac{\alpha-1}{\alpha}\Bigg)\Bigg)+\frac{u}{(x+c/\hat{r})\,\mathrm{e}^{\hat{r}(t-v)}-c/\hat{r}} \\ &\quad \times\tilde{f}'\Bigg(\ln\Bigg(\frac{u}{\alpha\big((x+c/\hat{r})\,\mathrm{e}^{\hat{r}(t-v)}-c/\hat{r}\big)} \\ &\quad +\frac{\alpha-1}{\alpha}\Bigg)\Bigg)\Bigg)\,\mathrm{d}u\Bigg)\,\mathrm{d}v\end{aligned} \qquad [6.39]$$

and

$$\begin{aligned}\frac{\partial\varphi(x,t)}{\partial t} =& -(\lambda+\tilde{\lambda})\,\varphi(x,t) + \lambda\int_0^x \varphi(u,t)\,f(x-u)\,\mathrm{d}u \\ &+ \tilde{\lambda}\int_{(1-\alpha)x}^{+\infty} \frac{\varphi(u,t)}{u+(\alpha-1)x}\,\tilde{f}\bigg(\ln\Big(\frac{u}{\alpha x}+\frac{\alpha-1}{\alpha}\Big)\bigg)\,\mathrm{d}u \\ &+ (\hat{r}x+c)\,\mathrm{e}^{(r-\lambda-\tilde{\lambda})t}\int_0^t \mathrm{e}^{(\lambda+\tilde{\lambda}-r)v} \\ &\times\bigg(\lambda f(0)\,\varphi\big((x+c/\hat{r})\,\mathrm{e}^{\hat{r}(t-v)}-c/\hat{r},\,v\big) \\ &+\lambda\int_0^{(x+c/\hat{r})\,\mathrm{e}^{\hat{r}(t-v)}-c/\hat{r}} \varphi(u,v) \\ &\times f'\big((x+c/\hat{r})\,\mathrm{e}^{\hat{r}(t-v)}-c/\hat{r}-u\big)\,\mathrm{d}u \\ &-\tilde{\lambda}\int_{(1-\alpha)\left((x+c/\hat{r})\,\mathrm{e}^{\hat{r}(t-v)}-c/\hat{r}\right)}^{+\infty} \varphi(u,v) \\ &\times\frac{1}{\big(u+(1-\alpha)\big((x+c/\hat{r})\,\mathrm{e}^{\hat{r}(t-v)}-c/\hat{r}\big)\big)^2} \\ &\times\bigg((\alpha-1)\,\tilde{f}\bigg(\ln\bigg(\frac{u}{\alpha\big((x+c/\hat{r})\,\mathrm{e}^{\hat{r}(t-v)}-c/\hat{r}\big)} \\ &+\frac{\alpha-1}{\alpha}\bigg)\bigg) + \frac{u}{(x+c/\hat{r})\,\mathrm{e}^{\hat{r}(t-v)}-c/\hat{r}} \\ &\times\tilde{f}'\bigg(\ln\bigg(\frac{u}{\alpha\big((x+c/\hat{r})\,\mathrm{e}^{\hat{r}(t-v)}-c/\hat{r}\big)} \\ &+\frac{\alpha-1}{\alpha}\bigg)\bigg)\bigg)\,\mathrm{d}u\bigg)\,\mathrm{d}v.\end{aligned} \qquad [6.40]$$

Combining [6.39] and [6.40], we have

$$\begin{aligned}\frac{\partial\varphi(x,t)}{\partial t} =& -(\lambda+\tilde{\lambda})\,\varphi(x,t) + \lambda\int_0^x \varphi(u,t)\,f(x-u)\,\mathrm{d}u \\ &+ \tilde{\lambda}\int_{(1-\alpha)x}^{+\infty} \frac{\varphi(u,t)}{u+(\alpha-1)x}\,\tilde{f}\bigg(\ln\Big(\frac{u}{\alpha x}+\frac{\alpha-1}{\alpha}\Big)\bigg)\,\mathrm{d}u\end{aligned}$$

$$+ (\hat{r}x + c)\,\frac{\partial \varphi(x,t)}{\partial x}$$

$$= (\hat{r}x + c)\,\frac{\partial \varphi(x,t)}{\partial x} - (\lambda + \tilde{\lambda})\,\varphi(x,t) + \lambda \int_0^x \varphi(x-y,\,t)\,\mathrm{d}F(y)$$

$$+ \tilde{\lambda} \int_{-\infty}^{+\infty} \varphi\big(\big((1-\alpha) + \alpha \mathrm{e}^{y}\big)\,x,\,t\big)\,\mathrm{d}\tilde{F}(y),$$

which gives [6.33].

From [6.39], it follows that

$$\left|\frac{\partial \varphi(x,t)}{\partial x}\right| \le \mathrm{e}^{(r-\lambda-\tilde{\lambda})t} \int_0^t \mathrm{e}^{(\lambda+\tilde{\lambda}-r)v} \Bigg(\lambda f(0) + \lambda \int_0^{+\infty} |f'(y)|\,\mathrm{d}y$$

$$+ \frac{\tilde{\lambda}(1-\alpha)}{\alpha\big((x + c/\hat{r})\,\mathrm{e}^{\hat{r}(t-v)} - c/\hat{r}\big)} \int_{-\infty}^{+\infty} \tilde{f}(y)\,\mathrm{e}^{-y}\,\mathrm{d}y$$

$$+ \frac{\tilde{\lambda}}{\alpha\big((x + c/\hat{r})\,\mathrm{e}^{\hat{r}(t-v)} - c/\hat{r}\big)}$$

$$\times \int_{-\infty}^{+\infty} |\tilde{f}'(y)|\,\big((1-\alpha)\,\mathrm{e}^{-y} + \alpha\big)\,\mathrm{d}y\Bigg)\,\mathrm{d}v$$

for all $x > 0$ and $t \ge 0$, which yields [6.34]. As in theorem 2.2, the boundary condition $\varphi(x,0) = 1$ is obvious and the condition $\lim_{x\to+\infty} \varphi(x,t) = 1$ follows immediately from the fact that there can be only a finite number of claims on any finite time interval a.s. This completes the proof. □

REMARK 6.3.– As in theorem 2.2, the requirements that we impose on the distributions of $\bar{Y}_i$, Y_i and $\tilde{Y}_i$, $i \ge 1$, in theorem 6.4 are not necessary for the continuity of $\varphi(x,t)$ or the existence of its partial derivatives (see also remark 2.4). On the other hand, $\varphi(x,t)$ may not have partial derivatives in some cases.

6.3. Risk model with stochastic premiums and investments in risk-free and risky assets

We will now deal with the generalization of the risk model with stochastic premiums considered in section 6.1. As in section 6.2.1, we introduce the discrete-time process $X_n^x = X_{\tau_n}(x)$, $n \ge 0$, where $X_0^x = x$. Applying arguments similar to those in section 6.2.1 shows that [6.3] is also true for this model, where $P_{1,n}$, $n \geq 1$, are given by [6.4] and

$$P_{2,n} = \sum_{i:\, \tau_{n-1}<\bar{\tau}_i\leq\tau_n} \left(\bar{Y}_i\, \mathrm{e}^{\hat{r}(\tau_n-\bar{\tau}_i)} \prod_{j:\, \bar{\tau}_i<\tilde{\tau}_j\leq\tau_n} (1-\alpha+\alpha\mathrm{e}^{\tilde{Y}_i}) \right), \quad n \geq 1.$$

Moreover, theorems 6.1 and 6.2 also hold in this case. They can be proved in the same manner.

We now formulate the corresponding analogues of theorems 6.3 and 6.4.

THEOREM 6.5.–

1) Let the surplus process $(X_t(x))_{t\geq 0}$ follow [6.2] under the above assumptions. Then, the function $\varphi(x)$ is continuous on $(0, +\infty)$.

2) Moreover, let there be $z_0 \in Z^*$, such that [6.15] holds. Then the following assertions hold:

i) the function $\varphi(x)$ is continuous on $\mathbb{R}_+$;

ii) the function $\varphi(x)$ is continuously differentiable on $(0, +\infty)$, except at positive points of discontinuity of $F(y)$;

iii) if $x > 0$ is a point of discontinuity of $F(y)$ and $F(x) - F(x_-) = p$, then $\varphi(x)$ has the left and right derivatives $\varphi'_-(x)$ and $\varphi'_+(x)$, respectively, and

$$\varphi'_-(x) - \varphi'_+(x) = \frac{\lambda p \varphi(0)}{\hat{r}x} > 0;$$

iv) the function $\varphi(x)$ satisfies the integro-differential equation

$$\hat{r}x\varphi'_+(x) = (\bar{\lambda} + \lambda + \tilde{\lambda})\varphi(x) - \bar{\lambda} \int_0^{+\infty} \varphi(x+y)\, \mathrm{d}\bar{F}(y)$$

$$- \lambda \int_0^{x} \varphi(x-y)\, \mathrm{d}F(y) - \tilde{\lambda} \int_{-\infty}^{+\infty} \varphi(((1-\alpha) + \alpha\mathrm{e}^{y})\, x)\, \mathrm{d}\tilde{F}(y)$$

on $(0, +\infty)$ with the boundary condition $\lim_{x\to+\infty} \varphi(x) = 1$ and its value at the point $x = 0$ is given by

$$\varphi(0) = \frac{\bar{\lambda}}{\bar{\lambda} + \lambda} \int_0^{+\infty} \varphi(y)\mathrm{d}\bar{F}(y);$$

v) for all $x_* > 0$, we have

$$\sup_{x\in[x_*,+\infty)} |\varphi'_+(x)| \leq \frac{\bar{\lambda} + \lambda + \tilde{\lambda}}{\hat{r}x_*}. \qquad [6.41]$$

THEOREM 6.6.– Let the surplus process $\left(X_t(x)\right)_{t\geq 0}$ follow [6.2] under the above assumptions.

1) If $\bar{Y}_i$ and Y_i, $i \geq 1$, have p.d.f.'s $\bar{f}(y)$ and $f(y)$, respectively, which are continuous on $\mathbb{R}_+$, and $\tilde{Y}_i$, $i \geq 1$, have a p.d.f. $\tilde{f}(y)$, which is continuous on $\mathbb{R}$, then $\varphi(x,t)$ is continuous on $(0,+\infty)\times[0,+\infty)$ as a function of two variables.

2) If the p.d.f.'s $\bar{f}(y)$ and $f(y)$ of $\bar{Y}_i$ and Y_i, $i \geq 1$, respectively, have the derivatives $\bar{f}'(y)$ and $f'(y)$ on $\mathbb{R}_+$, such that $|\bar{f}'(y)|$ and $|f'(y)|$ are integrable and bounded on $\mathbb{R}_+$, and the p.d.f. $\tilde{f}(y)$ of $\tilde{Y}_i$, $i \geq 1$, have the derivative $\tilde{f}'(y)$ on $\mathbb{R}$, such that $|\tilde{f}'(y)|$, $\tilde{f}(y)\,\mathrm{e}^{-y}$ and $|\tilde{f}'(y)|\,\mathrm{e}^{-y}$ are integrable and bounded on $\mathbb{R}$, then:

i) $\varphi(x,t)$ has partial derivatives w.r.t. x and t on $(0,+\infty)\times[0,+\infty)$, which are continuous as functions of two variables;

ii) $\varphi(x,t)$ satisfies the partial integro-differential equation

$$\begin{aligned}&\frac{\partial\varphi(x,t)}{\partial t} - \hat{r}x\,\frac{\partial\varphi(x,t)}{\partial x} + (\bar{\lambda}+\lambda+\tilde{\lambda})\,\varphi(x,t)\\ &\quad - \bar{\lambda}\int_0^{+\infty}\varphi(x+y,\,t)\,\mathrm{d}\bar{F}(y) - \lambda\int_0^{x}\varphi(x-y,\,t)\,\mathrm{d}F(y)\\ &\quad - \tilde{\lambda}\int_{-\infty}^{+\infty}\varphi\big(((1-\alpha)+\alpha\mathrm{e}^{y})\,x,\,t\big)\,\mathrm{d}\tilde{F}(y) = 0\end{aligned}$$

on $(0,+\infty)\times[0,+\infty)$ with the boundary conditions $\varphi(x,0) = 1$ and $\lim_{x\to+\infty}\varphi(x,t) = 1$;

iii) for all $x_* > 0$ and $T > 0$, we have

$$\sup_{\substack{x\in[x_*,+\infty),\\ t\in[0,T]}}\left|\frac{\partial\varphi(x,t)}{\partial x}\right| \leq C_1\left(C_2 + \frac{C_3}{x_*}\right), \qquad [6.42]$$

where

$$C_1 = \begin{cases}\dfrac{1-\mathrm{e}^{(\hat{r}-\bar{\lambda}-\lambda-\tilde{\lambda})T}}{\bar{\lambda}+\lambda+\tilde{\lambda}-\hat{r}} & \text{if } \ \bar{\lambda}+\lambda+\tilde{\lambda}\neq\hat{r},\\ T & \text{if } \ \bar{\lambda}+\lambda+\tilde{\lambda}=\hat{r},\end{cases}$$

$$C_2 = \left|\bar{\lambda}\bar{f}(0) - \lambda f(0)\right| + \bar{\lambda}\int_0^{+\infty}|\bar{f}'(y)|\,\mathrm{d}y + \lambda\int_0^{+\infty}|f'(y)|\,\mathrm{d}y,$$

$$C_3 = \frac{\tilde{\lambda}}{\alpha}\left((1-\alpha)\int_{-\infty}^{+\infty}\tilde{f}(y)\,\mathrm{e}^{-y}\,\mathrm{d}y + \int_{-\infty}^{+\infty}|\tilde{f}'(y)|\,((1-\alpha)\,\mathrm{e}^{-y}+\alpha)\,\mathrm{d}y\right).$$

We leave out proofs of these theorems. The proof of theorem 6.5 is analogous to the proofs of theorems 2.1, 3.1 and 6.3. The proof of theorem 6.6 is analogous to the proofs of theorems 2.2, 3.2 and 6.4.

6.4. Accuracy and reliability of uniform approximations of the survival probabilities by their statistical estimates

In this section, we obtain relations connecting accuracy and reliability of *uniform approximations* of the survival probabilities by their statistical estimates. We deal with the finite- and infinite-horizon survival probabilities in sections 6.4.1 and 6.4.2, respectively.

6.4.1. *Finite-horizon survival probability*

Let now the surplus process $\left(X_t(x)\right)_{t\geq 0}$ follow either [6.1] or [6.2] under the above assumptions. For any fixed $T > 0$, the function $\varphi(x, T)$ is defined on $\mathbb{R}_+$. Let x_* and x^* be fixed and such that $x^* > x_* > 0$. Our aim is to derive a relation connecting accuracy and reliability of a uniform approximation of the finite-horizon survival probability by its statistical estimate for all $x \in [x_*, x^*]$. To this end, we will use bounds [6.34] and [6.42]. Note that we can apply similar considerations to the risk models investigated in Chapters 2 and 3 and use the corresponding bounds for the partial derivative of $\varphi(x, t)$ w.r.t. x obtained in those chapters.

Fix $h > 0$, such that $h \ll x^* - x_*$. Next, let

$$h_1 = h, \quad x_1 = x_* + h_1;$$

$$h_i = h\,\frac{C_2\,x_*\,x_{i-1} + C_3\,x_{i-1}}{C_2\,x_*\,x_{i-1} + C_3\,x_*}, \quad x_i = x_{i-1} + h_i, \quad 2 \leq i \leq k-1,$$

where C_2 and C_3 are defined in theorems 6.4 and 6.6 and $k = k(x_*, x^*, h, C_2, C_3)$ is such that $x_{k-1} < x^* \leq x_k$. For simplicity of notation, we set $x_0 = x_*$, $x_k = x^*$ and $h_k = x^* - x_{k-1}$.

Let the piecewise-linear function $\bar{\varphi}(x, T)$ be defined on $[x_*,\, x^*]$ by

$$\bar{\varphi}(x,T) = \varphi(x_{i-1},\, T) + \big(\varphi(x_i,\, T) - \varphi(x_{i-1},\, T)\big)\, \frac{x - x_{i-1}}{x_i - x_{i-1}},$$

$$x \in [x_{i-1},\, x_i], \quad 1 \le i \le k.$$

We simulate paths of the surplus processes $\big(X_t(x_i)\big)_{t\ge 0}$, $0 \le i \le k$, N times on $[0, T]$ using Monte Carlo methods. In the end, ruin does not occur $\nu_{x_i}(N, T)$ times.

Let the piecewise-linear function $\hat{\varphi}(x, T)$ be defined on $[x_*,\, x^*]$ as follows. First, we define $\hat{\varphi}(x, T)$ at points x_i as a statistical estimate for $\varphi(x, T)$ by

$$\hat{\varphi}(x_i, T) = \frac{\nu_{x_i}(N, T)}{N}, \quad 0 \le i \le k.$$

Moreover, set

$$\hat{\varphi}(x,T) = \hat{\varphi}(x_{i-1},\, T) + \big(\hat{\varphi}(x_i,\, T) - \hat{\varphi}(x_{i-1},\, T)\big)\, \frac{x - x_{i-1}}{x_i - x_{i-1}},$$

$$x \in (x_{i-1},\, x_i), \quad 1 \le i \le k.$$

THEOREM 6.7.– Let the surplus process $\big(X_t(x)\big)_{t\ge 0}$ follow either [6.1] or [6.2] under the above assumptions. Moreover, let the conditions of either theorem 6.4 or theorem 6.6, respectively, hold that are sufficient for the existence of partial derivatives of $\varphi(x, t)$. Then, for any $x^* > x_* > 0$, $h \in (0,\, x^* - x_*)$ and $\varepsilon > 0$, we have

$$\mathbb{P}\left[\sup_{x\in[x_*,\, x^*]} \big|\varphi(x,T) - \hat{\varphi}(x,T)\big| \le \varepsilon + C_1 h \left(C_2 + \frac{C_3}{x_*}\right)\right] \qquad [6.43]$$

$$\ge 1 - 2(k+1)\,\mathrm{e}^{-2N\varepsilon^2}.$$

PROOF.– In what follows, $\varphi_1'(\cdot,\cdot)$ denotes the partial derivative of $\varphi(x, t)$ w.r.t. the first argument. By the definition of $\varphi(x, T)$, Lagrange's mean value theorem, bounds [6.34] or [6.42] and the definition of h_i, we get

$$\big|\varphi(x,T) - \bar{\varphi}(x,T)\big| \le \big|\varphi(x_{i+1},\, T) - \varphi(x_i,\, T)\big| = \varphi_1'(\hat{x}_{i+1},\, T)\, h_{i+1}$$

$$\le C_1 h_{i+1}\left(C_2 + \frac{C_3}{x_i}\right) = C_1 h\left(C_2 + \frac{C_3}{x_*}\right),$$

$$x \in [x_i,\, x_{i+1}], \quad 0 \le i \le k-1,$$

where $\hat{x}_{i+1} \in [x_i,\, x_{i+1}]$, $0 \le i \le k-1$.

Hence, we have

$$\sup_{x\in[x_*,x^*]} \left|\varphi(x,T)-\bar{\varphi}(x,T)\right| \le C_1 h\left(C_2+\frac{C_3}{x_*}\right). \qquad [6.44]$$

Next, we introduce the following events

$$E_{i,\varepsilon}=\left\{\left|\bar{\varphi}(x_i,T)-\hat{\varphi}(x_i,T)\right|\le\varepsilon\right\},\quad 0\le i\le k.$$

By Hoeffding's inequality (see section A.2), for any $\varepsilon>0$, we have

$$\mathbb{P}\left[\bar{E}_{i,\varepsilon}\right]\le 2\mathrm{e}^{-2N\varepsilon^2},\quad 0\le i\le k.$$

Consequently, we get

$$\begin{aligned}\mathbb{P}\left[\sup_{x\in[x_*,x^*]}\left|\bar{\varphi}(x,T)-\hat{\varphi}(x,T)\right|\le\varepsilon\right)\Bigg] &= \mathbb{P}\left[\bigcap_{i=0}^{k}E_{i,\varepsilon}\right]\\ =1-\mathbb{P}\left[\bigcup_{i=0}^{k}\bar{E}_{i,\varepsilon}\right]\ge 1-\sum_{i=0}^{k}\mathbb{P}\left[\bar{E}_{i,\varepsilon}\right]&\ge 1-2(k+1)\,\mathrm{e}^{-2N\varepsilon^2}.\end{aligned} \qquad [6.45]$$

Relation [6.43] follows immediately from [6.44] and [6.45], which completes the proof. □

REMARK 6.4.– Relation [6.43] connects accuracy and reliability of the uniform approximation of the finite-horizon survival probability by its statistical estimate for all $x\in[x_*,x^*]$. It is easily seen that any given accuracy of the approximation can be attained by small enough $\varepsilon>0$ and $h>0$ and for the given accuracy, any required reliability is provided by large enough N. Nevertheless, even pointwise accuracy of Monte Carlo methods is in fact limited. Indeed, if the required reliability is $1-10^{-l}$, where l is a positive integer, then $N=\left\lceil\frac{l\ln 10+\ln 2}{2\varepsilon^2}\right\rceil$, which grows rapidly as $\varepsilon\to 0$. For example, $N=2\,649\,159$ if $l=2$ and $\varepsilon=0.001$. Hence, it is difficult to estimate the survival probability appropriately when an initial surplus is large enough. In this case, we have to simulate too large number of paths.

REMARK 6.5.– We now give an upper bound for $k = k(x_*, x^*, h, C_2, C_3)$. Note that

$$h_i = h\,\frac{C_2\,x_*\,x_{i-1} + C_3\,x_{i-1}}{C_2\,x_*\,x_{i-1} + C_3\,x_*} = h\left(1 + \frac{C_3(x_{i-1} - x_*)}{x_*(C_2 x_{i-1} + C_3)}\right)$$
$$> h\left(1 + \frac{(i-1)hC_3}{x_*(C_2 x^* + C_3)}\right), \quad 1 \le i \le k-1.$$

Let

$$h_i' = h\left(1 + \frac{(i-1)hC_3}{x_*(C_2 x^* + C_3)}\right), \quad 1 \le i \le k',$$

where k' is the least integer number, such that $\sum_{i=1}^{k'} h_i' \ge x^* - x_*$. Since

$$\sum_{i=1}^{k'} h_i' = h \sum_{i=1}^{k'} \left(1 + \frac{(i-1)hC_3}{x_*(C_2 x^* + C_3)}\right) = h\left(k' + \frac{hC_3 k'(k'-1)}{2x_*(C_2 x^* + C_3)}\right),$$

we get

$$k' = \left\lceil \frac{hC_3 - 2x_*(C_2 x^* + C_3)}{2hC_3} + \frac{\sqrt{(2x_*(C_2 x^* + C_3) - hC_3)^2 + 8C_3 x_*(C_2 x^* + C_3)(x^* - x_*)}}{2hC_3} \right\rceil.$$

It is easily seen that $k \le k'$.

We will now give two examples.

EXAMPLE 6.3.– Let the claim sizes be exponentially distributed with mean μ and $\tilde{Y}_i$, $i \ge 1$, have the Laplace distribution with p.d.f. $\tilde{f}(y) = \frac{\mathrm{e}^{-|y|/\tilde{\mu}}}{2\tilde{\mu}}$, where $y \in \mathbb{R}$ and $\tilde{\mu} \in (0,1)$. In spite of the fact that $\tilde{f}(y)$ is not differentiable at the point $x = 0$, it is easy to show that theorems 6.4 and 6.6 are applicable in this case. Moreover, let the premium sizes be exponentially distributed with mean $\bar{\mu}$ when $\left(X_t(x)\right)_{t\ge0}$ follows [6.2].

A simple computation shows that $C_2 = 2\lambda/\mu$ if $\left(X_t(x)\right)_{t\ge0}$ follows [6.1] and $C_2 = \left|\bar{\lambda}/\bar{\mu} - \lambda/\mu\right| + \bar{\lambda}/\bar{\mu} + \lambda/\mu$ if $\left(X_t(x)\right)_{t\ge0}$ follows [6.2]. Since

$$\int_{-\infty}^{+\infty} \tilde{f}(y)\,\mathrm{e}^{-y}\,\mathrm{d}y = \frac{1}{1-\tilde{\mu}^2},$$

$$\int_{-\infty}^{+\infty} |\tilde{f}'(y)|\,\mathrm{d}y = \frac{1}{\tilde{\mu}},$$

$$\int_{-\infty}^{+\infty} |\tilde{f}'(y)|\,\mathrm{e}^{-y}\,\mathrm{d}y = \frac{1}{\mu(1-\tilde{\mu}^2)}$$

for $\tilde{\mu} \in (0,1)$, we get

$$C_3 = \frac{\tilde{\lambda}}{\tilde{\mu}} + \frac{\tilde{\lambda}(1-\alpha)}{\alpha\mu(1-\tilde{\mu})}.$$

Set $T = 1$, $x_* = 1$, $x^* = 8$, $h = 0.00001$, $\lambda = 1$, $\mu = 5$, $r = 0.001$, $\tilde{r} = 0.004$, $\tilde{\lambda} = 2$ and $\tilde{\mu} = 0.08$. Moreover, set $c = 6$ if $\left(X_t(x)\right)_{t\geq 0}$ follows [6.1] and set $\bar{\lambda} = 6$ and $\bar{\mu} = 1$ if $\left(X_t(x)\right)_{t\geq 0}$ follows [6.2]. Let the required accuracy and reliability be 0.001 and 0.99, respectively,i.e.

$$\mathbb{P}\left[\sup_{x\in[1,\,8]} \left|\varphi(x,1) - \hat{\varphi}(x,1)\right| \leq 0.001\right] \geq 0.99.$$

Tables 6.1 and 6.3 give comparisons of $\hat{\varphi}(x)$ for integer initial surpluses from $[1,\ 8]$ and different α if $\left(X_t(x)\right)_{t\geq 0}$ follows [6.1] and [6.2], respectively. The corresponding values of $\hat{r}$, ε, k and N, which depend on α, are given in Tables 6.2 and 6.4.

x	$\alpha = 0.25$	$\alpha = 0.50$	$\alpha = 0.75$	$\alpha = 0.99$
1	0.561270	0.561093	0.561981	0.559461
2	0.610120	0.609695	0.610136	0.609609
3	0.677954	0.678559	0.677589	0.676448
4	0.721247	0.720687	0.718702	0.717148
5	0.760956	0.758686	0.760075	0.759429
6	0.786624	0.788435	0.787550	0.787254
7	0.816830	0.817659	0.817152	0.815577
8	0.843835	0.844012	0.845231	0.845337

Table 6.1. *Comparison of $\hat{\varphi}(x)$ for different α if $\left(X_t(x)\right)_{t\geq 0}$ follows [6.1]. Exponentially distributed claim sizes*

EXAMPLE 6.4.– Let the claim sizes have the Pareto distribution with p.d.f.

$$f(y) = \frac{\gamma}{\mu(\gamma-1)}\left(\frac{\mu(\gamma-1)}{\mu(\gamma-1)+y}\right)^{\gamma+1},$$

where $y \in \mathbb{R}_+$, $\mu > 0$, $\gamma > 1$, and $\tilde{Y}_i$, $i \geq 1$, have the Laplace distribution with p.d.f. $\tilde{f}(y) = \frac{e^{-|y|/\tilde{\mu}}}{2\tilde{\mu}}$, where $y \in \mathbb{R}$ and $\tilde{\mu} \in (0, 1)$. Furthermore, let the premium sizes have the Pareto distribution with p.d.f.

$$\bar{f}(y) = \frac{\bar{\gamma}}{\bar{\mu}(\bar{\gamma}-1)} \left(\frac{\bar{\mu}(\bar{\gamma}-1)}{\bar{\mu}(\bar{\gamma}-1)+y} \right)^{\bar{\gamma}+1},$$

where $y \in \mathbb{R}_+$, $\bar{\mu} > 0$, $\bar{\gamma} > 1$, when $\left(X_t(x)\right)_{t\geq 0}$ follows [6.2]. Note that we write the p.d.f.'s of the Pareto distributions in such a form in order to get $\mathbb{E}[Y_i] = \mu$ and $\mathbb{E}[\bar{Y}_i] = \bar{\mu}$, $i \geq 1$.

α	$\hat{r}$	ε	k	N
0.25	0.00175	0.000661	209 786	20 075 749
0.50	0.00250	0.000833	211 689	12 643 204
0.75	0.00325	0.000891	213 658	11 072 168
0.99	0.00397	0.000919	215 612	10 416 840

Table 6.2. *Values of $\hat{r}$, ε, k and N for different α if $\left(X_t(x)\right)_{t\geq 0}$ follows [6.1]. Exponentially distributed claim sizes*

x	$\alpha = 0.25$	$\alpha = 0.50$	$\alpha = 0.75$	$\alpha = 0.99$
1	0.660976	0.660751	0.660778	0.661011
2	0.698463	0.699885	0.701007	0.703572
3	0.738687	0.740966	0.738998	0.739811
4	0.773619	0.773273	0.778589	0.777929
5	0.809001	0.809035	0.809151	0.810166
6	0.829187	0.830122	0.828774	0.830821
7	0.842306	0.844995	0.848979	0.849062
8	0.875158	0.873945	0.871285	0.874557

Table 6.3. *Comparison of $\hat{\varphi}(x)$ for different α if $\left(X_t(x)\right)_{t\geq 0}$ follows [6.2]. Exponentially distributed premium and claim sizes*

It is easy to check that $C_2 = \frac{2\lambda\gamma}{\mu(\gamma-1)}$ if $\left(X_t(x)\right)_{t\geq 0}$ follows [6.1] and

$$C_2 = \left| \frac{\bar{\lambda}\bar{\gamma}}{\bar{\mu}(\bar{\gamma}-1)} - \frac{\lambda\gamma}{\mu(\gamma-1)} \right| + \frac{\bar{\lambda}\bar{\gamma}}{\bar{\mu}(\bar{\gamma}-1)} + \frac{\lambda\gamma}{\mu(\gamma-1)}$$

if $\left(X_t(x)\right)_{t\geq 0}$ follows [6.2]. The constant C_3 is given in example 6.3.

Set $T = 1$, $x_* = 1$, $x^* = 8$, $h = 0.00001$, $\lambda = 1$, $\mu = 5$, $\gamma = 2$, $r = 0.001$, $\tilde{r} = 0.004$, $\tilde{\lambda} = 2$ and $\tilde{\mu} = 0.08$. Moreover, set $c = 6$ if $\left(X_t(x)\right)_{t\geq 0}$

follows [6.1] and set $\bar{\lambda} = 6$, $\bar{\mu} = 1$ and $\bar{\gamma} = 2$ if $\left(X_t(x)\right)_{t\geq 0}$ follows [6.2]. Let the required accuracy and reliability be 0.001 and 0.99, respectively.

α	$\hat{r}$	ε	k	N
0.25	0.00175	0.000868	257 765	11 776 810
0.50	0.00250	0.000929	299 956	10 382 981
0.75	0.00325	0.000949	336 146	10 010 400
0.99	0.00397	0.000959	366 356	9 854 452

Table 6.4. *Values of $\hat{r}$, ε, k and N for different α if $\left(X_t(x)\right)_{t\geq 0}$ follows [6.2]. Exponentially distributed premium and claim sizes*

Tables 6.5 and 6.7 give comparisons of $\hat{\varphi}(x)$ for integer initial surpluses from $[1,\ 8]$ and different α if $\left(X_t(x)\right)_{t\geq 0}$ follows [6.1] and [6.2], respectively. The corresponding values of $\hat{r}$, ε, k and N are given in Tables 6.6 and 6.8.

x	$\alpha = 0.25$	$\alpha = 0.50$	$\alpha = 0.75$	$\alpha = 0.99$
1	0.682065	0.681105	0.680689	0.679641
2	0.729443	0.729544	0.727982	0.728079
3	0.766450	0.765893	0.766246	0.766806
4	0.789548	0.789777	0.790657	0.791278
5	0.812283	0.813272	0.814265	0.814473
6	0.834393	0.834281	0.833948	0.833822
7	0.855646	0.855007	0.854873	0.854690
8	0.872234	0.871462	0.870616	0.870087

Table 6.5. *Comparison of $\hat{\varphi}(x)$ for different α if $\left(X_t(x)\right)_{t\geq 0}$ follows [6.1]. Pareto distributed claim sizes*

α	$\hat{r}$	ε	k	N
0.25	0.00175	0.000660	211 614	20 162 910
0.50	0.00250	0.000832	215 377	12 694 234
0.75	0.00325	0.000889	219 238	11 120 053
0.99	0.00397	0.000917	223 043	10 465 800

Table 6.6. *Values of $\hat{r}$, ε, k and N for different α if $\left(X_t(x)\right)_{t\geq 0}$ follows [6.1]. Pareto distributed claim sizes*

In the above examples, the survival probabilities do not depend considerably on α. If we take other values of parameters, e.g. larger T, then dependence can be more significant. Moreover, from these examples, we see that the accuracy of Monte Carlo methods is limited. We had to simulate a large number of paths to provide accuracy 0.001.

x	$\alpha = 0.25$	$\alpha = 0.50$	$\alpha = 0.75$	$\alpha = 0.99$
1	0.681357	0.681070	0.680748	0.680560
2	0.728993	0.729162	0.730631	0.730929
3	0.766374	0.765045	0.764757	0.762156
4	0.795067	0.792973	0.792025	0.788622
5	0.812547	0.811227	0.810828	0.809768
6	0.834758	0.835565	0.836113	0.835144
7	0.846817	0.846612	0.846782	0.847044
8	0.870882	0.870119	0.868036	0.868128

Table 6.7. *Comparison of $\hat{\varphi}(x)$ for different α if $\left(X_t(x)\right)_{t\geq 0}$ follows [6.2]. Pareto distributed premium and claim sizes*

α	$\hat{r}$	ε	k	N
0.25	0.00175	0.000855	298 423	12 247 208
0.50	0.00250	0.000915	362 976	10 801 518
0.75	0.00325	0.000935	411 351	10 413 216
0.99	0.00397	0.000945	447 609	10 246 613

Table 6.8. *Values of $\hat{r}$, ε, k and N for different α if $\left(X_t(x)\right)_{t\geq 0}$ follows [6.2]. Pareto distributed premium and claim sizes*

6.4.2. *Infinite-horizon survival probability*

Let the surplus process $\left(X_t(x)\right)_{t\geq 0}$ follow either [6.1] or [6.2] under the above assumptions. Our next aim is to get a relation connecting accuracy and reliability of a uniform approximation of the infinite-horizon survival probability by its statistical estimate for all $x \in [x_*, x^*]$. We assume that x_* and x^* are also such that $x^* > x_* > 0$. Nevertheless, we can take $x_* = 0$ if $\left(X_t(x)\right)_{t\geq 0}$ follows [6.1]. To get that relation, we use bounds [6.21] and [6.41]. Consequently, it is true for a much wider class of distributions than [6.43].

We also fix $h > 0$, such that $h \ll x^* - x_*$. Moreover, let

$$h_1 = h, \quad x_1 = x_* + h_1;$$

$$h_i = \frac{h(\hat{r}x_{i-1} + c)}{\hat{r}x_* + c} \quad \text{or} \quad h_i = \frac{hx_{i-1}}{x_*}$$

if $\left(X_t(x)\right)_{t\geq 0}$ follows [6.1] or [6.2], respectively, and

$$x_i = x_{i-1} + h_i, \quad 2 \leq i \leq k-1,$$

where $k = k(x_*, x^*, h)$ is such that $x_{k-1} < x^* \leq x_k$. To simplify notation, we set $x_0 = x_*$, $x_k = x^*$ and $h_k = x^* - x_{k-1}$. Let

$$C_0 = \frac{\lambda + \tilde{\lambda}}{\hat{r} x_* + c} \quad \text{or} \quad C_0 = \frac{\bar{\lambda} + \lambda + \tilde{\lambda}}{\hat{r} x_*}$$

if $\left(X_t(x)\right)_{t\geq 0}$ follows [6.1] or [6.2], respectively.

Next, we define the piecewise-linear function $\bar{\varphi}(x)$ on $[x_*, x^*]$ by

$$\bar{\varphi}(x) = \varphi(x_{i-1}) + \left(\varphi(x_i) - \varphi(x_{i-1})\right) \frac{x - x_{i-1}}{x_i - x_{i-1}},$$
$$x \in [x_{i-1}, x_i], \quad 1 \leq i \leq k.$$

When we simulate paths of the surplus processes $\left(X_t(x_i)\right)_{t\geq 0}$, $0 \leq i \leq k$, N times on the infinite time interval using Monte Carlo methods, ruin does not occur $\nu_{x_i}(N)$ times. Note that in this case we actually simulate paths on a sufficiently long finite time interval.

Let the piecewise-linear function $\hat{\varphi}(x)$ be defined on $[x_*, x^*]$ as follows:

$$\hat{\varphi}(x_i) = \frac{\nu_{x_i}(N)}{N}, \quad 0 \leq i \leq k;$$

$$\hat{\varphi}(x) = \hat{\varphi}(x_{i-1}) + \left(\hat{\varphi}(x_i) - \hat{\varphi}(x_{i-1})\right) \frac{x - x_{i-1}}{x_i - x_{i-1}},$$
$$x \in (x_{i-1}, x_i), \quad 1 \leq i \leq k.$$

THEOREM 6.8.– Let the surplus process $\left(X_t(x)\right)_{t\geq 0}$ follow either [6.1] or [6.2] under the above assumptions. Moreover, let there be $z_0 \in Z^*$, such that [6.15] holds. Then, for any $x^* > x_* > 0$, $h \in (0, x^* - x_*)$ and $\varepsilon > 0$, we have

$$\mathbb{P}\left[\sup_{x \in [x_*, x^*]} \left|\varphi(x) - \hat{\varphi}(x)\right| \leq \varepsilon + C_0 h\right] \geq 1 - 2(k+1)\,\mathrm{e}^{-2N\varepsilon^2}. \quad [6.46]$$

The proof of this theorem is analogous to the proof of theorem 6.7. We only note that we use bounds [6.21] and [6.41] to get the required analogue of [6.44]. Furthermore, to apply Lagrange's mean value theorem on $[x_i, x_{i+1}]$, $0 \leq i \leq k-1$, we can divide these intervals by points where $\varphi(x)$ is not differentiable, if any.

6.5. Bibliographical notes

Risk models where investment income is random are of great interest from the mathematical point of view. Since it is usually assumed that such income can be negative, the investment risk is added to the insurance risk. In 1993, Paulsen [PAU 93] introduced a risk model that allows a stochastic rate of return on investments as well as a stochastic level of inflation. Integro-differential equations for $\psi(x)$ were derived, but the differentiability of this function was not studied. These equations were used to obtain analytic expressions for $\psi(x)$ in some cases and inequalities for it in other cases. Examples showed that stochastic economic factors may have a serious impact on the infinite-horizon ruin probability. For some early results in this direction, see also [PAU 98, PAU 97]. Paulsen and Gjessing [PAU 97] considered a risk process with stochastic interest rate and showed that the infinite-horizon ruin probability can be found by solving a certain boundary value problem involving an integro-differential equation. In [PAU 98], the classical risk process compounded by another independent process was considered and both of these components were assumed to be Lévy processes. Sharp conditions were given on the parameters of these two components when ruin is certain.

The fact that risky investments can be dangerous was first justified mathematically by Kalashnikov and Norberg [KAL 02]. They modeled the basic surplus process due to insurance activity and the logarithm of the price of the risky asset by Lévy processes, and obtained upper and lower power bounds for the infinite-horizon ruin probability when the initial surplus is large enough. These bounds turned out to be power functions. Some similar results were obtained independently by Paulsen [PAU 02]. Yuen, Wang and Wu [YUE 06] generalized results of [KAL 02] for the renewal risk process with stochastic interest.

Frolova, Kabanov and Pergamenshchikov [FRO 02] used the bounds obtained in [KAL 02] to show that ruin occurs with probability 1 in the classical risk model if all surplus is invested in the risky asset, the price of which is modeled by a geometric Brownian motion, and some additional conditions for parameters of the geometric Brownian motion hold. They also showed that if these conditions are not fulfilled, a power asymptotics is true for $\psi(x)$ when the claim sizes are exponentially distributed. The power asymptotics for $\psi(x)$ was obtained by Cai and Xu [CAI 06] in the case where the classical risk process is perturbed by Brownian motion. Moreover, Pergamenshchikov and Zeitouny [PER 06] considered the risk model where the premium intensity is a bounded non-negative random function, and generalized results of [FRO 02].

On the other hand, numerous results indicate that risky investments can be used to improve the solvency of the insurance company. Gaier, Grandits and Schachermayer [GAI 03] considered the classical risk model under the additional assumptions that the company is allowed to borrow and invest in the risky asset, the price of which follows a geometric Brownian motion. They obtained an upper exponential bound for $\psi(x)$ when the claim sizes have exponential moments and a fixed quantity, which is independent of the current surplus, is invested in the risky asset. It appeared that this bound is better than the classical one. Mishura [MIS 06] obtained an exponential bound for the infinite-horizon ruin probability when the price of the risky asset is modeled by a semimartingale with absolutely continuous characteristics w.r.t. the Lebesgue measure. Ma and Sun [MA 03] introduced a new type of exponential martingale parameterized by a general rate function to get various exponential bounds for $\psi(x)$ in an extension of the classical risk model, which contains stochastic interest rate, reserve-dependent expense loading, diffusion perturbed models and many others as special cases.

The continuity and differentiability of the infinite-horizon survival probability in models with risky investments were first investigated by Wang and Wu [WAN 01]. They dealt with the classical risk model under the additional assumption that investment income follows a Brownian motion with drift. The authors established the continuity of $\varphi(x)$, obtained sufficient conditions when $\varphi(x)$ is twice continuously differentiable and derived an integro-differential equation for this function, which involves its second derivative. Those sufficient conditions turned out to be too limiting. In particular, it was assumed that the p.d.f. of claim sizes has the second derivative.

Yuen, Wang and Ng [YUE 04] generalized some results of [WAN 01] for the case when investment income follows a compound Poisson process plus a Brownian motion with positive drift. Grandits [GRA 04b] and Gaier and Grandits [GAI 04] extended results obtained in [WAN 01] to the classical risk model when the insurance company invests all surplus in risk-free and risky assets proportionally and the price of the risky asset follows a geometric Brownian motion. Moreover, they investigated the asymptotics of $\psi(x)$ for claim sizes with tails of regular variation. In all the works mentioned above, the sufficient conditions assumed that the p.d.f. of claim sizes has the second derivative.

Cai [CAI 04] considered the case when the classical risk process is invested in a stochastic interest process, which is a Lévy process, derived recursive and integral equations for the ruin probabilities and got lower and upper bounds for $\psi(x)$ using these equations. Cai and Yang [CAI 05] considered the perturbed classical risk process under stochastic or constant interest force and showed that $\psi(x)$ is twice continuously differentiable provided that claim sizes have continuous p.d.f. using stochastic control techniques.

Paulsen, Kasozi and Steigen [PAU 05] dealt with the classical risk process compounded by a linear Brownian motion. They gave sufficient conditions for $\varphi(x)$ to be four times continuously differentiable, which in particular implies that $\varphi(x)$ is a solution to a second order integro-differential equation. The authors also analyzed properties of the numerical solution to this equation, transforming it into an ordinary Volterra integral equation of the second kind. Gerber and Yang [GER 07] considered the classical risk model perturbed by diffusion, where the insurance company invests all surplus in risk-free and risky assets proportionally and the price of the risky asset follows a geometric Brownian motion. An integro-differential equation for the absolute ruin probability was derived and solved in some special cases, but the differentiability of this function was not investigated further.

For further references in this direction, see also the survey by Paulsen [PAU 08], which deals with the problem of ruin in risk models where assets earn investment income. Finally, numerous investigations are devoted to solving optimal investment problems (see references given in section 5.4).

Though a geometric Brownian motion is often used to model prices of risky assets, financial models based on jump processes have acquired increasing popularity recently. A self-contained overview of the theoretical, numerical and empirical aspects involved in using jump processes in financial modeling was provided in [CON 04]; see also, for example [GEM 02, MER 76] for motivation.

For general aspects of simulation and Monte Carlo methods, we refer the reader to, for example [ASM 07, FIS 96, KOR 10, RIP 87, RUB 08]. Since the ruin probabilities are too small for large initial surpluses, rare events simulation is of great interest when we evaluate the ruin probabilities. For a survey on rare events simulation, see, for example [ASM 07]. For evaluation of the ruin probabilities in the classical risk model by the Monte Carlo method, see references given in section 1.1.9.

Part 2

Supermartingale Approach to the Estimation of Ruin Probabilities

7

Risk Model with Variable Premium Intensity and Investments in One Risky Asset

In this chapter, we consider a generalization of the classical risk model where premium intensity depends on a current surplus of an insurance company. All surplus is invested in one risky asset, the price of which follows a geometric Brownian motion. Our main aim is to show that if the premium intensity grows rapidly with increasing surplus, then an upper exponential bound for the ruin probability holds under certain conditions in spite of the fact that all surplus is invested in the risky asset. To this end, we apply the supermartingale approach and allow the surplus process to explode. To be more precise, we let the premium intensity be a quadratic function. In addition, we investigate the question concerning the probability of explosion of the surplus process between claim arrivals in detail.

7.1. Description of the model

We assume that an insurance company has a non-negative initial surplus x and denote by $X_t(x)$ its surplus at time $t \geq 0$. For simplicity of notation, we also write X_t instead of $X_t(x)$ when no confusion can arise. Let $c\colon \mathbb{R} \to \mathbb{R}_+ \setminus \{0\}$ be a measurable function, such that $c(u) = c(0)$ for all $u < 0$, and let $c(X_t)$ be a premium intensity, which depends on the surplus at time t.

As for claim arrivals, we make the same assumptions and notation as for the classical risk model in section 1.1.1. Moreover, we suppose that there exists the shifted moment generating function $h\colon \mathbb{R}_+ \to \mathbb{R}_+$ of the claim sizes Y_i, $i \geq 1$, introduced in section 1.1.7 and satisfying the classical assumption that is also formulated in this section.

In addition, we assume that all surplus is invested in one *risky asset*, the price of which equals S_t at time t. We model the process $(S_t)_{t\geq 0}$ by a *geometric Brownian motion.* Thus,

$$\mathrm{d}S_t = S_t(a\,\mathrm{d}t + b\,\mathrm{d}W_t), \qquad [7.1]$$

where $a \in \mathbb{R}$, $b > 0$ and $(W_t)_{t\geq 0}$ is a standard Brownian motion. We suppose that the r.v.'s $(Y_i)_{i\geq 1}$ and the processes $(N_t)_{t\geq 0}$ and $(W_t)_{t\geq 0}$ are mutually independent.

Let $(\mathfrak{F}_t)_{t\geq 0}$ be the filtration generated by $(Y_i)_{i\geq 1}$, $(N_t)_{t\geq 0}$ and $(W_t)_{t\geq 0}$, i.e.

$$\mathfrak{F}_t = \sigma\big((N_s)_{0\leq s\leq t}, (W_s)_{0\leq s\leq t}, Y_1, Y_2, \ldots, Y_{N_t}\big).$$

Under the above assumptions, the *surplus process* $(X_t)_{t\geq 0}$ follows the equation

$$X_t = x + \int_0^t c(X_s)\,\mathrm{d}s + \int_0^t \frac{X_s}{S_s}\,\mathrm{d}S_s - \sum_{i=1}^{N_t} Y_i, \quad t \geq 0. \qquad [7.2]$$

Substituting [7.1] into [7.2] yields

$$X_t = x + \int_0^t \big(c(X_s) + aX_s\big)\,\mathrm{d}s + b\int_0^t X_s\,\mathrm{d}W_s - \sum_{i=1}^{N_t} Y_i, \quad t \geq 0. \qquad [7.3]$$

To simplify notation, we also let τ stand for $\tau(x)$ in this chapter.

The rest of the chapter is organized as follows. Section 7.2 contains some auxiliary results and deals with the detailed investigation of the question concerning the probability of explosion of the surplus process between two successive claim arrivals. In section 7.3, we formulate and prove the existence and uniqueness theorem for the stochastic differential equation that describes the surplus process. In section 7.4, we establish the supermartingale property for an auxiliary exponential process. This property allows us to obtain an exponential bound for the ruin probability under certain conditions. Finally, in section 7.5, we consider the case where the premium intensity is a quadratic function and obtain an upper exponential bound for the ruin probability. Section 7.6 contains some bibliographical notes concerning the problem considered.

7.2. Auxiliary results

Consider now the following stochastic differential equation:

$$X_t = x + \int_0^t q(X_s)\, \mathrm{d}s + b \int_0^t X_s\, \mathrm{d}W_s, \quad t \geq 0, \qquad [7.4]$$

where $x > 0$, $b > 0$, $(W_t)_{t\geq 0}$ is a standard Brownian motion, $q\colon \mathbb{R} \to \mathbb{R}_+$ is a locally Lipschitz continuous function, such that $q(u)$ is strictly increasing on $\mathbb{R}_+$ and $q(u) = q(0)$ for all $u < 0$.

Equation [7.4] describes the surplus process between two successive jumps of $(N_t)_{t\geq 0}$ up to the first exit time of $(X_t)_{t\geq 0}$ from $[0, +\infty)$ in the model considered above provided that we impose the corresponding restrictions on $c(u)$, set $q(u) = c(u) + au$ for $u \geq 0$ and take the surplus at the time of the last jump of $(N_t)_{t\geq 0}$ instead of x.

7.2.1. *Sufficient conditions for the solution to explode*

We now give some results which show that $(X_t)_{t\geq 0}$ goes to $+\infty$ either with probability 1 or with positive probability, which is less than 1, under certain conditions.

Let t^* be a possible *explosion time* of $(X_t)_{t\geq 0}$, i.e.

$$t^* = \inf\{t \geq 0\colon X_t \notin (-\infty, +\infty)\}.$$

Moreover, let $t^*_{(0,+\infty)}$ be the *first exit time* from $(0, +\infty)$ for $(X_t)_{t\geq 0}$, i.e.

$$t^*_{(0,+\infty)} = \inf\{t \geq 0\colon X_t \notin (0, +\infty)\}.$$

By theorem A.8, equation [7.4] has a unique strong solution up to the explosion time t^*. Note that here, and subsequently, we imply the pathwise uniqueness of solutions.

For $x > 0$, we define

$$I_1 = \int_x^{+\infty} \exp\left(-\frac{2}{b^2}\int_x^v \frac{q(u)}{u^2}\,du\right) dv \qquad [7.5]$$

and

$$I_2 = -\int_0^x \exp\left(\frac{2}{b^2}\int_v^x \frac{q(u)}{u^2}\,du\right) dv. \qquad [7.6]$$

The following lemmas provide sufficient conditions for I_1 being finite and I_2 being infinite.

LEMMA 7.1.– If

$$\limsup_{v\to+\infty}\left((1+\varepsilon)\ln v - \frac{2}{b^2}\int_x^v \frac{q(u)}{u^2}\,du\right) < +\infty \quad \text{for some} \quad \varepsilon > 0, \qquad [7.7]$$

then $I_1 < +\infty$.

PROOF.– Since $\int_x^{+\infty} \frac{1}{v^{1+\varepsilon}}\,dv < +\infty$ for all $\varepsilon > 0$ and $x > 0$, it suffices to show that

$$\limsup_{v\to+\infty} \frac{\exp\left(-\frac{2}{b^2}\int_x^v \frac{q(u)}{u^2}\,du\right)}{\exp(-(1+\varepsilon)\ln v)} < +\infty \quad \text{for some} \quad \varepsilon > 0 \qquad [7.8]$$

in order to obtain $I_1 < +\infty$.

We can rewrite [7.8] as

$$\limsup_{v\to+\infty} \exp\left((1+\varepsilon)\ln v - \frac{2}{b^2}\int_x^v \frac{q(u)}{u^2}\,du\right) < +\infty \quad \text{for some} \quad \varepsilon > 0,$$

which gives [7.7]. □

LEMMA 7.2.– If $q(0) > 0$, then $I_2 = -\infty$.

PROOF.– It is easily seen that

$$-I_2 \geq \int_0^x \exp\left(\frac{2q(0)}{b^2}\int_v^x \frac{1}{u^2}\,\mathrm{d}u\right)\mathrm{d}v$$

$$= \exp\left(-\frac{2q(0)}{b^2 x}\right)\int_0^x \exp\left(\frac{2q(0)}{b^2 v}\right)\mathrm{d}v$$

$$= \exp\left(-\frac{2q(0)}{b^2 x}\right)\int_{1/x}^{+\infty} \frac{1}{u^2}\exp\left(\frac{2q(0)u}{b^2}\right)\mathrm{d}u = +\infty,$$

which proves the lemma. □

PROPOSITION 7.1.– If $q(0) > 0$ and condition [7.7] holds, then

$$\mathbb{P}\left[\lim_{t\uparrow t^*_{(0,+\infty)}} X_t = +\infty\right] = 1.$$

PROOF.– Note that in this case $I_1 < +\infty$ and $I_2 = -\infty$ by lemmas 7.1 and 7.2. Thus, the assertion of the proposition follows immediately from theorem A.10. □

REMARK 7.1.– If $q(0) = 0$, then I_2 may be finite (see examples 7.1–7.3 below). By theorem A.10, if $I_1 < +\infty$ and $I_2 > -\infty$, then $\lim_{t\uparrow t^*_{(0,+\infty)}} X_t$ exists a.s., $0 < \mathbb{P}\left[\lim_{t\uparrow t^*_{(0,+\infty)}} X_t = +\infty\right] < 1$ and $\mathbb{P}\left[\lim_{t\uparrow t^*_{(0,+\infty)}} X_t = 0\right]$ $= 1 - \mathbb{P}\left[\lim_{t\uparrow t^*_{(0,+\infty)}} X_t = +\infty\right]$.

REMARK 7.2.– Proposition 7.1 does not give us whether the exit time $t^*_{(0,+\infty)}$ is finite. It is well known that Feller's test for explosions (see, for example, theorem 5.29 in [KAR 91, p. 348]) gives precise conditions for whether or not a one-dimensional diffusion process explodes in finite time. This test is very useful when we want to show that a diffusion process does not explode in finite time (see, for example [MIJ 10]), but it does not solve our problem.

We will now give a few examples.

EXAMPLE 7.1.– Let

$$q(u) = \begin{cases} q_1 u + q_0 & \text{if} \quad u \geq 0, \\ q_0 & \text{if} \quad u < 0. \end{cases}$$

The function $q(u)$ has the asserted properties provided that $q_0 \geq 0$ and $q_1 > 0$.

Since

$$I_1 = \int_x^{+\infty} \exp\left(-\frac{2}{b^2}\int_x^v \frac{q_1 u + q_0}{u^2}\,\mathrm{d}u\right)\mathrm{d}v$$
$$= \int_x^{+\infty} \left(\frac{x}{v}\right)^{2q_1/b^2} \exp\left(\frac{2q_0}{b^2}\left(\frac{1}{v}-\frac{1}{x}\right)\right)\mathrm{d}v,$$

we have $I_1 = +\infty$ for $2q_1 \leq b^2$ and $I_1 < +\infty$ for $2q_1 > b^2$.

We first consider the case $q_0 > 0$. From theorem A.10 and lemma 7.2, we conclude that $\mathbb{P}\big[t^*_{(0,+\infty)} = \infty\big] = 1$ if $2q_1 \leq b^2$ and $\mathbb{P}\big[\lim_{t\uparrow t^*_{(0,+\infty)}} X_t = +\infty\big] = 1$ if $2q_1 > b^2$.

Consider now the case $q_0 = 0$. Since

$$I_2 = -\int_0^x \exp\left(\frac{2}{b^2}\int_v^x \frac{q_1 u}{u^2}\,\mathrm{d}u\right)\mathrm{d}v = -\int_0^x \left(\frac{x}{v}\right)^{2q_1/b^2}\mathrm{d}v,$$

we obtain $I_2 > -\infty$ for $2q_1 < b^2$ and $I_2 = -\infty$ for $2q_1 \geq b^2$. Theorem A.10 yields $\mathbb{P}\big[\lim_{t\uparrow t^*_{(0,+\infty)}} X_t = 0\big] = 1$ if $2q_1 < b^2$, $\mathbb{P}\big[t^*_{(0,+\infty)} = \infty\big] = 1$ if $2q_1 = b^2$ and $\mathbb{P}\big[\lim_{t\uparrow t^*_{(0,+\infty)}} X_t = +\infty\big] = 1$ if $2q_1 > b^2$. □

EXAMPLE 7.2.– Let

$$q(u) = \begin{cases} q_1(u+q_2)^\beta & \text{if} \quad u \geq 0, \\ q_1 q_2^\beta & \text{if} \quad u < 0. \end{cases}$$

We impose the following restrictions on the parameters of $q(u)$: $\beta > 1$, $q_1 > 0$ and $q_2 \geq 0$.

Since

$$\limsup_{v\to+\infty}\left((1+\varepsilon)\ln v - \frac{2}{b^2}\int_x^v \frac{q_1(u+q_2)^\beta}{u^2}\,\mathrm{d}u\right)$$
$$\leq \limsup_{v\to+\infty}\left((1+\varepsilon)\ln v - \frac{2q_1}{b^2}\int_x^v u^{\beta-2}\,\mathrm{d}u\right)$$
$$= \lim_{v\to+\infty}\left((1+\varepsilon)\ln v - \frac{2q_1\left(v^{\beta-1}-x^{\beta-1}\right)}{b^2(\beta-1)}\right) = -\infty$$

for all $\varepsilon > 0$, lemma 7.1 gives $I_1 < +\infty$.

If $q_2 > 0$, then $\mathbb{P}\big[\lim_{t\uparrow t^*_{(0,+\infty)}} X_t = +\infty\big] = 1$ by proposition 7.1.

For $q_2 = 0$, we have

$$I_2 = -\int_0^x \exp\left(\frac{2q_1}{b^2}\int_x^v u^{\beta-2}\,du\right)dv$$
$$= -\int_0^x \exp\left(\frac{2q_1\left(v^{\beta-1}-x^{\beta-1}\right)}{b^2(\beta-1)}\right)dv > -\infty.$$

Hence, in this case $\lim_{t\uparrow t^*_{(0,+\infty)}} X_t$ exists a.s., $0 < \mathbb{P}\big[\lim_{t\uparrow t^*_{(0,+\infty)}} X_t = +\infty\big] < 1$ and $\mathbb{P}\big[\lim_{t\uparrow t^*_{(0,+\infty)}} X_t = 0\big] = 1 - \mathbb{P}\big[\lim_{t\uparrow t^*_{(0,+\infty)}} X_t = +\infty\big]$ by theorem A.10. □

EXAMPLE 7.3.– Let

$$q(u) = \begin{cases} q_2u^2 + q_1u + q_0 & \text{if} \quad u \geq 0, \\ q_0 & \text{if} \quad u < 0. \end{cases} \qquad [7.9]$$

If $q_0 \geq 0$, $q_1 \geq 0$ and $q_2 > 0$, then $q(u)$ has all the properties required.

For all $\varepsilon > 0$, we have

$$\limsup_{v\to+\infty}\left((1+\varepsilon)\ln v - \frac{2}{b^2}\int_x^v \frac{q_2u^2+q_1u+q_0}{u^2}\,du\right)$$
$$\leq \limsup_{v\to+\infty}\left((1+\varepsilon)\ln v - \frac{2}{b^2}\int_x^v q_2\,du\right)$$
$$= \lim_{v\to+\infty}\left((1+\varepsilon)\ln v - \frac{2q_2(v-x)}{b^2}\right) = -\infty.$$

Hence, $I_1 < +\infty$ by lemma 7.1.

If $q_0 > 0$, then $\mathbb{P}\big[\lim_{t\uparrow t^*_{(0,+\infty)}} X_t = +\infty\big] = 1$ by proposition 7.1.

For $q_0 = 0$, we obtain

$$I_2 = -\int_0^x \exp\left(\frac{2}{b^2}\int_v^x \frac{q_2u^2+q_1u}{u^2}\,du\right)dv$$
$$= -\int_0^x \left(\frac{x}{v}\right)^{2q_1/b^2} \exp\left(\frac{2q_2(x-v)}{b^2}\right)dv.$$

This gives $I_2 > -\infty$ for $2q_1 < b^2$ and $I_2 = -\infty$ for $2q_1 \geq b^2$. Consequently, if $2q_1 < b^2$, then $\lim_{t\uparrow t^*_{(0,+\infty)}} X_t$ exists a.s., $0 < \mathbb{P}\big[\lim_{t\uparrow t^*_{(0,+\infty)}} X_t = +\infty\big] < 1$ and $\mathbb{P}\big[\lim_{t\uparrow t^*_{(0,+\infty)}} X_t = 0\big] = 1 - \mathbb{P}\big[\lim_{t\uparrow t^*_{(0,+\infty)}} X_t = +\infty\big]$; if $2q_1 \geq b^2$, then $\mathbb{P}\big[\lim_{t\uparrow t^*_{(0,+\infty)}} X_t = +\infty\big] = 1$. □

7.2.2. *Finiteness of the first exit time*

One question which is still unanswered is whether $t^*_{(0,+\infty)}$ is finite. We will now study it under the conditions of example 7.3.

THEOREM 7.1.– Let $(X_t)_{t\geq 0}$ be a strong solution to [7.4] and $q(u)$ be defined by [7.9] with $q_0 \geq 0$, $q_1 \geq 0$ and $q_2 > 0$.

i) If $q_0 = 0$ and $\frac{2q_1}{b^2} < 1$, then

$$\mathbb{P}\big[t^*_{(0,+\infty)} < \infty,\ \lim_{t\uparrow t^*_{(0,+\infty)}} X_t = +\infty\big]$$
$$= \frac{\int_0^x v^{-2q_1/b^2} \exp\left(-\frac{2q_2 v}{b^2}\right) dv}{\int_0^{+\infty} v^{-2q_1/b^2} \exp\left(-\frac{2q_2 v}{b^2}\right) dv}. \tag{7.10}$$

ii) If either $q_0 = 0$ and $\frac{2q_1}{b^2} \geq 1$ or $q_0 > 0$, then

$$\mathbb{P}\big[t^*_{(0,+\infty)} < \infty,\ \lim_{t\uparrow t^*_{(0,+\infty)}} X_t = +\infty\big] = 1. \tag{7.11}$$

PROOF.– Let $n_0 = \min\{n \in \mathbb{N}\colon 1/n < x\}$. For all integer n, such that $n \geq n_0$, we denote by $t^*_{(1/n,+\infty)}$ the first exit time from $(1/n, +\infty)$ for $(X_t)_{t\geq 0}$, i.e.

$$t^*_{(1/n,+\infty)} = \inf\{t \geq 0\colon X_t \notin (1/n, +\infty)\}.$$

Note that the sequence of events

$$\Big(\big\{\omega \in \Omega\colon t^*_{(1/n,+\infty)}(\omega) < \infty,\ \lim_{t\uparrow t^*_{(1/n,+\infty)}} X_t(\omega) = +\infty\big\}\Big)_{n\geq n_0}$$

is non-decreasing. Hence,

$$\lim_{n\to\infty} \big\{\omega \in \Omega\colon t^*_{(1/n,+\infty)}(\omega) < \infty,\ \lim_{t\uparrow t^*_{(1/n,+\infty)}} X_t(\omega) = +\infty\big\}$$
$$= \bigcup_{n=n_0}^{\infty} \big\{\omega \in \Omega\colon t^*_{(1/n,+\infty)}(\omega) < \infty,\ \lim_{t\uparrow t^*_{(1/n,+\infty)}} X_t(\omega) = +\infty\big\}.$$

Furthermore,

$$\bigcup_{n=n_0}^{\infty} \{\omega \in \Omega \colon t^*_{(1/n,+\infty)}(\omega) < \infty,\ \lim_{t \uparrow t^*_{(1/n,+\infty)}} X_t(\omega) = +\infty\}$$
$$= \{\omega \in \Omega \colon t^*_{(0,+\infty)}(\omega) < \infty,\ \lim_{t \uparrow t^*_{(0,+\infty)}} X_t(\omega) = +\infty\}.$$

Therefore, by the continuity of probability measures, we conclude that

$$\begin{aligned}
&\mathbb{P}\big[t^*_{(0,+\infty)} < \infty,\ \lim_{t \uparrow t^*_{(0,+\infty)}} X_t = +\infty\big] \\
&= \mathbb{P}\Big[\lim_{n\to\infty} \{t^*_{(1/n,+\infty)} < \infty,\ \lim_{t \uparrow t^*_{(1/n,+\infty)}} X_t = +\infty\}\Big] \\
&= \lim_{n\to\infty} \mathbb{P}\big[t^*_{(1/n,+\infty)} < \infty,\ \lim_{t \uparrow t^*_{(1/n,+\infty)}} X_t = +\infty\big].
\end{aligned} \quad [7.12]$$

From the considerations given in [KAR 91, pp. 343–344] it follows that $\mathbb{E}[t^*_{(1/n,+\infty)}] = M_n(x)$ for all $n \geq n_0$, where $M_n(x)$ is a solution to the boundary value problem

$$\begin{aligned}
&\frac{1}{2} b^2 x^2 M_n''(x) + (q_2 x^2 + q_1 x + q_0) M_n'(x) = -1, \\
&M_n\left(\frac{1}{n}\right) = 0, \quad M_n(+\infty) = 0,
\end{aligned} \quad [7.13]$$

which can be solved by the usual technique (see, for example [AGA 08]). Here and subsequently, the value of a function at $+\infty$ stands for its limit as the value of the argument tends to $+\infty$.

Boundary value problem [7.13] has the unique solution

$$\begin{aligned}
M_n(x) = {} & \frac{2 m_n(x)}{b^2 m_n(+\infty)} \int_{1/n}^{+\infty} \frac{m_n(+\infty) - m_n(z)}{z^2 m_n'(z)}\, \mathrm{d}z \\
& - \frac{2}{b^2} \int_{1/n}^{x} \frac{m_n(x) - m_n(z)}{z^2 m_n'(z)}\, \mathrm{d}z,
\end{aligned}$$

where

$$m_n(x) = \int_{1/n}^{x} \exp\left(-\frac{2}{b^2} \int_{1/n}^{v} \frac{q_2 u^2 + q_1 u + q_0}{u^2}\, \mathrm{d}u\right) \mathrm{d}v.$$

Note that $m_n(+\infty) < +\infty$. Furthermore,

$$\lim_{z\to+\infty} \frac{m_n(+\infty) - m_n(z)}{m_n'(z)}$$

$$= \lim_{z\to+\infty} \frac{\int_z^{+\infty} \exp\left(-\frac{2}{b^2}\int_{1/n}^{v} \frac{q_2u^2+q_1u+q_0}{u^2}\,\mathrm{d}u\right)\mathrm{d}v}{\exp\left(-\frac{2}{b^2}\int_{1/n}^{z} \frac{q_2u^2+q_1u+q_0}{u^2}\,\mathrm{d}u\right)} = \frac{b^2}{2q_2} < +\infty$$

(here we applied L'Hopital's rule) and $\int_{1/n}^{+\infty} \frac{1}{z^2}\,\mathrm{d}z < +\infty$. Therefore, we obtain

$$\int_{1/n}^{+\infty} \frac{m_n(+\infty) - m_n(z)}{z^2 m_n'(z)}\,\mathrm{d}z < +\infty.$$

Thus, $\mathbb{E}[t^*_{(1/n,+\infty)}] < \infty$ for all $n \geq n_0$. This gives $\mathbb{P}[t^*_{(1/n,+\infty)} < \infty] = 1$ for all $n \geq n_0$. Moreover, by the considerations given in [KAR 91, pp. 343–344], we have

$$\mathbb{P}\left[\lim_{t\uparrow t^*_{(1/n,+\infty)}} X_t = +\infty\right]$$

$$= \frac{\int_{1/n}^{x} \exp\left(-\frac{2}{b^2}\int_{1/n}^{v} \frac{q_2u^2+q_1u+q_0}{u^2}\,\mathrm{d}u\right)\mathrm{d}v}{\int_{1/n}^{+\infty} \exp\left(-\frac{2}{b^2}\int_{1/n}^{v} \frac{q_2u^2+q_1u+q_0}{u^2}\,\mathrm{d}u\right)\mathrm{d}v} \qquad [7.14]$$

$$= \frac{\int_{1/n}^{x} v^{-2q_1/b^2} \exp\left(\frac{2q_0}{b^2 v} - \frac{2q_2 v}{b^2}\right)\mathrm{d}v}{\int_{1/n}^{+\infty} v^{-2q_1/b^2} \exp\left(\frac{2q_0}{b^2 v} - \frac{2q_2 v}{b^2}\right)\mathrm{d}v}.$$

Consequently, [7.12] and [7.14] yield

$$\mathbb{P}\left[t^*_{(0,+\infty)} < \infty,\ \lim_{t\uparrow t^*_{(0,+\infty)}} X_t = +\infty\right]$$

$$= \lim_{n\to\infty} \frac{\int_{1/n}^{x} v^{-2q_1/b^2} \exp\left(\frac{2q_0}{b^2 v} - \frac{2q_2 v}{b^2}\right)\mathrm{d}v}{\int_{1/n}^{+\infty} v^{-2q_1/b^2} \exp\left(\frac{2q_0}{b^2 v} - \frac{2q_2 v}{b^2}\right)\mathrm{d}v}. \qquad [7.15]$$

Consider the following two cases:

1) If $q_0 = 0$ and $\frac{2q_1}{b^2} < 1$, then both of the integrals in the right-hand side of [7.15] are finite as $n \to \infty$. This yields [7.10]. Note that in this case $0 < \mathbb{P}\big[t^*_{(0,+\infty)} < \infty,\ \lim_{t \uparrow t^*_{(0,+\infty)}} X_t = +\infty\big] < 1$.

2) If either $q_0 = 0$ and $\frac{2q_1}{b^2} \geq 1$ or $q_0 > 0$, then both of the integrals in the right-hand side of [7.15] are infinite as $n \to \infty$. Applying L'Hopital's rule, we obtain [7.11].

The theorem is thus proved. □

REMARK 7.3.– Since $c(u)$ is positive by our assumption, the surplus of the insurance company becomes infinitely large in finite time a.s. if the premium intensity is a quadratic function and the claims do not arrive. Note that the time interval between two successive claims can be large enough with positive probability. Hence, the process $\big(X_t(x)\big)_{t \geq 0}$ that follows [7.3] goes to $+\infty$ with positive probability. It is clear that ruin does not occur in this case. Consequently, from now on we can consider $\big(X_t(x)\big)_{t \geq 0}$ up to the minimum from the ruin time and its possible explosion.

7.3. Existence and uniqueness theorem

Consider now equation [7.3]. Let $t^*(x)$ be a possible *explosion time* of $\big(X_t(x)\big)_{t \geq 0}$, i.e.

$$t^*(x) = \inf\{t \geq 0 \colon X_t(x) \notin (-\infty, +\infty)\}.$$

To shorten notation, we let t^* stand for $t^*(x)$.

THEOREM 7.2.– If $c(u)$ is a locally Lipschitz continuous function on $\mathbb{R}$, then [7.3] has a unique strong solution up to the time $\tau \wedge t^*$.

PROOF.– Since the process $(N_t)_{t \geq 0}$ is homogeneous, it has only a finite number of jumps on any finite time interval a.s. To prove the theorem, we study [7.3] between two successive jumps of $(N_t)_{t \geq 0}$.

Let us first consider [7.3] on the time interval $[\tau_0, \tau_1)$, where $\tau_0 = 0$. It can be rewritten as

$$X_t = X_{\tau_0} + \int_{\tau_0}^{t} \big(c(X_s) + aX_s\big)\,\mathrm{d}s + b\int_{\tau_0}^{t} X_s\,\mathrm{d}W_s, \quad \tau_0 \leq t < \tau_1. \quad [7.16]$$

By theorem A.8, the locally Lipschitz continuity of $c(u) + au$ and bu on $\mathbb{R}$ implies the existence of a unique strong solution to [7.16] on $[\tau_0, \tau_1 \wedge t^*)$. Moreover, the comparison theorem (see section A.3.2) shows that this solution is not less than the solution to the equation

$$X_t = X_{\tau_0} + a \int_{\tau_0}^{t} X_s \, \mathrm{d}s + b \int_{\tau_0}^{t} X_s \, \mathrm{d}W_s, \quad \tau_0 \leq t \leq \tau_1 \wedge t^*, \qquad [7.17]$$

a.s. Since the solution to [7.17] is positive, so is the solution to [7.16] on $[\tau_0, \tau_1 \wedge t^*)$. Hence, $\lim_{t \uparrow t^*} X_t = +\infty$ if $t^* \leq \tau_1$. Thus, ruin does not occur up to the time $\tau_1 \wedge t^*$.

If $t^* \leq \tau_1$, then the theorem follows. Otherwise $X_{\tau_1 -} < +\infty$ and we set $X_{\tau_1} = X_{\tau_1 -} - Y_1$. Next, if $X_{\tau_1} < 0$, then $\tau = \tau_1$, which completes the proof. Otherwise we consider [7.3] on the time interval $[\tau_1, \tau_2)$. We rewrite it as

$$X_t = X_{\tau_1} + \int_{\tau_1}^{t} \big(c(X_s) + aX_s\big) \, \mathrm{d}s + b \int_{\tau_1}^{t} X_s \, \mathrm{d}W_s, \quad \tau_1 \leq t < \tau_2. \qquad [7.18]$$

Repeating the same arguments, we conclude that [7.18] has a unique strong solution on $[\tau_1, \tau_2 \wedge t^*)$ and ruin does not occur up to the time $\tau_2 \wedge t^*$.

Thus, we have proved that [7.3] has a unique strong solution on $[0, \tau_2 \wedge t^*)$, which is our assertion if $t^* \leq \tau_2$. For the case $t^* > \tau_2$, we set $X_{\tau_2} = X_{\tau_2 -} - Y_2$. Next, if $X_{\tau_2} < 0$, then $\tau = \tau_2$, which proves the theorem. Otherwise we continue in this fashion and prove the theorem by induction. □

REMARK 7.4.– Note that if $t^* < \infty$, then the proof of theorem 7.2 implies $\lim_{t \uparrow t^*} X_t = +\infty$ and [7.3] also holds for $t = t^*$ provided that we let both of its sides be formally equal to $+\infty$. In this case, we formally set $X_{t^*} = +\infty$. In addition, if $\tau < \infty$, then we set $X_\tau = X_{\tau_i -} - Y_i$, where i is the number of the claim that caused the ruin, and [7.3] also holds for $t = \tau$.

7.4. Supermartingale property for the exponential process

Let the stopped process $\big(\tilde{X}_t(x)\big)_{t \geq 0}$ be defined by $\tilde{X}_t(x) = X_{t \wedge \tau \wedge t^*}(x)$. Note that $\big(\tilde{X}_t(x)\big)_{t \geq 0}$ is a solution to [7.3] provided that so is $\big(X_t(x)\big)_{0 \leq t < \tau \wedge t^*}$. In particular, if $t^* < \infty$, we formally let $\tilde{X}_t(x) = +\infty$ for all $t \geq t^*$ and [7.3] holds.

For all $R \geq 0$, we define the processes $\big(U_t(x,R)\big)_{t\geq 0}$ and $\big(V_t(x,R)\big)_{t\geq 0}$ by

$$U_t(x,R) = -R\tilde{X}_t(x) \quad \text{and} \quad V_t(x,R) = \mathrm{e}^{U_t(x,R)}.$$

In what follows, we write $\tilde{X}_t$, U_t and V_t instead of $\tilde{X}_t(x)$, $U_t(x,R)$ and $V_t(x,R)$, respectively, when no confusion can arise. Recall that the function $h(R)$ and the constant $R_\infty \in (0,+\infty]$, which are used below, were defined in section 1.1.7.

THEOREM 7.3.– If [7.3] has a unique strong solution up to the time $\tau \wedge t^*$ and there is $\hat{R} \in (0, R_\infty)$, such that

$$\frac{\hat{R}^2 b^2}{2} u^2 - \hat{R}\big(c(u) + au\big) + \lambda h(\hat{R}) \leq 0 \quad \text{for all} \quad u \geq 0, \qquad [7.19]$$

then $\big(V_t(x,\hat{R})\big)_{t\geq 0}$ is an $(\mathfrak{F}_t)$-supermartingale.

PROOF.– Let $R \in (0, R_\infty)$. Since $(\tilde{X}_t)_{t\geq 0}$ is a solution to [7.3], we have

$$\begin{aligned} U_t = &- Rx - R\int_0^{t\wedge\tau\wedge t^*} \big(c(X_s) + aX_s\big)\,\mathrm{d}s \\ &- Rb\int_0^{t\wedge\tau\wedge t^*} X_s\,\mathrm{d}W_s + R\sum_{i=1}^{N_{t\wedge\tau\wedge t^*}} Y_i, \quad t \geq 0. \end{aligned} \qquad [7.20]$$

Set

$$\hat{\tau}_n = \inf\{t \geq 0\colon X_t \geq n\} \wedge n, \quad n \geq 1.$$

The process $(\tilde{X}_t)_{t\geq 0}$ is a sum of local martingales and càdlàg processes of locally bounded variation. Indeed, as $\mathbb{E}\left[\left|\int_0^{t\wedge\tau\wedge t^*\wedge\hat{\tau}_n} X_s\,\mathrm{d}W_s\right|\right] < +\infty$ for all $t \geq 0$, the process $\left(\int_0^{t\wedge\tau\wedge t^*} X_s\,\mathrm{d}W_s\right)_{t\geq 0}$ is a local $(\mathfrak{F}_t)$-martingale with the localizing sequence $(\hat{\tau}_n)_{n\geq 1}$. Similarly, $\left(\int_0^{t\wedge\tau\wedge t^*} X_s\,\mathrm{d}s\right)_{t\geq 0}$ and

$\left(\int_0^{t\wedge\tau\wedge t^*} c(X_s)\,\mathrm{d}s\right)_{t\geq 0}$ are càdlàg processes of locally bounded variation with the localizing sequence $(\hat{\tau}_n)_{n\geq 1}$. Next, the process

$$\left(\sum_{i=1}^{N_{t\wedge\tau\wedge t^*}} Y_i - \lambda\mu(t\wedge\tau\wedge t^*)\right)_{t\geq 0}$$

is a compensated process with independent increments. Hence, it is an $(\mathfrak{F}_t)$-martingale. Thus, $(\tilde{X}_t)_{t\geq 0}$ is an $(\mathfrak{F}_t)$-semimartingale as a sum of local martingales and càdlàg processes of locally bounded variation and so are $(U_t)_{t\geq 0}$ and $(V_t)_{t\geq 0}$ (see section A.4).

Let us denote by $(U_t^{\mathrm{c}})_{t\geq 0}$ the continuous part of the local martingale in the decomposition of $(U_t)_{t\geq 0}$. By [7.20], applying Itô's formula (see section A.4), we obtain

$$\begin{aligned} V_t = {} & \mathrm{e}^{-Rx} + \int_{0+}^{t\wedge\tau\wedge t^*} \mathrm{e}^{U_{s-}}\,\mathrm{d}U_s + \frac{1}{2}\int_{0+}^{t\wedge\tau\wedge t^*} \mathrm{e}^{U_{s-}}\,\mathrm{d}\langle U^{\mathrm{c}}, U^{\mathrm{c}}\rangle_s \\ & + \sum_{0<s\leq t\wedge\tau\wedge t^*} \left(\mathrm{e}^{U_s} - \mathrm{e}^{U_{s-}} - \mathrm{e}^{U_{s-}}(U_s - U_{s-})\right), \quad t\geq 0, \end{aligned} \qquad [7.21]$$

where

$$\begin{aligned} U_{t_-} = {} & -Rx - R\int_0^{t\wedge\tau\wedge t^*} \big(c(X_s) + aX_s\big)\,\mathrm{d}s \\ & - Rb\int_0^{t\wedge\tau\wedge t^*} X_s\,\mathrm{d}W_s + R\sum_{0<s\leq t_-\wedge\tau\wedge t^*} Y_{N_s}\,\mathbb{1}_{\{\Delta N_s\neq 0\}}, \end{aligned}$$

$$\begin{aligned} \mathrm{d}U_s &= \mathbb{1}_{\{s\leq\tau\wedge t^*\}}\left(-R\big(c(X_s)+aX_s\big)\,\mathrm{d}s - RbX_s\,\mathrm{d}W_s + RY_{N_s}\,\mathbb{1}_{\{\Delta N_s\neq 0\}}\right), \\ \mathrm{d}\langle U^{\mathrm{c}}, U^{\mathrm{c}}\rangle_s &= R^2b^2X_s^2\,\mathbb{1}_{\{s\leq\tau\wedge t^*\}}\,\mathrm{d}s, \\ \mathrm{e}^{U_s} - \mathrm{e}^{U_{s-}} &= \mathrm{e}^{U_{s-}}\,\mathbb{1}_{\{s\leq\tau\wedge t^*\}}\,(\mathrm{e}^{RY_{N_s}\mathbb{1}_{\{\Delta N_s\neq 0\}}} - 1), \\ U_s - U_{s_-} &= RY_{N_s}\,\mathbb{1}_{\{\Delta N_s\neq 0\}}\,\mathbb{1}_{\{s\leq\tau\wedge t^*\}}, \\ \Delta N_s &= N_s - N_{s_-}. \end{aligned}$$

Substituting the above equalities into [7.21] yields

$$\begin{aligned} V_t = \mathrm{e}^{-Rx} &- R \int_{0+}^{t \wedge \tau \wedge t^*} \mathrm{e}^{U_{s-}} \big(c(X_s) + aX_s\big)\, \mathrm{d}s \\ &- Rb \int_{0+}^{t \wedge \tau \wedge t^*} \mathrm{e}^{U_{s-}} X_s \,\mathrm{d}W_s \\ &+ R \sum_{0<s\le t\wedge\tau\wedge t^*} \mathrm{e}^{U_{s-}} Y_{N_s} \mathbb{1}_{\{\Delta N_s \neq 0\}} \\ &+ \frac{1}{2} R^2 b^2 \int_{0+}^{t \wedge \tau \wedge t^*} \mathrm{e}^{U_{s-}} X_s^2 \,\mathrm{d}s \\ &+ \sum_{0<s\le t\wedge\tau\wedge t^*} \mathrm{e}^{U_{s-}} \big(\mathrm{e}^{RY_{N_s} \mathbb{1}_{\{\Delta N_s \neq 0\}}} - 1 - RY_{N_s} \mathbb{1}_{\{\Delta N_s \neq 0\}}\big), \\ t \ge 0. \end{aligned} \qquad [7.22]$$

Simplifying [7.22] gives

$$\begin{aligned} V_t = \mathrm{e}^{-Rx} &+ \int_{0+}^{t \wedge \tau \wedge t^*} \mathrm{e}^{U_{s-}} \left(\frac{1}{2} R^2 b^2 X_s^2 - R\big(c(X_s) + aX_s\big)\right) \mathrm{d}s \\ &- Rb \int_{0+}^{t \wedge \tau \wedge t^*} \mathrm{e}^{U_{s-}} X_s \,\mathrm{d}W_s \\ &+ \sum_{0<s\le t\wedge\tau\wedge t^*} \mathrm{e}^{U_{s-}} \big(\mathrm{e}^{RY_{N_s} \mathbb{1}_{\{\Delta N_s \neq 0\}}} - 1\big), \quad t \ge 0. \end{aligned} \qquad [7.23]$$

Next, the process

$$\left(\sum_{0<s\le t\wedge\tau\wedge t^*} \mathrm{e}^{U_{s-}} \big(\mathrm{e}^{RY_{N_s} \mathbb{1}_{\{\Delta N_s \neq 0\}}} - 1\big) \right)_{t\ge 0}$$

is non-decreasing and can be written in the integral form

$$\sum_{0<s\le t\wedge\tau\wedge t^*} \mathrm{e}^{U_{s-}} \big(\mathrm{e}^{RY_{N_s} \mathbb{1}_{\{\Delta N_s \neq 0\}}} - 1\big) = \int_{0+}^{t \wedge \tau \wedge t^*} \mathrm{e}^{U_{s-}} \,\mathrm{d}Q_s, \quad t \ge 0,$$

where

$$Q_t = \sum_{0<s\leq t} \left(\mathrm{e}^{RY_{N_s}\mathbb{1}_{\{\Delta N_s \neq 0\}}} - 1\right).$$

By Wald's identity, $\mathbb{E}[Q_t] = \lambda t h(R)$. Hence, $\mathbb{E}[Q_t] < \infty$ for all $t \geq 0$ and $R \in (0, R_\infty)$. Furthermore, since $(Q_t)_{t\geq 0}$ is a process with independent increments, the compensated process $\big(Q_t - \mathbb{E}[Q_t]\big)_{t\geq 0}$ is an $(\mathfrak{F}_t)$-martingale. Thus,

$$\left(\sum_{0<s\leq t\wedge\tau\wedge t^*} \left(\mathrm{e}^{RY_{N_s}\mathbb{1}_{\{\Delta N_s \neq 0\}}} - 1\right) - \lambda h(R)\int_{0+}^{t\wedge\tau\wedge t^*} \mathrm{e}^{U_{s-}}\,\mathrm{d}s\right)_{t\geq 0}$$

is a local $(\mathfrak{F}_t)$-martingale with the localizing sequence $(\hat{\tau}_n)_{n\geq 1}$ defined above.

Since we have already justified at the beginning of the proof that

$$\left(-Rb\int_{0+}^{t\wedge\tau\wedge t^*} \mathrm{e}^{U_{s-}} X_s\,\mathrm{d}W_s\right)_{t\geq 0}$$

is also a local $(\mathfrak{F}_t)$-martingale with the localizing sequence $(\hat{\tau}_n)_{n\geq 1}$, so is

$$\Bigg(-Rb\int_{0+}^{t\wedge\tau\wedge t^*} \mathrm{e}^{U_{s-}} X_s\,\mathrm{d}W_s + \sum_{0<s\leq t\wedge\tau\wedge t^*} \left(\mathrm{e}^{RY_{N_s}\mathbb{1}_{\{\Delta N_s \neq 0\}}} - 1\right)$$
$$- \lambda h(R)\int_{0+}^{t\wedge\tau\wedge t^*} \mathrm{e}^{U_{s-}}\,\mathrm{d}s\Bigg)_{t\geq 0}.$$

We define the process $(P_t)_{t\geq 0}$ by

$$P_t = V_t - V_0 + Rb\int_{0+}^{t\wedge\tau\wedge t^*} \mathrm{e}^{U_{s-}} X_s\,\mathrm{d}W_s$$
$$- \sum_{0<s\leq t\wedge\tau\wedge t^*} \left(\mathrm{e}^{RY_{N_s}\mathbb{1}_{\{\Delta N_s \neq 0\}}} - 1\right) + \lambda h(R)\int_{0+}^{t\wedge\tau\wedge t^*} \mathrm{e}^{U_{s-}}\,\mathrm{d}s.$$

Substituting V_t from [7.23], we obtain

$$P_t = \int_{0+}^{t\wedge\tau\wedge t^*} \mathrm{e}^{-RX_{s-}}\left(\frac{1}{2}R^2 b^2 X_s^2 - R\big(c(X_s) + aX_s\big) + \lambda h(R)\right)\mathrm{d}s,$$
$$t \geq 0.$$

Note that $(V_t)_{t\geq 0}$ is a local $(\mathfrak{F}_t)$-supermartingale with the localizing sequence $(\hat{\tau}_n)_{n\geq 1}$ provided that $(P_t)_{t\geq 0}$ is a measurable non-increasing process, i.e.

$$\int_{t_1\wedge\tau\wedge t^*}^{t_2\wedge\tau\wedge t^*} \mathrm{e}^{-RX_{s-}} \left(\frac{1}{2}R^2b^2X_s^2 - R\big(c(X_s)+aX_s\big)+\lambda h(R)\right)\mathrm{d}s \qquad [7.24]$$
$$\leq 0 \quad \text{a.s.}$$

for all $t_2 \geq t_1 \geq 0$.

By the assumption of the theorem, there is $\hat{R} \in (0, R_\infty)$, such that [7.19] holds. Therefore, [7.24] is true with $R = \hat{R}$ and $\big(V_t(x,\hat{R})\big)_{t\geq 0}$ is a non-negative local $(\mathfrak{F}_t)$-supermartingale with the localizing sequence $(\hat{\tau}_n)_{n\geq 1}$.

By Fatou's lemma, for all $t_2 \geq t_1 \geq 0$, we obtain

$$0 \leq \mathbb{E}\big[V_{t_2}(x,\hat{R})\,/\,\mathfrak{F}_{t_1}\big] = \mathbb{E}\Big[\lim_{n\to\infty} V_{t_2\wedge\hat{\tau}_n}(x,\hat{R})\,/\,\mathfrak{F}_{t_1}\Big]$$
$$= \mathbb{E}\Big[\liminf_{n\to\infty} V_{t_2\wedge\hat{\tau}_n}(x,\hat{R})\,/\,\mathfrak{F}_{t_1}\Big] \leq \liminf_{n\to\infty} \mathbb{E}\big[V_{t_2\wedge\hat{\tau}_n}(x,\hat{R})\,/\,\mathfrak{F}_{t_1}\big]$$
$$\leq \liminf_{n\to\infty} V_{t_1\wedge\hat{\tau}_n}(x,\hat{R}) = V_{t_1}(x,\hat{R}).$$

Hence, $\big(V_t(x,\hat{R})\big)_{t\geq 0}$ is an $(\mathfrak{F}_t)$-supermartingale, which completes the proof. □

Theorem 7.3 allows us to obtain an upper exponential bound for the ruin probability under certain conditions.

7.5. Upper exponential bound for the ruin probability

Let the premium intensity $c(u)$ be a quadratic function for $u \geq 0$, i.e.

$$c(u) = \begin{cases} c_2u^2 + c_1u + c_0 & \text{if} \quad u \geq 0, \\ c_0 & \text{if} \quad u < 0, \end{cases} \qquad [7.25]$$

where $c_2 \neq 0$. The function $c(u)$ is strictly increasing and positive on $\mathbb{R}_+$ if and only if $c_0 > 0$, $c_1 \geq 0$ and $c_2 > 0$. This model implies that the premium intensity grows rapidly with increasing surplus.

THEOREM 7.4.– Let the surplus process $\left(X_t(x)\right)_{t\geq 0}$ follow [7.3] under the above assumptions, the premium intensity $c(u)$ be defined by [7.25] with $c_0 > 0$, $c_1 \geq 0$ and $c_2 > 0$. Moreover, let $a+c_1 > 0$ and at least one of the following two conditions hold:

i) $\frac{2c_2}{b^2} < R_\infty$ and $h\left(\frac{2c_2}{b^2}\right) \leq \frac{2c_0c_2}{b^2\lambda}$;

ii) $\lambda\mu < c_0$.

Then, for all $x \geq 0$, we have

$$\psi(x) \leq \mathrm{e}^{-\hat{R}x}, \qquad [7.26]$$

where $\hat{R} = \frac{2c_2}{b^2}$ if condition i) holds and $\hat{R} = \min\left\{R_0, \frac{2c_2}{b^2}\right\}$ if condition ii) holds. Here R_0 stands for the unique positive solution to

$$h(R) = \frac{c_0 R}{\lambda}. \qquad [7.27]$$

LEMMA 7.3.– Let the conditions of theorem 7.4 hold. Then, the process $\left(V_t(x,\hat{R})\right)_{t\geq 0}$ is an $(\mathfrak{F}_t)$-supermartingale, where $\hat{R} = \frac{2c_2}{b^2}$ if condition i) of theorem 7.4 is true and $\hat{R}$ is an arbitrary number from $(0, R_0]$ for $R_0 < \frac{2c_2}{b^2}$ or from $\left(0, \frac{2c_2}{b^2}\right)$ for $R_0 \geq \frac{2c_2}{b^2}$ if condition ii) of theorem 7.4 is true. Here R_0 stands for the unique positive solution to [7.27].

PROOF (Proof of lemma 7.3).– Since $c(u)$ defined by [7.25] is a locally Lipschitz continuous function on $\mathbb{R}$, equation [7.3] has a unique strong solution up to the time $\tau \wedge t^*$ by theorem 7.2. According to theorem 7.3, if there is $\hat{R} \in (0, R_\infty)$, such that

$$\left(\frac{\hat{R}^2 b^2}{2} - \hat{R}c_2\right) u^2 - \hat{R}(a+c_1)u - \hat{R}c_0 + \lambda h(\hat{R}) \leq 0 \quad \text{for all} \quad u \geq 0, \qquad [7.28]$$

then $\left(V_t(x,\hat{R})\right)_{t\geq 0}$ is an $(\mathfrak{F}_t)$-supermartingale.

Condition [7.28] holds in one of the following two cases:

1) The coefficient of u^2 is equal to 0, i.e. $\hat{R} = \frac{2c_2}{b^2}$.

Then [7.28] is true if and only if

$$\frac{2c_2}{b^2} < R_\infty \quad \text{and} \quad -\frac{2c_2}{b^2}c_0 + \lambda h\left(\frac{2c_2}{b^2}\right) \leq 0,$$

which coincides with condition i) of the theorem;

2) The coefficient of u^2 is negative, i.e. $\hat{R} \in \left(0, \frac{2c_2}{b^2}\right)$.

Since $u = \frac{a+c_1}{\hat{R}b^2-2c_2}$, which is negative, maximizes the left-hand side of [7.28], the last one is true if and only if

$$\hat{R} \in \left(0, \min\left\{\frac{2c_2}{b^2}, R_\infty\right\}\right) \qquad [7.29]$$

and

$$\lambda h(\hat{R}) \le c_0 \hat{R}. \qquad [7.30]$$

Consider the functions $g_1(R) = \lambda h(R)$ and $g_2(R) = c_0 R$ on $[0, R_\infty)$. Note that $g_1(0) = 0$, $g_2(0) = 0$, $g_1'(0) = \lambda\mu$ and $g_2'(0) = c_0$. Moreover, since $h(R)$ is increasing and convex, so is $g_1(R)$. Thus, we obtain the following:

– if $\lambda\mu \ge c_0$, then $g_2(R) < g_1(R)$ for all $R \in (0, R_\infty)$. Hence, for no $\hat{R} \in (0, R_\infty)$ does [7.28] hold;

– if $\lambda\mu < c_0$, then the equation $g_1(R) = g_2(R)$ has a unique solution $R_0 \in (0, R_\infty)$. Therefore, [7.27] has a unique positive solution and [7.30] is true for all $\hat{R} \in (0, R_0]$. Taking into account the condition [7.29] we conclude that [7.28] holds for all $\hat{R} \in (0, R_0]$ if $R_0 < \frac{2c_2}{b^2}$ and for all $\hat{R} \in \left(0, \frac{2c_2}{b^2}\right)$ if $R_0 \ge \frac{2c_2}{b^2}$.

The lemma is thus proved. □

PROOF (of theorem 7.4).– Let $\hat{R}$ be defined in the assertion of lemma 7.3. Then, $\left(V_t(x, \hat{R})\right)_{t\ge 0}$ is an $(\mathfrak{F}_t)$-supermartingale by this lemma. Therefore, for all $t \ge 0$, we obtain

$$\begin{aligned} e^{-\hat{R}x} &= V_0(x, \hat{R}) \ge \mathbb{E}\left[V_t(x, \hat{R}) \,/\, \mathfrak{F}_0\right] = \mathbb{E}\left[e^{-\hat{R}X_{t\wedge\tau\wedge t^*}(x)}\right] \\ &= \mathbb{E}\left[e^{-\hat{R}X_\tau(x)} \mathbb{1}_{\{\tau(x)<t\wedge t^*(x)\}}\right] \\ &\quad + \mathbb{E}\left[e^{-\hat{R}X_{t\wedge t^*}(x)} \mathbb{1}_{\{\tau(x)\ge t\wedge t^*(x)\}}\right] \\ &\ge \mathbb{E}\left[e^{-\hat{R}X_\tau(x)} \mathbb{1}_{\{\tau(x)<t\wedge t^*(x)\}}\right]. \end{aligned} \qquad [7.31]$$

Letting $t \to \infty$ in [7.31] gives

$$\mathbb{E}\left[e^{-\hat{R}X_\tau(x)} \mathbb{1}_{\{\tau(x)<t^*(x)\}}\right] \le e^{-\hat{R}x}. \qquad [7.32]$$

Since the surplus becomes infinitely large at the explosion time, ruin does not occur after $t^*(x)$. Hence,

$$\{\omega \in \Omega\colon \tau(x,\omega) < t^*(x,\omega)\} = \{\omega \in \Omega\colon \tau(x,\omega) < \infty\}$$

and [7.32] can be rewritten as

$$\mathbb{E}\big[\mathrm{e}^{-\hat{R}X_{\tau}(x)}\,\mathbb{1}_{\{\tau(x)<\infty\}}\big] \leq \mathrm{e}^{-\hat{R}x}. \qquad [7.33]$$

Furthermore,

$$\mathbb{E}\big[\mathrm{e}^{-\hat{R}X_{\tau}(x)}\,\mathbb{1}_{\{\tau(x)<\infty\}}\big] = \mathbb{E}\big[\mathrm{e}^{-\hat{R}X_{\tau}(x)} \,/\, \tau(x) < \infty\big]\,\mathbb{P}[\tau(x) < \infty],$$

and

$$\mathbb{E}\big[\mathrm{e}^{-\hat{R}X_{\tau}(x)} \,/\, \tau(x) < \infty\big] \geq 1$$

by the definition of the ruin time. Therefore, from [7.33], we conclude that

$$\mathbb{P}[\tau(x) < \infty] \leq \frac{\mathrm{e}^{-\hat{R}x}}{\mathbb{E}\big[\mathrm{e}^{-\hat{R}X_{\tau}(x)} \,/\, \tau(x) < \infty\big]} \leq \mathrm{e}^{-\hat{R}x},$$

which yields [7.26].

What is left is to note that the larger $\hat{R}$ we choose, the better bound in [7.26] we obtain. Thus, if condition i) of the theorem holds and $R_0 < \frac{2c_2}{b^2}$, then we set $\hat{R} = R_0$. If condition ii) of the theorem holds and $R_0 \geq \frac{2c_2}{b^2}$, then [7.26] is true for all $\hat{R} \in \big(0, \frac{2c_2}{b^2}\big)$. Hence, it is also true for $\hat{R} = \frac{2c_2}{b^2}$. This completes the proof. □

7.6. Bibliographical notes

To solve our problem, we used the supermartingale approach, which generalizes the martingale approach introduced by Gerber [GER 73]. In particular, Gerber used the martingale approach to find the famous exponential bound, which is known as the Lundberg inequality, for $\psi(x)$ in the classical risk model (see section 1.1.8). Sørensen [SØR 96] attracted attention to the semimartingale methods that can be applied to very general types of risk models to obtain local martingales or martingales, which can then be used in the classical way to evaluate the ruin probabilities.

The fact that risky investments can be dangerous was first justified mathematically by Kalashnikov and Norberg [KAL 02]. They modeled the basic surplus process due to insurance activity and the logarithm of the price of the risky asset by Lévy processes and obtained upper and lower power bounds for the infinite-horizon ruin probability when the initial surplus is large enough. These bounds turned out to be power functions. For some similar results, see [PAU 02].

Frolova, Kabanov and Pergamenshchikov [FRO 02] used the bounds obtained in [KAL 02] to show that ruin occurs with probability 1 in the classical risk model if all surplus is invested in the risky asset, the price of which is modeled by a geometric Brownian motion and $2a/b^2 < 1$. They also showed that if $2a/b^2 > 1$, a power asymptotics is true for $\psi(x)$ when the claim sizes are exponentially distributed. Pergamenshchikov and Zeitouny [PER 06] considered the risk model where the premium intensity is a bounded non-negative random function and generalized results of [FRO 02].

Nevertheless, numerous results show that risky investments can improve the solvency of the insurance company (see the references given in section 6.5 and the survey by Paulsen [PAU 08]). For the references concerning the problem of optimal control by investments in the classical risk model and some its generalizations, see section 5.4.

Finally, note that the geometric Brownian motion as a model for the price of risky asset was introduced by Samuelson in [SAM 65]. An important property of this model is the absence of arbitrage. For more information in this direction, see textbooks on mathematical finance [DOT 90, KAR 98, ROL 99, SHR 04].

8

Risk Model with Variable Premium Intensity and Investments in One Risky Asset up to the Stopping Time of Investment Activity

In this chapter, we consider a generalization of the classical risk model where a premium intensity depends on a current surplus of an insurance company. All surplus is invested in one risky asset, the price of which follows a geometric Brownian motion, but an insurance company stops its investment activity when the price of the risky asset goes down below some fixed level. Our main result asserts that an upper exponential bound for the ruin probability holds under certain conditions. To this end, we give another representation for the surplus process, redefine the ruin time and establish the supermartingale property for an auxiliary exponential process. Then, we concentrate on the case of exponentially distributed claim sizes. Moreover, we consider the modification of the model where the insurance company stops its investment activity when the price of the risky asset exits from some fixed interval and extend the results obtained to this modification.

8.1. Description of the model

We assume that an insurance company has a non-negative initial surplus x and denote by $X_t(x)$ its surplus at time $t \geq 0$. In what follows, we also write X_t instead of $X_t(x)$ for brevity when no confusion can arise. Let $c\colon \mathbb{R} \to$

$\mathbb{R}_+ \setminus \{0\}$ be a measurable function, such that $\inf_{u \in \mathbb{R}} c(u) > 0$, and let $c(X_t)$ be a premium intensity, which depends on the surplus at time t.

As for claim arrivals, we make the same assumptions and notation as for the classical risk model in section 1.1.1. Moreover, we suppose that there exists the shifted moment generating function $h\colon \mathbb{R}_+ \to \mathbb{R}_+$ of the claim sizes Y_i, $i \geq 1$, introduced in section 1.1.7 and satisfying the classical assumption that is also formulated in this section.

Furthermore, the insurance company can invest the surplus in a *risky asset,* the price of which equals S_t at time t. We model the process $(S_t)_{t\geq 0}$ by a *geometric Brownian motion.* Thus,

$$\mathrm{d}S_t = S_t(a\,\mathrm{d}t + b\,\mathrm{d}W_t), \qquad [8.1]$$

where $a \in \mathbb{R}$, $b > 0$ and $(W_t)_{t\geq 0}$ is a standard Brownian motion. We suppose that the r.v.'s $(Y_i)_{i\geq 1}$ and the processes $(N_t)_{t\geq 0}$ and $(W_t)_{t\geq 0}$ are mutually independent.

Let $(\mathfrak{F}_t)_{t\geq 0}$ be the filtration generated by $(Y_i)_{i\geq 1}$, $(N_t)_{t\geq 0}$ and $(W_t)_{t\geq 0}$, i.e.

$$\mathfrak{F}_t = \sigma\big((N_s)_{0\leq s\leq t}, (W_s)_{0\leq s\leq t}, Y_1, Y_2, \ldots, Y_{N_t}\big).$$

We assume that all surplus is invested in the risky asset only up to the *stopping time*

$$\tau^* = \inf\{t \geq 0\colon S_t < S_*\},$$

where $S_* \in (0, S_0)$. This means that the insurance company stops its investment activity when the price of the risky asset goes down below S_*. Since $(S_t)_{t\geq 0}$ is continuous a.s., it is clear that $S_{\tau^*} = S_*$ a.s. We suppose that $\tau^* = \infty$ if $S_t \geq S_*$ for all $t \geq 0$.

Under the above assumptions, the *surplus process* $(X_t)_{t\geq 0}$ follows the equation

$$X_t = x + \int_0^t c(X_s)\,\mathrm{d}s + \int_0^t \frac{X_s\, \mathbb{1}_{\{s<\tau^*\}}}{S_s}\,\mathrm{d}S_s - \sum_{i=1}^{N_t} Y_i, \quad t \geq 0. \qquad [8.2]$$

Substituting [8.1] into [8.2] yields

$$X_t = x + \int_0^t \left(c(X_s) + aX_s \, \mathbb{1}_{\{s<\tau^*\}}\right) \mathrm{d}s + b\int_0^t X_s \, \mathbb{1}_{\{s<\tau^*\}} \, \mathrm{d}W_s - \sum_{i=1}^{N_t} Y_i, \quad t \ge 0. \qquad [8.3]$$

The rest of the chapter is organized as follows. In section 8.2, we formulate and prove the existence and uniqueness theorem for the stochastic differential equation that describes the surplus process. In section 8.3, we give another representation for the surplus process and redefine the ruin time. In section 8.4, we establish the supermartingale property for an auxiliary exponential process, which allows us to get an upper exponential bound for the ruin probability in section 8.5. Section 8.6 deals with the case of exponentially distributed claim sizes. In section 8.7, we consider the modification of the model described above where the insurance company stops its investment activity when the price of the risky asset exits from some fixed interval and extend results obtained in sections 8.2–8.6 to this modification. For some bibliographical notes concerning the problem considered, see section 7.6.

8.2. Existence and uniqueness theorem

In what follows, we imply the pathwise uniqueness of solutions.

THEOREM 8.1.– If $c(u)$ is a Lipschitz continuous function on $\mathbb{R}$, then [8.3] has a unique strong solution on $\mathbb{R}_+$.

PROOF.– The process $(N_t)_{t\ge 0}$ is homogeneous, so it has only a finite number of jumps on any finite time interval a.s. Consequently, we study [8.3] between two successive jumps of $(N_t)_{t\ge 0}$ and prove the theorem by induction.

We first consider [8.3] on the time interval $[\tau_0, \tau_1)$, where $\tau_0 = 0$. It can be rewritten as

$$X_t = X_{\tau_0} + \int_{\tau_0}^t \left(c(X_s) + aX_s \, \mathbb{1}_{\{s<\tau^*\}}\right) \mathrm{d}s + b\int_{\tau_0}^t X_s \, \mathbb{1}_{\{s<\tau^*\}} \, \mathrm{d}W_s, \quad \tau_0 \le t < \tau_1. \qquad [8.4]$$

Note that [8.4] is a stochastic differential equation up to the time τ^*, which has a unique strong solution by theorem A.7, and if $\tau^* \in [\tau_0, \tau_1)$, it is an ordinary differential equation after the time τ^*, which also has a unique solution by the Lipschitz continuity of $c(u)$. Thus, [8.3] has a unique strong solution on $[\tau_0, \tau_1)$. Next, we set $X_{\tau_1} = X_{\tau_1 -} - Y_1$ and consider [8.3] on $[\tau_1, \tau_2)$.

In general, to show that [8.3] has a unique strong solution on $[\tau_i, \tau_{i+1})$, $i \geq 1$, we rewrite it as

$$X_t = X_{\tau_i} + \int_{\tau_i}^{t} \left(c(X_s) + aX_s \, \mathbb{1}_{\{s<\tau^*\}}\right) \mathrm{d}s + b \int_{\tau_i}^{t} X_s \, \mathbb{1}_{\{s<\tau^*\}} \, \mathrm{d}W_s, \quad \tau_i \leq t < \tau_{i+1},$$

and apply the same arguments. Then we set $X_{\tau_{i+1}} = X_{\tau_{i+1}-} - Y_{i+1}$ and continue in the same manner to prove the theorem by induction.

Therefore, [8.3] has a unique strong solution on $\mathbb{R}_+$, which is our assertion. □

8.3. Redefinition of the ruin time

In what follows, let $\Delta N_s = N_s - N_{s-}$, $s > 0$.

THEOREM 8.2.– If $(X_t)_{t\geq 0}$ is a unique strong solution to [8.3] on $\mathbb{R}_+$, then

$$X_t = S_{t\wedge\tau^*} \left(\frac{x}{S_0} + \int_0^t \frac{c(X_s)}{S_{s\wedge\tau^*}} \, \mathrm{d}s - \sum_{0<s\leq t} \frac{Y_{N_s} \, \mathbb{1}_{\{\Delta N_s \neq 0\}}}{S_{s\wedge\tau^*}} \right), \quad t \geq 0. \qquad [8.5]$$

PROOF.– To get [8.5], we apply considerations similar to those in [GĨH 72, KLO 92] when we solve linear stochastic differential equations. We now make the temporary assumption that $x > 0$ and solve the equation

$$X'_t = x + a \int_0^t X'_s \, \mathbb{1}_{\{s<\tau^*\}} \, \mathrm{d}s + b \int_0^t X'_s \, \mathbb{1}_{\{s<\tau^*\}} \, \mathrm{d}W_s, \quad t \geq 0, \qquad [8.6]$$

where $(X'_t)_{t\geq 0}$ is some new process.

Since [8.6] has a unique strong solution on $\mathbb{R}_+$, which is continuous a.s., we have $X'_t > 0$ in some right semi-neighborhood of the point $t = 0$. Applying Itô's formula, we get

$$\begin{aligned}\ln X'_t &= \ln x + \int_0^t \left(\frac{aX'_s \,\mathbb{1}_{\{s<\tau^*\}}}{X'_s} - \frac{b^2 (X'_s)^2 \,\mathbb{1}_{\{s<\tau^*\}}}{2(X'_s)^2} \right) \mathrm{d}s \\ &\quad + \int_0^t \frac{bX'_s \,\mathbb{1}_{\{s<\tau^*\}}}{X'_s} \,\mathrm{d}W_s \\ &= \ln x + (a - b^2/2)(t \wedge \tau^*) + bW_{t\wedge\tau^*},\end{aligned}$$

which yields

$$X'_t = x \exp\big((a - b^2/2)(t \wedge \tau^*) + bW_{t\wedge\tau^*}\big). \qquad [8.7]$$

It is easily seen that [8.7] is a solution to [8.6] while $X'_t \neq 0$. Since the right-hand side of [8.7] is positive for all $t \in \mathbb{R}_+$, we conclude that [8.7] is a solution to [8.6] on $\mathbb{R}_+$.

Next, let now $x \geq 0$ and

$$X''_t = \exp\big((b^2/2 - a)(t \wedge \tau^*) - bW_{t\wedge\tau^*}\big), \quad t \geq 0. \qquad [8.8]$$

Applying Itô's formula to $(X''_t)_{t\geq 0}$ yields

$$\begin{aligned}X''_t &= 1 + (b^2 - a) \int_0^t X''_s \,\mathbb{1}_{\{s<\tau^*\}} \,\mathrm{d}s \\ &\quad - b \int_0^t X''_s \,\mathbb{1}_{\{s<\tau^*\}} \,\mathrm{d}W_s, \quad t \geq 0.\end{aligned} \qquad [8.9]$$

Introduce the new process $(X'''_t)_{t\geq 0}$, such that $X'''_t = X''_t X_t$, $t \geq 0$. By [8.3] and [8.9], we get

$$\begin{aligned}X'''_t &= x + \int_0^t X''_s \,\mathrm{d}X_s + \int_0^t X_s \,\mathrm{d}X''_s - b^2 \int_0^t X''_s X_s \,\mathbb{1}_{\{s<\tau^*\}} \,\mathrm{d}s \\ &= x + \int_0^t X''_s \left(c(X_s) + aX_s \,\mathbb{1}_{\{s<\tau^*\}} \right) \mathrm{d}s \\ &\quad + b \int_0^t X''_s X_s \,\mathbb{1}_{\{s<\tau^*\}} \,\mathrm{d}W_s\end{aligned}$$

$$
\begin{aligned}
&- \sum_{0<s\le t} X''_s Y_{N_s} \mathbb{1}_{\{\Delta N_s \neq 0\}} + (b^2 - a) \int_0^t X_s X''_s \mathbb{1}_{\{s<\tau^*\}} \, \mathrm{d}s \\
&- b \int_0^t X_s X''_s \mathbb{1}_{\{s<\tau^*\}} \, \mathrm{d}W_s - b^2 \int_0^t X''_s X_s \mathbb{1}_{\{s<\tau^*\}} \, \mathrm{d}s \qquad [8.10] \\
&= x + \int_0^t c(X_s) X''_s \, \mathrm{d}s - \sum_{0<s\le t} X''_s Y_{N_s} \mathbb{1}_{\{\Delta N_s \neq 0\}}, \quad t \ge 0.
\end{aligned}
$$

It follows immediately from [8.8] that $X''_t = \frac{S_0}{S_{t\wedge\tau^*}}$. From this and [8.10], we obtain

$$
\begin{aligned}
X_t &= \frac{X'''_t}{X''_t} = \frac{X'''_t \, S_{t\wedge\tau^*}}{S_0} \\
&= \frac{S_{t\wedge\tau^*}}{S_0} \left(x + \int_0^t \frac{c(X_s)\, S_0}{S_{s\wedge\tau^*}} \, \mathrm{d}s - \sum_{0<s\le t} \frac{S_0 \, Y_{N_s} \mathbb{1}_{\{\Delta N_s \neq 0\}}}{S_{s\wedge\tau^*}} \right) \\
&= S_{t\wedge\tau^*} \left(\frac{x}{S_0} + \int_0^t \frac{c(X_s)}{S_{s\wedge\tau^*}} \, \mathrm{d}s - \sum_{0<s\le t} \frac{Y_{N_s} \mathbb{1}_{\{\Delta N_s \neq 0\}}}{S_{s\wedge\tau^*}} \right), \quad t \ge 0,
\end{aligned}
$$

which is the desired conclusion. □

REMARK 8.1.– By [8.5], we can redefine the *ruin time* as

$$
\tau(x) = \inf\{t \ge 0 \colon \tilde{X}_t(x) < 0\},
$$

where the process $\big(\tilde{X}_t(x)\big)_{t\ge0}$, which is briefly denoted by $\big(\tilde{X}_t\big)_{t\ge0}$ when no confusion can arise, is given by

$$
\tilde{X}_t = \frac{x}{S_0} + \int_0^t \frac{c(X_s)}{S_{s\wedge\tau^*}} \, \mathrm{d}s - \sum_{0<s\le t} \frac{Y_{N_s} \mathbb{1}_{\{\Delta N_s \neq 0\}}}{S_{s\wedge\tau^*}}, \quad t \ge 0. \qquad [8.11]
$$

8.4. Supermartingale property for the exponential process

For all $R \ge 0$, we define the processes $\big(U_t(x,R)\big)_{t\ge0}$ and $\big(V_t(x,R)\big)_{t\ge0}$ by

$$
U_t(x,R) = -R\tilde{X}_t(x) \quad \text{and} \quad V_t(x,R) = \mathrm{e}^{U_t(x,R)},
$$

where $\big(\tilde{X}_t(x)\big)_{t\ge0}$ is given by [8.11]. In what follows, we write $\tilde{X}_t$, U_t and V_t instead of $\tilde{X}_t(x)$, $U_t(x,R)$ and $V_t(x,R)$, respectively, when no confusion can arise. Recall that the function $h(R)$ and the constant $R_\infty \in (0,+\infty]$, which are used below, were defined in section 1.1.7.

THEOREM 8.3.– If [8.3] has a unique strong solution on $\mathbb{R}_+$ and there is $\hat{R} \in (0,\ S_* R_\infty)$, such that

$$\inf_{u\in\mathbb{R}} c(u) \geq \lambda h'\left(\frac{\hat{R}}{S_*}\right), \qquad [8.12]$$

then $\big(V_t(x,\hat{R})\big)_{t\geq 0}$ is an $(\mathfrak{F}_t)$-supermartingale.

PROOF.– Let $\big(X_t(x)\big)_{t\geq 0}$ be a unique strong solution to [8.3] on $\mathbb{R}_+$ and $R \in (0,\ S_* R_\infty)$. By theorem 8.2 and remark 8.1, we have $X_t = S_{t\wedge\tau^*}\tilde{X}_t$ for all $t \geq 0$, where $(\tilde{X}_t)_{t\geq 0}$ is given by [8.11]. Consequently, we get

$$U_t = -\frac{Rx}{S_0} - R\int_0^t \frac{c(X_s)}{S_{s\wedge\tau^*}}\,\mathrm{d}s + R\sum_{0<s\leq t}\frac{Y_{N_s}\mathbb{1}_{\{\Delta N_s\neq 0\}}}{S_{s\wedge\tau^*}}, \quad t\geq 0. \qquad [8.13]$$

From [8.11] it can be easily seen that the process $(\tilde{X}_t)_{t\geq 0}$ is a sum of a martingale and càdlàg processes of bounded variation. Indeed, the process

$$\left(\sum_{0<s\leq t}\frac{Y_{N_s}\mathbb{1}_{\{\Delta N_s\neq 0\}} - \lambda\mu}{S_{s\wedge\tau^*}}\right)_{t\geq 0}$$

is a compensated process with independent increments. Therefore, it is an $(\mathfrak{F}_t)$-martingale. The processes $\int_0^t \frac{c(X_s)}{S_{s\wedge\tau^*}}\,\mathrm{d}s$ and $\int_0^t \frac{\lambda\mu}{S_{s\wedge\tau^*}}\,\mathrm{d}s$ are increasing. Hence, they are càdlàg processes of bounded variation.

Thus, $(\tilde{X}_t)_{t\geq 0}$ is an $(\mathfrak{F}_t)$-semimartingale as a sum of a martingale and càdlàg processes of bounded variation and so are $(U_t)_{t\geq 0}$ and $(V_t)_{t\geq 0}$ (see section A.4). By [8.13], applying It's formula, we get

$$V_t = \exp\left(-\frac{Rx}{S_0}\right) + \int_{0_+}^t \mathrm{e}^{U_{s-}}\,\mathrm{d}U_s + \sum_{0<s\leq t}\left(\mathrm{e}^{U_s} - \mathrm{e}^{U_{s-}} - \mathrm{e}^{U_{s-}}(U_s - U_{s-})\right), \quad t\geq 0, \qquad [8.14]$$

where

$$U_{t-} = -\frac{Rx}{S_0} - R\int_0^t \frac{c(X_s)}{S_{s\wedge\tau^*}}\,\mathrm{d}s + R\sum_{0<s\leq t_-}\frac{Y_{N_s}\mathbb{1}_{\{\Delta N_s\neq 0\}}}{S_{s\wedge\tau^*}},$$

$$\mathrm{d}U_s = -\frac{Rc(\tilde{X}_s)}{S_{s\wedge\tau^*}}\,\mathrm{d}s + \frac{RY_{N_s}\,\mathbb{1}_{\{\Delta N_s \neq 0\}}}{S_{s\wedge\tau^*}},$$

$$\mathrm{e}^{U_s} - \mathrm{e}^{U_{s-}} = \mathrm{e}^{U_{s-}}\left(\exp\left(\frac{RY_{N_s}\,\mathbb{1}_{\{\Delta N_s \neq 0\}}}{S_{s\wedge\tau^*}}\right) - 1\right),$$

$$U_s - U_{s_-} = \frac{RY_{N_s}\,\mathbb{1}_{\{\Delta N_s \neq 0\}}}{S_{s\wedge\tau^*}},$$

$$\Delta N_s = N_s - N_{s_-}.$$

Substituting the above equalities into [8.14] gives

$$\begin{aligned} V_t = \exp\left(-\frac{Rx}{S_0}\right) &- R\int_{0+}^{t} \frac{\mathrm{e}^{U_{s-}}\,c(X_s)}{S_{s\wedge\tau^*}}\,\mathrm{d}s \\ &+ R\sum_{0<s\leq t} \frac{\mathrm{e}^{U_{s-}}\,Y_{N_s}\,\mathbb{1}_{\{\Delta N_s \neq 0\}}}{S_{s\wedge\tau^*}} \\ &+ \sum_{0<s\leq t} \mathrm{e}^{U_{s-}}\left(\exp\left(\frac{RY_{N_s}\,\mathbb{1}_{\{\Delta N_s \neq 0\}}}{S_{s\wedge\tau^*}}\right) - 1\right. \\ &\left. - \frac{RY_{N_s}\,\mathbb{1}_{\{\Delta N_s \neq 0\}}}{S_{s\wedge\tau^*}}\right), \quad t \geq 0. \end{aligned} \qquad [8.15]$$

Simplifying [8.15] yields

$$\begin{aligned} V_t = \exp\left(-\frac{Rx}{S_0}\right) &- R\int_{0+}^{t} \frac{\mathrm{e}^{U_{s-}}\,c(X_s)}{S_{s\wedge\tau^*}}\,\mathrm{d}s \\ &+ \sum_{0<s\leq t} \mathrm{e}^{U_{s-}}\left(\exp\left(\frac{RY_{N_s}\,\mathbb{1}_{\{\Delta N_s \neq 0\}}}{S_{s\wedge\tau^*}}\right) - 1\right), \quad t \geq 0. \end{aligned} \qquad [8.16]$$

We now show that $\left(V_t\right)_{t\geq 0}$ is integrable. Taking into account [8.13], the positivity of the premium intensity, the fact that $S_{t\wedge\tau^*} \geq S_*$ for all $t \geq 0$ and the definition of $h(R)$, for all $R \in (0,\, S_* R_\infty)$ and $t \geq 0$, we get

$$\mathbb{E}\big[|V_t|\big] = \mathbb{E}\big[\mathrm{e}^{U_t}\big] \leq \mathbb{E}\left[\exp\left(-\frac{Rx}{S_0} + R\sum_{0<s\leq t} \frac{Y_{N_s}\,\mathbb{1}_{\{\Delta N_s \neq 0\}}}{S_{s\wedge\tau^*}}\right)\right]$$

$$\leq \mathbb{E}\Big[\exp\Big(-\frac{Rx}{S_0} + \frac{R}{S_*}\sum_{i=1}^{N_t} Y_i\Big)\Big]$$

$$= \exp\Big(-\frac{Rx}{S_0}\Big) \sum_{n=0}^{\infty} \mathbb{E}\Big[\exp\Big(\frac{R}{S_*}\sum_{i=1}^{n} Y_i\Big)\Big] \frac{(\lambda t)^n}{n!} \,\mathrm{e}^{-\lambda t}$$

$$= \exp\Big(-\frac{Rx}{S_0} - \lambda t\Big) \sum_{n=0}^{\infty} \Big(h\Big(\frac{R}{S_*}\Big) + 1\Big)^n \frac{(\lambda t)^n}{n!}$$

$$= \exp\Big(-\frac{Rx}{S_0} - \lambda t h\Big(\frac{R}{S_*}\Big)\Big) < \infty,$$

which proves the integrability of $(V_t)_{t\geq 0}$.

What is left is to show that

$$\mathbb{E}\big[V_{t_2}(x,\hat{R}) - V_{t_1}(x,\hat{R}) \,/\, \mathfrak{F}_{t_1}\big] \leq 0 \qquad [8.17]$$

for all $t_2 \geq t_1 \geq 0$, where $\hat{R} \in (0,\ S_* R_\infty)$ is such that [8.12] holds.

By [8.16], for all $t_2 \geq t_1 \geq 0$, we have

$$V_{t_2} - V_{t_1} = -\, R \int_{t_1}^{t_2} \frac{\mathrm{e}^{U_{s-}}\, c(X_s)}{S_{s\wedge\tau^*}} \,\mathrm{d}s + \sum_{t_1 < s \leq t_2} \mathrm{e}^{U_{s-}} \Big(\exp\Big(\frac{RY_{N_s} \mathbb{1}_{\{\Delta N_s \neq 0\}}}{S_{s\wedge\tau^*}}\Big) - 1\Big). \qquad [8.18]$$

Since $\mathrm{e}^u - 1 \leq u\mathrm{e}^u$ for all $u \geq 0$ and $S_{t\wedge\tau^*} \geq S_*$ for all $t \geq 0$ a.s., from [8.18] we get

$$V_{t_2} - V_{t_1} \leq -\, R \int_{t_1}^{t_2} \frac{\mathrm{e}^{U_{s-}}\, c(X_s)}{S_{s\wedge\tau^*}} \,\mathrm{d}s + \sum_{t_1 < s \leq t_2} \frac{RY_{N_s} \mathbb{1}_{\{\Delta N_s \neq 0\}}\, \mathrm{e}^{U_{s-}}}{S_{s\wedge\tau^*}} \exp\Big(\frac{RY_{N_s}}{S_*}\Big),$$

which can be rewritten as

$$V_{t_2} - V_{t_1} \leq -R \int_{t_1}^{t_2} \frac{\mathrm{e}^{U_{s-}}\, c(X_s)}{S_{s\wedge\tau^*}} \,\mathrm{d}s + \int_{t_1}^{t_2} \frac{\mathrm{e}^{U_{s-}}}{S_{s\wedge\tau^*}} \,\mathrm{d}Q_s, \qquad [8.19]$$

where

$$Q_t = R \sum_{0<s\leq t} Y_{N_s}\, \mathbb{1}_{\{\Delta N_s \neq 0\}} \exp\biggl(\frac{RY_{N_s}}{S_*}\biggr).$$

It is evident that we can rewrite the last equality as

$$Q_t = R \sum_{i=1}^{N_t} Y_i \exp\biggl(\frac{RY_i}{S_*}\biggr).$$

Next, since $h(R) = \mathbb{E}\bigl[\mathrm{e}^{RY_i}\bigr] < \infty$ for all $R \in [0, R_\infty)$, we conclude that $h'(R) = \mathbb{E}\bigl[Y_i\, \mathrm{e}^{RY_i}\bigr] < \infty$ for all $R \in [0, R_\infty)$. By Wald's identity, we obtain

$$\mathbb{E}[Q_t] = \lambda t R\, \mathbb{E}\biggl[Y_i \exp\biggl(\frac{RY_i}{S_*}\biggr)\biggr] = \lambda t R\, h'\biggl(\frac{R}{S_*}\biggr)$$

for all $R \in (0, S_* R_\infty)$ and $t \geq 0$. Therefore, $\mathbb{E}[Q_t] < \infty$ for all $R \in (0, S_* R_\infty)$ and $t \geq 0$.

Since $(N_t)_{t\geq 0}$ has independent increments and r.v.'s Y_i, $i \geq 1$, are mutually independent, $(Q_t)_{t\geq 0}$ is a process with independent increments. Hence, the compensated process $\bigl(Q_t - \mathbb{E}[Q_t]\bigr)_{t\geq 0}$ is an $(\mathfrak{F}_t)$-martingale.

Moreover, we can rewrite [8.19] as

$$\begin{aligned} V_{t_2} - V_{t_1} \leq &- R \int_{t_1}^{t_2} \frac{\mathrm{e}^{U_{s-}}}{S_{s\wedge\tau^*}} \biggl(c(X_s) - \lambda\, h'\biggl(\frac{R}{S_*}\biggr)\biggr)\, \mathrm{d}s \\ &+ \int_{t_1}^{t_2} \frac{\mathrm{e}^{U_{s-}}}{S_{s\wedge\tau^*}}\, \mathrm{d}(Q_s - \mathbb{E}[Q_s]). \end{aligned} \qquad [8.20]$$

Since $\hat{R} \in (0, S_* R_\infty)$ is such that [8.12] holds, taking conditional expectation w.r.t. $\mathfrak{F}_{t_1}$ in [8.20] shows that [8.17] is true, which completes the proof. □

8.5. Upper exponential bound for the ruin probability

THEOREM 8.4.– Let the surplus process $\bigl(X_t(x)\bigr)_{t\geq 0}$ follow [8.3] under the above assumptions, the premium intensity $c(u)$ be a Lipschitz continuous function on $\mathbb{R}$ and there be $\hat{R} \in (0, S_* R_\infty)$, such that

$$\inf_{u\in\mathbb{R}} c(u) = \lambda h'\biggl(\frac{\hat{R}}{S_*}\biggr). \qquad [8.21]$$

Then, for all $x \geq 0$, we have

$$\psi(x) \leq \mathrm{e}^{-\hat{R}x/S_0}. \qquad [8.22]$$

PROOF.– Since $c(u)$ is a Lipschitz continuous function on $\mathbb{R}$, equation [8.3] has a unique strong solution on $\mathbb{R}_+$ by theorem 8.1. From theorem 8.2 and remark 8.1, it follows that we can define $\big(\tilde{X}_t(x)\big)_{t\geq 0}$ by [8.11] and introduce $\big(V_t(x,R)\big)_{t\geq 0}$.

Next, since there is $\hat{R} \in (0, S_* R_\infty)$, such that [8.21] holds, the process $\big(V_t(x,\hat{R})\big)_{t\geq 0}$ is an $(\mathfrak{F}_t)$-supermartingale by theorem 8.3. Recall that $\tau(x)$ is the ruin time. Thus, for all $t \geq 0$, we get

$$\begin{aligned}
\mathrm{e}^{-\hat{R}x/S_0} &= V_0(x,\hat{R}) \geq \mathbb{E}\big[V_{t\wedge\tau(x)}(x,\hat{R}) \,/\, \mathfrak{F}_0\big] \\
&= \mathbb{E}\big[V_{\tau(x)}(x,\hat{R})\; \mathbb{1}_{\{\tau(x)<t\}}\big] + \mathbb{E}\big[V_t(x,\hat{R})\; \mathbb{1}_{\{\tau(x)\geq t\}}\big] \qquad [8.23] \\
&\geq \mathbb{E}\big[V_{\tau(x)}(x,\hat{R})\; \mathbb{1}_{\{\tau(x)<t\}}\big].
\end{aligned}$$

Letting $t \to \infty$ in [8.23] yields

$$\mathbb{E}\big[V_{\tau(x)}(x,\hat{R})\; \mathbb{1}_{\{\tau(x)<\infty\}}\big] \leq \mathrm{e}^{-\hat{R}x/S_0}. \qquad [8.24]$$

Moreover,

$$\mathbb{E}\big[V_{\tau(x)}(x,\hat{R})\; \mathbb{1}_{\{\tau(x)<\infty\}}\big] = \mathbb{E}\big[V_{\tau(x)}(x,\hat{R}) \,/\, \tau(x)<\infty\big]\; \mathbb{P}[\tau(x)<\infty],$$

and

$$\mathbb{E}\big[V_{\tau(x)}(x,\hat{R}) \,/\, \tau(x)<\infty\big] \geq 1$$

by the definition of the ruin time. Consequently, from [8.24] we conclude that

$$\mathbb{P}[\tau(x)<\infty] \leq \frac{\mathrm{e}^{-\hat{R}x/S_0}}{\mathbb{E}\big[V_{\tau(x)}(x,\hat{R}) \,/\, \tau(x)<\infty\big]} \leq \mathrm{e}^{-\hat{R}x/S_0},$$

which gives [8.22]. The proof is complete. □

REMARK 8.2.– In theorem 8.4, we took $\hat{R} \in (0, S_* R_\infty)$, such that [8.21] holds, i.e. equality is attained in [8.12], to get the better bound in [8.22].

REMARK 8.3.– In the proof of theorem 8.3, we used the inequality $e^u - 1 \leq ue^u$ for all $u \geq 0$ and that is why we obtained non-classical condition [8.21] for finding $\hat{R}$. We now show that in the classical risk model without any investments the classical exponential bound for the ruin probability is better than our one.

To get the classical risk model without any investments from our model, we set $c(u) = c > 0$ for all $u \in \mathbb{R}$, $S_* = 1$ and $S_0 \in (0, S_*)$. Though our assumption was $S_* \in (0, S_0)$ to exclude the degenerate case when the insurance company does not start its investment activity, it is easy to check that all our considerations are true if $S_0 \in (0, S_*)$.

We assume that there is $\hat{R} \in (0, R_\infty)$, such that [8.21] holds, i.e. $\hat{R}$ is a unique solution to the equation

$$h'(\hat{R}) = \frac{c}{\lambda}.$$

Furthermore, by lemma 1.2, the equation

$$\frac{h(\tilde{R})}{\tilde{R}} = \frac{c}{\lambda}$$

has a unique solution $\tilde{R} \in (0, R_\infty)$ (see also, for example [ROL 99, p. 170]).

We now show that $\hat{R} < \tilde{R}$, which implies that the classical bound is better. To this end, we introduce the functions $g_1(R) = h'(R)$ and $g_2(R) = \frac{h(R)}{R}$ on $(0, R_\infty)$.

First of all, note that $g_1(R) > g_2(R)$ for all $R \in (0, R_\infty)$, which is equivalent to

$$\int_0^{+\infty} y e^{Ry} \, dF(y) > \frac{1}{R} \left(\int_0^{+\infty} e^{Ry} \, dF(y) - 1 \right). \qquad [8.25]$$

Indeed, to establish [8.25], we rewrite it as

$$\int_0^{+\infty} (1 - Ry) \, e^{Ry} \, dF(y) < 1. \qquad [8.26]$$

Since $e^u - 1 < ue^u$ for all $u > 0$, which is equivalent to $(1-u)\,e^u < 1$, we get $(1 - Ry)\,e^{Ry} < 1$ for all $y > 0$ and $R \in (0, R_\infty)$. Moreover, we do not deal with the case when $\mathbb{P}[Y_i = 0] = 1$. Hence, it is easily seen that [8.26] is true and so is [8.25].

Next, the functions $g_1(R)$ and $g_2(R)$ are increasing on $(0, R_\infty)$. This assertion is evident for $g_1(R)$. For $g_2(R)$, it follows immediately from the fact that

$$g_2'(R) = \frac{h'(R)R - h(R)}{R^2} = \frac{g_1(R) - g_2(R)}{R} > 0$$

for all $R \in (0, R_\infty)$.

From the above we conclude that $g_1(R)$ reaches c/λ faster than $g_2(R)$, which yields $\hat{R} < \tilde{R}$. □

8.6. Exponentially distributed claim sizes

In what follows, let $c_0 = \inf_{u \in \mathbb{R}} c(u) > 0$.

THEOREM 8.5.– Let the surplus process $\big(X_t(x)\big)_{t \geq 0}$ follow [8.3] under the above assumptions, the premium intensity $c(u)$ be a Lipschitz continuous function on $\mathbb{R}$, the claim sizes be exponentially distributed with mean μ and $c_0 > \lambda\mu$. Then, for all $x \geq 0$, we have

$$\psi(x) \leq \exp\left(-\frac{S_* x}{\mu S_0}\left(1 - \sqrt{\frac{\lambda\mu}{c_0}}\right)\right). \qquad [8.27]$$

PROOF.– For all $R \in [0,\ 1/\mu)$, we have

$$h(R) = \frac{\mu R}{1 - \mu R}$$

and

$$h'(R) = \frac{\mu}{(1 - \mu R)^2}.$$

We now show that there is $\hat{R} \in (0,\ S_*/\mu)$, such that [8.21] holds, i.e.

$$c_0 = \frac{\lambda\mu}{(1 - \mu\hat{R}/S_*)^2}. \qquad [8.28]$$

Solving [8.28] gives

$$\hat{R}_1 = \frac{S_*}{\mu}\left(1 - \sqrt{\frac{\lambda\mu}{c_0}}\right)$$

and

$$\hat{R}_2 = \frac{S_*}{\mu}\left(1 + \sqrt{\frac{\lambda\mu}{c_0}}\right).$$

It is obvious that $\hat{R}_2 \notin (0, S_*/\mu)$. Since $c_0 > \lambda\mu$, we conclude that $\hat{R}_1 \in (0, S_*/\mu)$. Therefore, by theorem 8.4, [8.22] holds with $\hat{R} = \hat{R}_1$, which yields [8.27]. The proof is complete. □

8.7. Modification of the model

We will now consider a modification of the model described in section 8.1. We assume that all surplus is invested in the risky asset only up to the *stopping time*

$$\bar{\tau}^* = \inf\{t \geq 0 \colon S_t \notin [S_*, S^*]\},$$

where $0 < S_* < S_0 < S^* < +\infty$. This means that the insurance company stops its investment activity when the price of the risky asset exits from the interval $[S_*, S^*]$. By the continuity of $(S_t)_{t\geq 0}$ a.s., we conclude that either $S_{\bar{\tau}^*} = S_*$ or $S_{\bar{\tau}^*} = S^*$ a.s.

All other assumptions are the same as in section 8.1. Hence, the *surplus process* $(X_t)_{t\geq 0}$ follows the equation

$$\begin{aligned} X_t = x + &\int_0^t \left(c(X_s) + aX_s\, \mathbb{1}_{\{s<\bar{\tau}^*\}}\right) \mathrm{d}s \\ &+ b\int_0^t X_s\, \mathbb{1}_{\{s<\bar{\tau}^*\}}\, \mathrm{d}W_s - \sum_{i=1}^{N_t} Y_i, \quad t \geq 0. \end{aligned} \qquad [8.29]$$

As in section 8.2, it is easy to prove that if $c(u)$ is a Lipschitz continuous function on $\mathbb{R}$, then [8.29] has a unique strong solution on $\mathbb{R}_+$. Moreover, applying arguments similar to those in section 8.3 shows that if $(X_t)_{t\geq 0}$ is a unique strong solution to [8.29] on $\mathbb{R}_+$, then

$$X_t = S_{t\wedge\bar{\tau}^*}\left(\frac{x}{S_0} + \int_0^t \frac{c(X_s)}{S_{s\wedge\bar{\tau}^*}}\,\mathrm{d}s - \sum_{0<s\leq t}\frac{Y_{N_s}\,\mathbb{1}_{\{\Delta N_s\neq 0\}}}{S_{s\wedge\bar{\tau}^*}}\right), \quad t\geq 0.$$

This allows us to redefine the ruin time as in remark 8.1 with

$$\tilde{X}_t = \frac{x}{S_0} + \int_0^t \frac{c(X_s)}{S_{s\wedge\bar{\tau}^*}}\,\mathrm{d}s - \sum_{0<s\leq t}\frac{Y_{N_s}\,\mathbb{1}_{\{\Delta N_s\neq 0\}}}{S_{s\wedge\bar{\tau}^*}}, \quad t\geq 0,$$

and introduce the processes $\big(U_t(x,R)\big)_{t\geq 0}$ and $\big(V_t(x,R)\big)_{t\geq 0}$ as in section 8.4.

THEOREM 8.6.– If [8.29] has a unique strong solution on $\mathbb{R}_+$ and there is $\hat{R}\in(0,\, S_*R_\infty)$, such that

$$\inf_{u\in\mathbb{R}} c(u) \geq \frac{\lambda S^*}{\hat{R}}\, h\left(\frac{\hat{R}}{S_*}\right), \qquad [8.30]$$

then $\big(V_t(x,\hat{R})\big)_{t\geq 0}$ is an $(\mathfrak{F}_t)$-supermartingale.

PROOF.– The proof of this theorem is similar to the proof of theorem 8.3. In particular, we can show that

$$\begin{aligned} V_t = {} & \exp\left(-\frac{Rx}{S_0}\right) - R\int_{0+}^t \frac{\mathrm{e}^{U_{s-}}\,c(X_s)}{S_{s\wedge\bar{\tau}^*}}\,\mathrm{d}s \\ & + \sum_{0<s\leq t}\mathrm{e}^{U_{s-}}\left(\exp\left(\frac{RY_{N_s}\,\mathbb{1}_{\{\Delta N_s\neq 0\}}}{S_{s\wedge\bar{\tau}^*}}\right) - 1\right), \quad t\geq 0, \end{aligned} \qquad [8.31]$$

and $\big(V_t\big)_{t\geq 0}$ is integrable.

We now only prove that [8.17] is true for all $t_2 \geq t_1 \geq 0$, where $\hat{R}\in(0,\, S_*R_\infty)$ is such that [8.30] holds. By [8.31], for all $t_2\geq t_1\geq 0$, we have

$$\begin{aligned} V_{t_2} - V_{t_1} = {} & -R\int_{t_1}^{t_2} \frac{\mathrm{e}^{U_{s-}}\,c(X_s)}{S_{s\wedge\bar{\tau}^*}}\,\mathrm{d}s \\ & + \sum_{t_1<s\leq t_2}\mathrm{e}^{U_{s-}}\left(\exp\left(\frac{RY_{N_s}\,\mathbb{1}_{\{\Delta N_s\neq 0\}}}{S_{s\wedge\bar{\tau}^*}}\right) - 1\right). \end{aligned} \qquad [8.32]$$

Since $S_* \leq S_{t\wedge\tau^*} \leq S^*$ for all $t \geq 0$ a.s., from [8.32] we obtain

$$V_{t_2} - V_{t_1} \leq - R \int_{t_1}^{t_2} \frac{\mathrm{e}^{U_{s-}}\, c(X_s)}{S^*}\, \mathrm{d}s + \sum_{t_1<s\leq t_2} \mathrm{e}^{U_{s-}} \left(\exp\left(\frac{RY_{N_s}\, \mathbb{1}_{\{\Delta N_s\neq 0\}}}{S_*}\right) - 1\right),$$

which is equivalent to

$$V_{t_2} - V_{t_1} \leq -\frac{R}{S^*} \int_{t_1}^{t_2} \mathrm{e}^{U_{s-}}\, c(X_s)\, \mathrm{d}s + \int_{t_1}^{t_2} \mathrm{e}^{U_{s-}}\, \mathrm{d}Q_s, \qquad [8.33]$$

where

$$Q_t = \sum_{0<s\leq t} \left(\exp\left(\frac{RY_{N_s}\, \mathbb{1}_{\{\Delta N_s\neq 0\}}}{S_*}\right) - 1\right).$$

By Wald's identity, we get $\mathbb{E}[Q_t] = \lambda t\, h\left(\frac{R}{S_*}\right)$ for all $R \in (0,\, S_* R_\infty)$ and $t \geq 0$. Hence, $\mathbb{E}[Q_t] < \infty$ for all $R \in (0,\, S_* R_\infty)$ and $t \geq 0$. Since $(Q_t)_{t\geq 0}$ is a process with independent increments, the compensated process $\big(Q_t - \mathbb{E}[Q_t]\big)_{t\geq 0}$ is an $(\mathfrak{F}_t)$-martingale. Moreover, we can rewrite [8.33] as

$$V_{t_2} - V_{t_1} \leq - \int_{t_1}^{t_2} \mathrm{e}^{U_{s-}} \left(\frac{Rc(X_s)}{S^*} - \lambda\, h\left(\frac{R}{S_*}\right)\right) \mathrm{d}s + \int_{t_1}^{t_2} \mathrm{e}^{U_{s-}}\, \mathrm{d}(Q_s - \mathbb{E}[Q_s]). \qquad [8.34]$$

Since $\hat{R} \in (0,\, S_* R_\infty)$ is such that [8.30] holds, taking conditional expectation w.r.t. $\mathfrak{F}_{t_1}$ in [8.34] shows that [8.17] is true, which completes the proof. □

THEOREM 8.7.– Let the surplus process $\big(X_t(x)\big)_{t\geq 0}$ follow [8.29] under the above assumptions, the premium intensity $c(u)$ be a Lipschitz continuous function on $\mathbb{R}$ and there be $\hat{R} \in (0,\, S_* R_\infty)$, such that

$$\inf_{u\in\mathbb{R}} c(u) = \frac{\lambda S^*}{\hat{R}}\, h\left(\frac{\hat{R}}{S_*}\right). \qquad [8.35]$$

Then, for all $x \geq 0$, we have

$$\psi(x) \leq \mathrm{e}^{-\hat{R}x/S_0}. \qquad [8.36]$$

The proof of theorem 8.7 is similar to the proof of theorem 8.4.

REMARK 8.4.– Note that in the classical risk model without any investments, theorem 8.7 gives the classical exponential bound for the ruin probability. This fact follows immediately if we set $c(u) = c > 0$ for all $u \in \mathbb{R}$ and $S_0 = S_* = S^* = 1$. It is easy to check that all our considerations are also true for this degenerate case.

We now deal with the case of exponentially distributed claim sizes. Recall that $c_0 = \inf_{u \in \mathbb{R}} c(u) > 0$.

THEOREM 8.8.– Let the surplus process $\big(X_t(x)\big)_{t\geq 0}$ follow [8.29] under the above assumptions, the premium intensity $c(u)$ be a Lipschitz continuous function on $\mathbb{R}$, the claim sizes be exponentially distributed with mean μ and $c_0 S_* > \lambda\mu S^*$. Then, for all $x \geq 0$, we have

$$\psi(x) \leq \exp\left(-\frac{(c_0 S_* - \lambda\mu S^*)x}{\mu c_0 S_0}\right). \qquad [8.37]$$

PROOF.– For all $R \in [0,\, 1/\mu)$, we have

$$h(R) = \frac{\mu R}{1 - \mu R}.$$

It is easy to check that since $c_0 S_* > \lambda\mu S^*$, there is $\hat{R} \in (0,\, S_*/\mu)$, such that [8.35] holds, i.e.

$$c_0 = \frac{\lambda S^* \mu \hat{R}}{\hat{R} S_* (1 - \mu\hat{R}/S_*)}. \qquad [8.38]$$

Indeed, from [8.38], we get

$$\hat{R} = \frac{c_0 S_* - \lambda\mu S^*}{\mu c_0},$$

which is in $(0,\, S_*/\mu)$ provided that $c_0 S_* > \lambda\mu S^*$.

Thus, by theorem 8.7, [8.36] holds with this $\hat{R}$, which yields [8.37]. The proof is complete. □

9

Risk Model with Variable Premium Intensity and Investments in One Risk-Free and a Few Risky Assets

In this chapter, we consider a generalization of the classical risk model where a premium intensity depends on a current surplus of an insurance company. The surplus is invested in one risk-free and a few risky assets. In addition, borrowing is also possible. The prices of the risky assets follow geometric Brownian motions, which are not necessarily independent. Our main aim is to obtain an upper exponential bound for the ruin probability under some constant investment strategy. To this end, we establish the supermartingale property for an auxiliary exponential process. Finally, we deal with exponentially distributed claim sizes and constant premium intensity. For this case, we consider some examples and compare results obtained.

9.1. Description of the model

We deal with a risk model where an insurance company invests its surplus in one risk-free and n risky assets. In what follows, we suppose that $n \geq 2$. The case $n = 1$ is considered separately in section 9.5.

We assume that an insurance company has a non-negative initial surplus x and denote by $X_t^{(K_t^1, K_t^2, \dots, K_t^n)}(x)$ its surplus at time $t \geq 0$ provided that the investment strategy $(K_t^1, K_t^2, \dots, K_t^n)_{t \geq 0}$ is used. The set of all admissible investment strategies will be defined later. For simplicity of notation, we also write X_t instead of $X_t^{(K_t^1,K_t^2,\dots,K_t^n)}(x)$ when no confusion can arise. Let $c\colon \mathbb{R} \to \mathbb{R}_+ \setminus \{0\}$ be a measurable function, such that $c(u) = c(0)$ for all $u < 0$ and $\inf_{u\in\mathbb{R}_+} c(u) > 0$, and let $c(X_t)$ be a premium intensity, which depends on the surplus at time t.

As for claim arrivals, we make the same assumptions and notation as for the classical risk model in section 1.1.1. Moreover, we suppose that there exists the shifted moment generating function $h\colon \mathbb{R}_+ \to \mathbb{R}_+$ of the claim sizes Y_i, $i \geq 1$, introduced in section 1.1.7 and satisfying the classical assumption that is also formulated in this section.

Next, the price of the jth *risky asset* equals S_t^j at time t. We model the processes $(S_t^j)_{t\geq 0}$ by *geometric Brownian motions*. To be more precise,

$$\mathrm{d}S_t^j = S_t^j(a_j\,\mathrm{d}t + b_j\,\mathrm{d}W_t^j), \quad 1 \leq j \leq n, \qquad [9.1]$$

where $a_j \in \mathbb{R}$, $b_j > 0$ and $(W_t^j)_{t\geq 0}$ are standard Brownian motions, such that $\mathrm{Cov}[W_t^j, W_t^k] = \rho_{jk}t$, $1 \leq j \leq n$ and $1 \leq k \leq n$. We denote by $\mathbf{C}$ the *correlation matrix* of the random vector $(W_t^1, W_t^2, \dots, W_t^n)$, $t \geq 0$. It is easily seen that

$$\mathbf{C} = \begin{pmatrix} \rho_{11} & \rho_{12} & \rho_{13} & \cdots & \rho_{1n} \\ \rho_{21} & \rho_{22} & \rho_{23} & \cdots & \rho_{2n} \\ \vdots & \vdots & \vdots & \ddots & \vdots \\ \rho_{n1} & \rho_{n2} & \rho_{n3} & \cdots & \rho_{nn} \end{pmatrix}.$$

Hence, the following conditions must be satisfied for all $1 \leq j \leq n$ and $1 \leq k \leq n$:

$$\rho_{jk} = \rho_{kj}, \quad \rho_{jk} \in [-1, 1], \quad \rho_{jj} = 1.$$

Moreover, the matrix $\mathbf{C}$ must be positive semi-definite (see, for example [ROU 94]).

We make some additional restrictions. Throughout this chapter, we assume that $\rho_{jk} \in (-1, 1)$ for all $j \neq k$ and the correlation matrix $\mathbf{C}$ is positive definite. By Sylvester's criterion, a real symmetric matrix is positive definite if and only if all its principal minors are positive (see, for example [GIL 91, HOR 85, STR 88]).

We suppose that the processes $(W_t^j)_{t\geq 0}$, $1 \leq j \leq n$, are independent of $(Y_i)_{i\geq 1}$ and $(N_t)_{t\geq 0}$. Let $(\mathfrak{F}_t^n)_{t\geq 0}$ be the filtration generated by $(Y_i)_{i\geq 1}$, $(N_t)_{t\geq 0}$ and $(W_t^j)_{t\geq 0}$, $1 \leq j \leq n$, i.e.

$$\mathfrak{F}_t^n = \sigma\big((N_s)_{0\leq s\leq t}, (W_s^1)_{0\leq s\leq t}, (W_s^2)_{0\leq s\leq t}, \dots, \\ (W_s^n)_{0\leq s\leq t}, Y_1, Y_2, \dots, Y_{N_t}\big).$$

We assume that the amount K_t^j is invested in the jth risky asset at time t and the rest of the surplus $X_t^{(K_t^1, K_t^2, \dots, K_t^n)}(x) - \sum_{j=1}^n K_t^j$ is invested in the *risk-free asset,* the price of which equals B_t at time t. The process $(B_t)_{t\geq 0}$ follows the equation

$$\mathrm{d}B_t = rB_t\,\mathrm{d}t, \qquad [9.2]$$

where $r > 0$ is a risk-free interest rate.

Any *admissible investment strategy* $(K_t^1, K_t^2, \dots, K_t^n)_{t\geq 0}$ is an $\mathfrak{F}_t^n$-predictable process, such that $K_t^j \geq 0$ and $\mathbb{P}\left[\int_0^t (K_s^j)^2\,\mathrm{d}s < +\infty\right] = 1$ for all $1 \leq j \leq n$ and $t \geq 0$. We impose the last constraint for the existence of the integrals w.r.t. $(W_t^j)_{t\geq 0}$ that are used below. Note that we do not put the restriction $K_t^j \leq X_t^{(K_t^1, K_t^2, \dots, K_t^n)}(x)$. Hence, if $\sum_{j=1}^n K_t^j > X_t^{(K_t^1, K_t^2, \dots, K_t^n)}(x)$, then the insurance company borrows the amount $\sum_{j=1}^n K_t^j - X_t^{(K_t^1, K_t^2, \dots, K_t^n)}(x)$ at time t with the same interest rate $r > 0$. In what follows, we will denote any admissible strategy $(K_t^1, K_t^2, \dots, K_t^n)_{t\geq 0}$ briefly by $(K_t^1, K_t^2, \dots, K_t^n)$.

Under the above assumptions, the *surplus process* $(X_t)_{t\geq 0}$ follows the equation

$$X_t = x + \int_0^t c(X_s)\,\mathrm{d}s + \sum_{j=1}^n \int_0^t \frac{K_s^j}{S_s^j}\,\mathrm{d}S_s^j \\ + \int_0^t \frac{1}{B_s}\left(X_s - \sum_{j=1}^n K_s^j\right)\mathrm{d}B_s - \sum_{i=1}^{N_t} Y_i, \quad t \geq 0. \qquad [9.3]$$

Substituting [9.1] and [9.2] into [9.3] gives

$$X_t = x + \int_0^t \left(c(X_s) + rX_s + \sum_{j=1}^{n} (a_j - r) K_s^j \right) \mathrm{d}s \\ + \sum_{j=1}^{n} b_j \int_0^t K_s^j \, \mathrm{d}W_s^j - \sum_{i=1}^{N_t} Y_i, \quad t \geq 0. \tag{9.4}$$

The *ruin time* under the admissible strategy $(K_t^1, K_t^2, \ldots, K_t^n)$ is defined by

$$\tau^{(K_t^1, K_t^2, \ldots, K_t^n)}(x) = \inf\{t \geq 0 \colon X_t^{(K_t^1, K_t^2, \ldots, K_t^n)}(x) < 0\}.$$

We suppose that $\tau^{(K_t^1, K_t^2, \ldots, K_t^n)}(x) = \infty$ if $X_t^{(K_t^1, K_t^2, \ldots, K_t^n)}(x) \geq 0$ for all $t \geq 0$. To simplify notation, we also let τ stand for $\tau^{(K_t^1, K_t^2, \ldots, K_t^n)}(x)$. The corresponding *infinite-horizon ruin probability* is given by

$$\psi^{(K_t^1, K_t^2, \ldots, K_t^n)}(x) = \mathbb{P}\big[\inf_{t \geq 0} \; X_t^{(K_t^1, K_t^2, \ldots, K_t^n)}(x) < 0\big],$$

which is equivalent to

$$\psi^{(K_t^1, K_t^2, \ldots, K_t^n)}(x) = \mathbb{P}[\tau^{(K_t^1, K_t^2, \ldots, K_t^n)}(x) < \infty].$$

The rest of the chapter is organized as follows. In section 9.2, we formulate the existence and uniqueness theorem for the stochastic differential equation that describes the surplus process. In section 9.3, we establish the supermartingale property for an auxiliary exponential process. The main result of this chapter is an upper exponential bound for the ruin probability under some constant investment strategy. We will obtain it in section 9.4. Next, in section 9.5, we concentrate on the case of one risky asset. In section 9.6, we will deal with exponentially distributed claim sizes and constant premium intensity and consider some examples. For some bibliographical notes regarding the problem considered, see section 7.6.

9.2. Existence and uniqueness theorem

In what follows, we imply the pathwise uniqueness of solutions.

THEOREM 9.1.– If $c(u)$ is a Lipschitz continuous function on $\mathbb{R}$, then [9.4] has a unique strong solution up to the time τ for any admissible strategy $(K_t^1, K_t^2, \ldots, K_t^n)$.

The proof of theorem 9.1 is analogous to the proof of theorems 7.2 and 8.1.

9.3. Supermartingale property for the exponential process

Let the stopped process $\left(\tilde{X}_t^{(K_t^1,K_t^2,\ldots,K_t^n)}(x)\right)_{t\geq 0}$ be defined by $\tilde{X}_t^{(K_t^1,K_t^2,\ldots,K_t^n)}(x) = X_{t\wedge\tau}^{(K_t^1,K_t^2,\ldots,K_t^n)}(x)$. Note that $\left(\tilde{X}_t^{(K_t^1,K_t^2,\ldots,K_t^n)}(x)\right)_{t\geq 0}$ is a solution to [9.4] provided that the process $\left(X_t^{(K_t^1,K_t^2,\ldots,K_t^n)}(x)\right)_{0\leq t<\tau}$ has the same property.

For all $R \geq 0$, we define the processes $\left(U_t^{(K_t^1,K_t^2,\ldots,K_t^n)}(x, R)\right)_{t\geq 0}$ and $\left(V_t^{(K_t^1,K_t^2,\ldots,K_t^n)}(x, R)\right)_{t\geq 0}$ by

$$U_t^{(K_t^1,K_t^2,\ldots,K_t^n)}(x, R) = -R\tilde{X}_t^{(K_t^1,K_t^2,\ldots,K_t^n)}(x)$$

and

$$V_t^{(K_t^1,K_t^2,\ldots,K_t^n)}(x, R) = \exp\Big(U_t^{(K_t^1,K_t^2,\ldots,K_t^n)}(x, R)\Big).$$

In what follows, we will write $\tilde{X}_t$, U_t and V_t instead of $\tilde{X}_t^{(K_t^1,K_t^2,\ldots,K_t^n)}(x)$, $U_t^{(K_t^1,K_t^2,\ldots,K_t^n)}(x, R)$ and $V_t^{(K_t^1,K_t^2,\ldots,K_t^n)}(x, R)$, respectively, when no confusion can arise. Recall that the function $h(R)$ and the constant $R_\infty \in (0, +\infty]$, which are used below, were defined in section 1.1.7.

THEOREM 9.2.– Let [9.4] have a unique strong solution up to the time τ for any admissible strategy $(K_t^1, K_t^2, \dots, K_t^n)$. If there are $\hat{R} \in (0, R_\infty)$ and a strategy $(\hat{K}_t^1, \hat{K}_t^2, \dots, \hat{K}_t^n)$, such that

$$\mathbb{P}\Bigg[\hat{R}^2\bigg(\frac{1}{2}\sum_{j=1}^{n} b_j^2(\hat{K}_t^j)^2 + \sum_{j=1}^{n-1}\sum_{k=j+1}^{n} b_j b_k \rho_{jk} \hat{K}_t^j \hat{K}_t^k\bigg) - \hat{R}\bigg(\inf_{u\in\mathbb{R}_+}(c(u)+ru) + \sum_{j=1}^{n}(a_j - r)\hat{K}_t^j\bigg) + \lambda h(\hat{R}) \le 0\Bigg] = 1 \quad \text{for all} \quad t \ge 0, \qquad [9.5]$$

then $\big(V_t^{(\hat{K}_t^1, \hat{K}_t^2, \dots, \hat{K}_t^n)}(x, \hat{R})\big)_{t\ge 0}$ is an $(\mathfrak{F}_t^n)$-supermartingale.

PROOF.– Let $R \in (0, R_\infty)$. Since $(\tilde{X}_t)_{t\ge0}$ is a solution to [9.4], we get

$$U_t = -Rx - R\int_0^{t\wedge\tau}\bigg(c(X_s) + rX_s + \sum_{j=1}^{n}(a_j - r)K_s^j\bigg)\,\mathrm{d}s - R\sum_{j=1}^{n} b_j \int_0^{t\wedge\tau} K_s^j\,\mathrm{d}W_s^j + R\sum_{i=1}^{N_{t\wedge\tau}} Y_i, \quad t \ge 0. \qquad [9.6]$$

Note that $(\tilde{X}_t)_{t\ge0}$ is a sum of local martingales and càdlàg processes of locally bounded variation. Indeed, it is easily seen that the processes $\left(\int_0^{t\wedge\tau} K_s^j\,\mathrm{d}W_s^j\right)_{t\ge0}$, $1 \le j \le n$, are local $(\mathfrak{F}_t^n)$-martingales. The process

$$\left(\sum_{i=1}^{N_{t\wedge\tau}} Y_i - \lambda\mu(t\wedge\tau)\right)_{t\ge0}$$

is a compensated process with independent increments. So, it is an $(\mathfrak{F}_t^n)$-martingale. Moreover, $(t \wedge \tau)_{t\ge0}$, $\left(\int_0^{t\wedge\tau} c(X_s)\,\mathrm{d}s\right)_{t\ge0}$, $\left(\int_0^{t\wedge\tau} X_s\,\mathrm{d}s\right)_{t\ge0}$ and $\left(\int_0^{t\wedge\tau} K_s^j\,\mathrm{d}s\right)_{t\ge0}$, $1 \le j \le n$, are càdlàg processes of locally bounded variation. Therefore, $(\tilde{X}_t)_{t\ge0}$ is an $(\mathfrak{F}_t^n)$-semimartingale as a sum of local martingales and càdlàg processes of locally bounded variation and so are $(U_t)_{t\ge0}$ and $(V_t)_{t\ge0}$ (see section A.4).

Let $(U_t^{\mathrm{c}})_{t\geq 0}$ denote the continuous part of the local martingale in the decomposition of $(U_t)_{t\geq 0}$. By [9.6], applying Itô's formula (see section A.4) we obtain

$$\begin{aligned} V_t = {} & \mathrm{e}^{-Rx} + \int_{0+}^{t\wedge\tau} \mathrm{e}^{U_{s-}}\,\mathrm{d}U_s + \frac{1}{2}\int_{0+}^{t\wedge\tau} \mathrm{e}^{U_{s-}}\,\mathrm{d}\langle U^{\mathrm{c}}, U^{\mathrm{c}}\rangle_s \\ & + \sum_{0<s\leq t\wedge\tau} \left(\mathrm{e}^{U_s} - \mathrm{e}^{U_{s-}} - \mathrm{e}^{U_{s-}}(U_s - U_{s-})\right), \quad t\geq 0, \end{aligned} \qquad [9.7]$$

where

$$\begin{aligned} U_{t-} = {} & -Rx - R\int_0^{t\wedge\tau}\Big(c(X_s) + rX_s + \sum_{j=1}^n (a_j - r)K_s^j\Big)\,\mathrm{d}s \\ & - R\sum_{j=1}^n b_j \int_0^{t\wedge\tau} K_s^j\,\mathrm{d}W_s^j + R\sum_{0<s\leq t_-\wedge\tau} Y_{N_s}\,\mathbb{1}_{\{\varDelta N_s\neq 0\}}, \end{aligned}$$

$$\begin{aligned} \mathrm{d}U_s = {} & \mathbb{1}_{\{s\leq\tau\}}\Big(-R\Big(c(X_s) + rX_s + \sum_{j=1}^n (a_j - r)K_s^j\Big)\,\mathrm{d}s \\ & - R\sum_{j=1}^n b_j K_s^j\,\mathrm{d}W_s^j + RY_{N_s}\,\mathbb{1}_{\{\varDelta N_s\neq 0\}}\Big), \end{aligned}$$

$$\mathrm{d}\langle U^{\mathrm{c}}, U^{\mathrm{c}}\rangle_s = R^2\,\mathbb{1}_{\{s\leq\tau\}}\Big(\sum_{j=1}^n b_j^2 (K_s^j)^2 + 2\sum_{j=1}^{n-1}\sum_{k=j+1}^n b_j b_k \rho_{jk} K_s^j K_s^k\Big)\,\mathrm{d}s,$$

$$\begin{aligned} \mathrm{e}^{U_s} - \mathrm{e}^{U_{s-}} &= \mathrm{e}^{U_{s-}}\,\mathbb{1}_{\{s\leq\tau\}}\left(\mathrm{e}^{RY_{N_s}\mathbb{1}_{\{\varDelta N_s\neq 0\}}} - 1\right), \\ U_s - U_{s-} &= RY_{N_s}\,\mathbb{1}_{\{\varDelta N_s\neq 0\}}\,\mathbb{1}_{\{s\leq\tau\}}, \\ \varDelta N_s &= N_s - N_{s-}. \end{aligned}$$

Substituting the above equalities into [9.7] yields

$$\begin{aligned} V_t = {} & \mathrm{e}^{-Rx} - R\int_{0+}^{t\wedge\tau} \mathrm{e}^{U_{s-}}\Big(c(X_s) + rX_s + \sum_{j=1}^n (a_j - r)K_s^j\Big)\,\mathrm{d}s \\ & - R\sum_{j=1}^n b_j \int_{0+}^{t\wedge\tau} \mathrm{e}^{U_{s-}} K_s^j\,\mathrm{d}W_s^j \end{aligned}$$

$$
\begin{aligned}
&+ R \sum_{0<s\leq t\wedge\tau} \mathrm{e}^{U_{s-}} Y_{N_s} \mathbb{1}_{\{\Delta N_s \neq 0\}} \\
&+ \frac{1}{2} R^2 \int_{0+}^{t\wedge\tau} \mathrm{e}^{U_{s-}} \Bigg(\sum_{j=1}^{n} b_j^2 (K_s^j)^2 \\
&+ 2 \sum_{j=1}^{n-1} \sum_{k=j+1}^{n} b_j b_k \rho_{jk} K_s^j K_s^k \Bigg) \,\mathrm{d}s \qquad [9.8] \\
&+ \sum_{0<s\leq t\wedge\tau} \mathrm{e}^{U_{s-}} \big(\mathrm{e}^{RY_{N_s} \mathbb{1}_{\{\Delta N_s \neq 0\}}} - 1 \\
&- RY_{N_s} \mathbb{1}_{\{\Delta N_s \neq 0\}} \big), \quad t \geq 0.
\end{aligned}
$$

Simplifying [9.8] gives

$$
\begin{aligned}
V_t = \mathrm{e}^{-Rx} &+ \int_{0+}^{t\wedge\tau} \mathrm{e}^{U_{s-}} \left(c(X_s) + rX_s + \sum_{j=1}^{n} (a_j - r) K_s^j \right) \mathrm{d}s \\
&+ R^2 \int_{0+}^{t\wedge\tau} \mathrm{e}^{U_{s-}} \Bigg(\frac{1}{2} \sum_{j=1}^{n} b_j^2 (K_s^j)^2 \\
&+ \sum_{j=1}^{n-1} \sum_{k=j+1}^{n} b_j b_k \rho_{jk} K_s^j K_s^k \Bigg) \,\mathrm{d}s \qquad [9.9] \\
&- R \sum_{j=1}^{n} b_j \int_{0+}^{t\wedge\tau} \mathrm{e}^{U_{s-}} K_s^j \,\mathrm{d}W_s^j \\
&+ \sum_{0<s\leq t\wedge\tau} \mathrm{e}^{U_{s-}} \big(\mathrm{e}^{RY_{N_s} \mathbb{1}_{\{\Delta N_s \neq 0\}}} - 1 \big), \quad t \geq 0.
\end{aligned}
$$

Arguments similar to those in the proof of theorem 7.3 show that

$$
\left(\sum_{0<s\leq t\wedge\tau} \big(\mathrm{e}^{RY_{N_s} \mathbb{1}_{\{\Delta N_s \neq 0\}}} - 1 \big) - \lambda h(R) \int_{0+}^{t\wedge\tau} \mathrm{e}^{U_{s-}} \,\mathrm{d}s \right)_{t\geq 0}
$$

is a local $(\mathfrak{F}_t^n)$-martingale. Moreover, it is easy to check that the processes

$$
\left(-Rb_j \int_{0+}^{t\wedge\tau} \mathrm{e}^{U_{s-}} K_s^j \,\mathrm{d}W_s^j \right)_{t\geq 0}, \quad 0 \leq j \leq n,
$$

are also local $(\mathfrak{F}_t^n)$-martingales. Thus, the process

$$\Big(-R\sum_{j=1}^{n} b_j \int_{0_+}^{t\wedge\tau} \mathrm{e}^{U_{s-}} K_s^j \,\mathrm{d}W_s^j + \sum_{0<s\le t\wedge\tau} \big(\mathrm{e}^{RY_{N_s}\mathbb{1}_{\{\Delta N_s\neq 0\}}} - 1\big) - \lambda h(R) \int_{0_+}^{t\wedge\tau} \mathrm{e}^{U_{s-}} \,\mathrm{d}s\Big)_{t\ge 0}.$$

is a local $(\mathfrak{F}_t^n)$-martingale.

Defining the process $(P_t)_{t\ge 0}$ by

$$P_t = V_t - V_0 + R\sum_{j=1}^{n} b_j \int_{0_+}^{t\wedge\tau} \mathrm{e}^{U_{s-}} K_s^j \,\mathrm{d}W_s^j - \sum_{0<s\le t\wedge\tau} \big(\mathrm{e}^{RY_{N_s}\mathbb{1}_{\{\Delta N_s\neq 0\}}} - 1\big) + \lambda h(R) \int_{0_+}^{t\wedge\tau} \mathrm{e}^{U_{s-}} \,\mathrm{d}s$$

and substituting V_t from [9.9], we get

$$P_t = \int_{0_+}^{t\wedge\tau} \mathrm{e}^{-RX_{s-}} \Big(R^2\Big(\frac{1}{2}\sum_{j=1}^{n} b_j^2 (K_s^j)^2 + \sum_{j=1}^{n-1}\sum_{k=j+1}^{n} b_j b_k \rho_{jk} K_s^j K_s^k\Big) - R\Big(c(X_s) + rX_s + \sum_{j=1}^{n} (a_j - r) K_s^j\Big) + \lambda h(R)\Big) \,\mathrm{d}s, \quad t\ge 0.$$

Next, $(V_t)_{t\ge 0}$ is a local $(\mathfrak{F}_t^n)$-supermartingale provided that $(P_t)_{t\ge 0}$ is a measurable non-increasing process, i.e.

$$\int_{t_1\wedge\tau}^{t_2\wedge\tau} \mathrm{e}^{-RX_{s-}} \Big(R^2\Big(\frac{1}{2}\sum_{j=1}^{n} b_j^2 (K_s^j)^2 + \sum_{j=1}^{n-1}\sum_{k=j+1}^{n} b_j b_k \rho_{jk} K_s^j K_s^k\Big) - R\Big(c(X_s) + rX_s + \sum_{j=1}^{n} (a_j - r) K_s^j\Big) + \lambda h(R)\Big) \,\mathrm{d}s \le 0 \quad \text{a.s.} \tag{9.10}$$

for all $t_2 \ge t_1 \ge 0$.

Since there are $\hat{R} \in (0, R_\infty)$ and the strategy $(\hat{K}_t^1, \hat{K}_t^2, \ldots, \hat{K}_t^n)$, such that [9.5] holds, [9.10] is true with $R = \hat{R}$ and $K_t^j = \hat{K}_t^j$, $1 \le j \le n$. Hence,

$\left(V_t^{(\hat{K}_t^1,\hat{K}_t^2,\dots,\hat{K}_t^n)}(x,\hat{R})\right)_{t\geq 0}$ is a non-negative local $(\mathfrak{F}_t^n)$-supermartingale. Applying Fatou's lemma and arguments similar to those in the proof of theorem 7.3 shows that $\left(V_t^{(\hat{K}_t^1,\hat{K}_t^2,\dots,\hat{K}_t^n)}(x,\hat{R})\right)_{t\geq 0}$ is an $(\mathfrak{F}_t^n)$-supermartingale, which completes the proof. □

9.4. Upper exponential bound for the ruin probability

Let the matrix $\mathbf{C}_j$, $1 \leq j \leq n$, be given by

$$\mathbf{C}_j = \begin{pmatrix} 1 & \rho_{12} & \rho_{13} & \cdots & \rho_{1(j-1)} & \frac{a_1-r}{b_1} & \rho_{1(j+1)} & \cdots & \rho_{1n} \\ \rho_{21} & 1 & \rho_{23} & \cdots & \rho_{2(j-1)} & \frac{a_2-r}{b_2} & \rho_{2(j+1)} & \cdots & \rho_{2n} \\ \vdots & \vdots & \vdots & \ddots & \vdots & \vdots & \vdots & \ddots & \vdots \\ \rho_{n1} & \rho_{n2} & \rho_{n3} & \cdots & \rho_{n(j-1)} & \frac{a_n-r}{b_n} & \rho_{n(j+1)} & \cdots & 1 \end{pmatrix},$$

i.e. $\mathbf{C}_j$ is the matrix obtained from the correlation matrix $\mathbf{C}$ by replacing the jth column by the column

$$\begin{pmatrix} \frac{a_1-r}{b_1} \\ \frac{a_2-r}{b_2} \\ \vdots \\ \frac{a_n-r}{b_n} \end{pmatrix}.$$

Recall that the correlation matrix $\mathbf{C}$ is positive definite by our assumption made in section 9.1. In particular, this means that $\det(\mathbf{C}) > 0$.

Moreover, let $g(K_t^1, K_t^2, \dots, K_t^n)$ be given by

$$g(K_t^1, K_t^2, \dots, K_t^n) = R^2\left(\frac{1}{2}\sum_{j=1}^{n} b_j^2 (K_t^j)^2 + \sum_{j=1}^{n-1}\sum_{k=j+1}^{n} b_j b_k \rho_{jk} K_t^j K_t^k\right)$$
$$- R\left(\inf_{u\in\mathbb{R}_+}(c(u)+ru) + \sum_{j=1}^{n}(a_j - r)K_t^j\right) + \lambda h(R)$$

for all $R \in (0, R_\infty)$ on the set of all admissible strategies $(K_t^1, K_t^2, \dots, K_t^n)$.

PROPOSITION 9.1.– If $\det(\mathbf{C}_j) \geq 0$ for all $1 \leq j \leq n$, then the point

$$\left(\frac{\det(\mathbf{C}_1)}{b_1 R \det(\mathbf{C})}, \frac{\det(\mathbf{C}_2)}{b_2 R \det(\mathbf{C})}, \dots, \frac{\det(\mathbf{C}_n)}{b_n R \det(\mathbf{C})}\right)$$

is a local minimum of $g(K_t^1, K_t^2, \dots, K_t^n)$ for all $t \geq 0$ and $R \in (0, R_\infty)$.

PROOF.– It is easily seen from the definition of $g(K_t^1, K_t^2, \ldots, K_t^n)$ that its local minimum, if any, does not depend on t. Since

$$\frac{\partial g(K_t^1, K_t^2, \ldots, K_t^n)}{\partial K_t^j} = R^2\left(b_j^2 K_t^j + b_j \sum_{\substack{1\le k\le n\\ k\ne j}} b_k \rho_{jk} K_t^k\right) - R(a_j - r), \quad 1 \le j \le n,$$

we get the system of linear equations

$$K_t^j + \frac{1}{b_j} \sum_{\substack{1\le k\le n\\ k\ne j}} b_k \rho_{jk} K_t^k = \frac{a_j - r}{b_j^2 R}, \quad 1 \le j \le n, \qquad [9.11]$$

to find all critical points of $g(K_t^1, K_t^2, \ldots, K_t^n)$.

System [9.11] can be rewritten in the matrix form

$$\begin{pmatrix} 1 & \frac{b_2}{b_1}\rho_{12} & \frac{b_3}{b_1}\rho_{13} & \cdots & \frac{b_n}{b_1}\rho_{1n} \\ \frac{b_1}{b_2}\rho_{21} & 1 & \frac{b_3}{b_2}\rho_{23} & \cdots & \frac{b_n}{b_2}\rho_{2n} \\ \vdots & \vdots & \vdots & \ddots & \vdots \\ \frac{b_1}{b_n}\rho_{n1} & \frac{b_2}{b_n}\rho_{n2} & \frac{b_3}{b_n}\rho_{n3} & \cdots & 1 \end{pmatrix} \begin{pmatrix} K_t^1 \\ K_t^2 \\ \vdots \\ K_t^n \end{pmatrix} = \begin{pmatrix} \frac{a_1-r}{b_1^2 R} \\ \frac{a_2-r}{b_2^2 R} \\ \vdots \\ \frac{a_n-r}{b_n^2 R} \end{pmatrix}.$$

By Cramer's formulas, we get

$$K_t^j = \frac{\det(\mathbf{C}'_j)}{\det(\mathbf{C}')}, \quad 1 \le j \le n,$$

where

$$\mathbf{C}' = \begin{pmatrix} 1 & \frac{b_2}{b_1}\rho_{12} & \frac{b_3}{b_1}\rho_{13} & \cdots & \frac{b_n}{b_1}\rho_{1n} \\ \frac{b_1}{b_2}\rho_{21} & 1 & \frac{b_3}{b_2}\rho_{23} & \cdots & \frac{b_n}{b_2}\rho_{2n} \\ \vdots & \vdots & \vdots & \ddots & \vdots \\ \frac{b_1}{b_n}\rho_{n1} & \frac{b_2}{b_n}\rho_{n2} & \frac{b_3}{b_n}\rho_{n3} & \cdots & 1 \end{pmatrix}$$

and

$$\mathbf{C}'_j = \begin{pmatrix} 1 & \frac{b_2}{b_1}\rho_{12} & \cdots & \frac{b_{j-1}}{b_1}\rho_{1(j-1)} & \frac{a_1-r}{b_1^2 R} & \frac{b_{j+1}}{b_1}\rho_{1(j+1)} & \cdots & \frac{b_n}{b_1}\rho_{1n} \\ \frac{b_1}{b_2}\rho_{21} & 1 & \cdots & \frac{b_{j-1}}{b_2}\rho_{2(j-1)} & \frac{a_2-r}{b_2^2 R} & \frac{b_{j+1}}{b_2}\rho_{2(j+1)} & \cdots & \frac{b_n}{b_2}\rho_{2n} \\ \vdots & \vdots & \ddots & \vdots & \vdots & \vdots & \ddots & \vdots \\ \frac{b_1}{b_n}\rho_{n1} & \frac{b_2}{b_n}\rho_{n2} & \cdots & \frac{b_{j-1}}{b_n}\rho_{n(j-1)} & \frac{a_n-r}{b_n^2 R} & \frac{b_{j+1}}{b_n}\rho_{n(j+1)} & \cdots & 1 \end{pmatrix},$$

provided that $\det(\mathbf{C}') \ne 0$.

We see at once that

$$\det(\mathbf{C}') = \frac{1}{\prod_{j=1}^{n} b_j} \det \begin{pmatrix} b_1 & b_2\rho_{12} & b_3\rho_{13} & \dots & b_n\rho_{1n} \\ b_1\rho_{21} & b_2 & b_3\rho_{23} & \dots & b_n\rho_{2n} \\ \vdots & \vdots & \vdots & \ddots & \vdots \\ b_1\rho_{n1} & b_2\rho_{n2} & b_3\rho_{n3} & \dots & b_n \end{pmatrix} = \det(\mathbf{C}),$$

which is positive since the correlation matrix $\mathbf{C}$ is positive definite by our assumption.

Moreover, it is evident that

$$\det(\mathbf{C}'_j) = \frac{\det(\mathbf{C}_j)}{b_j R}, \quad 1 \leq j \leq n,$$

which are non-negative.

Thus, $\left(\frac{\det(\mathbf{C}_1)}{b_1 R \det(\mathbf{C})}, \frac{\det(\mathbf{C}_2)}{b_2 R \det(\mathbf{C})}, \dots, \frac{\det(\mathbf{C}_n)}{b_n R \det(\mathbf{C})}\right)$ is the unique critical point of $g(K_t^1, K_t^2, \dots, K_t^n)$. We now show that it is a local minimum of $g(K_t^1, K_t^2, \dots, K_t^n)$.

Since

$$\frac{\partial g(K_t^1, K_t^2, \dots, K_t^n)}{\partial K_t^j \, \partial K_t^k} = \begin{cases} b_j^2 R^2 & \text{if} \quad j = k, \\ b_j b_k \rho_{jk} R^2 & \text{if} \quad j \neq k, \end{cases}$$

for all $1 \leq j \leq n$ and $1 \leq k \leq n$, we get the Hessian matrix of $g(K_t^1, K_t^2, \dots, K_t^n)$:

$$\begin{pmatrix} b_1^2 R^2 & b_1 b_2 \rho_{12} R^2 & b_1 b_3 \rho_{13} R^2 & \dots & b_1 b_n \rho_{1n} R^2 \\ b_2 b_1 \rho_{21} R^2 & b_2^2 R^2 & b_2 b_3 \rho_{23} R^2 & \dots & b_2 b_n \rho_{2n} R^2 \\ \vdots & \vdots & \vdots & \ddots & \vdots \\ b_n b_1 \rho_{n1} R^2 & b_n b_2 \rho_{n2} R^2 & b_n b_3 \rho_{n3} R^2 & \dots & b_n^2 R^2 \end{pmatrix}.$$

Since the correlation matrix $\mathbf{C}$ is positive definite, we conclude that so is the Hessian matrix. Therefore, $\left(\frac{\det(\mathbf{C}_1)}{b_1 R \det(\mathbf{C})}, \frac{\det(\mathbf{C}_2)}{b_2 R \det(\mathbf{C})}, \dots, \frac{\det(\mathbf{C}_n)}{b_n R \det(\mathbf{C})}\right)$ is indeed a local minimum of $g(K_t^1, K_t^2, \dots, K_t^n)$, which is the desired conclusion. □

REMARK 9.1.– If $\rho_{jk} = \rho_j \rho_k$ for all $1 \le j \le n$, $1 \le k \le n$ and $j \ne k$, i.e.

$$\mathbf{C} = \begin{pmatrix} 1 & \rho_1\rho_2 & \rho_1\rho_3 & \cdots & \rho_1\rho_n \\ \rho_2\rho_1 & 1 & \rho_2\rho_3 & \cdots & \rho_2\rho_n \\ \vdots & \vdots & \vdots & \ddots & \vdots \\ \rho_n\rho_1 & \rho_n\rho_2 & \rho_n\rho_3 & \cdots & 1 \end{pmatrix},$$

then $\det(\mathbf{C})$ can be easily computed by formula [9.12] below.

PROPOSITION 9.2.– If $\rho_{jk} = \rho_j \rho_k$ for all $1 \le j \le n$, $1 \le k \le n$ and $j \ne k$, then

$$\det(\mathbf{C}) = \prod_{j=1}^{n}(1-\rho_j^2)\left(1 + \sum_{k=1}^{n} \frac{\rho_k^2}{1-\rho_k^2}\right), \quad n \ge 2. \qquad [9.12]$$

PROOF.– The proof is by induction on n. An easy computation shows that the assertion holds for $n = 2$. We now assume that the assertion is true for $n - 1$ and show that it is true for n.

Let

$$D_n = \det \begin{pmatrix} 1 & \rho_1\rho_2 & \rho_1\rho_3 & \cdots & \rho_1\rho_n \\ \rho_2\rho_1 & 1 & \rho_2\rho_3 & \cdots & \rho_2\rho_n \\ \vdots & \vdots & \vdots & \ddots & \vdots \\ \rho_n\rho_1 & \rho_n\rho_2 & \rho_n\rho_3 & \cdots & 1 \end{pmatrix}, \quad n \ge 2,$$

and

$$D_n' = \det \begin{pmatrix} \frac{1}{\rho_1^2} & 1 & 1 & \cdots & 1 & 1 \\ 1 & \frac{1}{\rho_2^2} & 1 & \cdots & 1 & 1 \\ \vdots & \vdots & \vdots & \ddots & \vdots & \vdots \\ 1 & 1 & 1 & \cdots & \frac{1}{\rho_{n-1}^2} & 1 \\ 1 & 1 & 1 & \cdots & 1 & 1 \end{pmatrix}, \quad n \ge 2.$$

It is easily seen that

$$D'_n = \det \begin{pmatrix} \frac{1}{\rho_1^2} & 1 & 1 \dots & 1 & 1 \\ 1 & \frac{1}{\rho_2^2} & 1 \dots & 1 & 1 \\ \vdots & \vdots & \vdots \ddots & \vdots & \vdots \\ 1 & 1 & 1 \dots & \frac{1}{\rho_{n-1}^2} & 1 \\ 0 & 0 & 0 \dots 1 - & \frac{1}{\rho_{n-1}^2} & 0 \end{pmatrix}$$

$$= \left(\frac{1}{\rho_{n-1}^2} - 1\right) D'_{n-1}, \quad n \geq 3,$$

and

$$D'_2 = \frac{1}{\rho_1^2} - 1.$$

Hence, we obtain

$$D'_n = \prod_{j=1}^{n-1} \left(\frac{1}{\rho_j^2} - 1\right), \quad n \geq 2. \qquad [9.13]$$

Moreover, we have

$$D_n = \prod_{j=1}^{n} \rho_j^2 \det \begin{pmatrix} \frac{1}{\rho_1^2} & 1 & 1 \dots & 1 & 1 \\ 1 & \frac{1}{\rho_2^2} & 1 \dots & 1 & 1 \\ \vdots & \vdots & \vdots \ddots & \vdots & \vdots \\ 1 & 1 & 1 \dots & \frac{1}{\rho_{n-1}^2} & 1 \\ 1 & 1 & 1 \dots & 1 & \frac{1}{\rho_n^2} \end{pmatrix}$$

$$= \prod_{j=1}^{n} \rho_j^2 \det \begin{pmatrix} \frac{1}{\rho_1^2} & 1 & 1 \dots & 1 & 1 \\ 1 & \frac{1}{\rho_2^2} & 1 \dots & 1 & 1 \\ \vdots & \vdots & \vdots \ddots & \vdots & \vdots \\ 1 & 1 & 1 \dots & \frac{1}{\rho_{n-1}^2} & 1 \\ 0 & 0 & 0 \dots 1 - & \frac{1}{\rho_{n-1}^2} & \frac{1}{\rho_n^2} - 1 \end{pmatrix}$$

$$= \prod_{j=1}^{n} \rho_j^2 \left(\frac{1}{\rho_{n-1}^2} - 1\right) D'_{n-1} + \rho_n^2 \left(\frac{1}{\rho_n^2} - 1\right) D_{n-1}.$$

By the induction hypothesis and [9.13], we get

$$\det(\mathbf{C}) = D_n = \prod_{j=1}^{n} \rho_j^2 \prod_{k=1}^{n-1}\left(\frac{1}{\rho_k^2} - 1\right)$$

$$+ \rho_n^2\left(\frac{1}{\rho_n^2} - 1\right)\prod_{j=1}^{n-1}(1-\rho_j^2)\left(1 + \sum_{k=1}^{n-1}\frac{\rho_k^2}{1-\rho_k^2}\right)$$

$$= \rho_n^2 \prod_{j=1}^{n-1}(1-\rho_j^2) + \prod_{j=1}^{n}(1-\rho_j^2)\left(1 + \sum_{k=1}^{n-1}\frac{\rho_k^2}{1-\rho_k^2}\right)$$

$$= \prod_{j=1}^{n}(1-\rho_j^2)\left(1 + \sum_{k=1}^{n}\frac{\rho_k^2}{1-\rho_k^2}\right),$$

which completes the proof. □

REMARK 9.2.– In proposition 9.1 we impose the restriction $\det(\mathbf{C}_j) \geq 0$, $1 \leq j \leq n$, only because $g(K_t^1, K_t^2, \ldots, K_t^n)$ is defined on the set of all admissible strategies $(K_t^1, K_t^2, \ldots, K_t^n)$ and $K_t^j \geq 0$, $1 \leq j \leq n$, by the assumption of the model. Nevertheless, if we leave out the assumption $K_t^j \geq 0$, we do not need the condition $\det(\mathbf{C}_j) \geq 0$ in proposition 9.1.

We will now use the result of proposition 9.1 to get an upper exponential bound for the ruin probability under a constant strategy.

THEOREM 9.3.– Let $\det(\mathbf{C}_j) \geq 0$ for all $1 \leq j \leq n$ and $\sum_{j=1}^{n} \frac{(a_j - r)\det(\mathbf{C}_j)}{b_j} > 0$. Moreover, let $c(u)$ be a Lipschitz continuous function on $\mathbb{R}$ and the surplus process $\left(X_t^{(K_t^1, K_t^2, \ldots, K_t^n)}(x)\right)_{t\geq 0}$ follows [9.4] under the above assumptions with $K_t^j = \hat{K}^j$ for all $t \geq 0$ and $1 \leq j \leq n$, where $\hat{K}^j = \frac{\det(\mathbf{C}_j)}{b_j \hat{R} \det(\mathbf{C})}$ and $\hat{R} \in (0, R_\infty)$ is the unique solution to the equation

$$\lambda h(R) = \frac{1}{2\det(\mathbf{C})}\sum_{j=1}^{n}\frac{(a_j - r)\det(\mathbf{C}_j)}{b_j} + R \inf_{u\in\mathbb{R}_+}(c(u) + ru). \quad [9.14]$$

Then, for all $x \geq 0$, we have

$$\psi^{(\hat{K}^1, \hat{K}^2, \ldots, \hat{K}^n)}(x) \leq \mathrm{e}^{-\hat{R}x}. \quad [9.15]$$

REMARK 9.3.– Since $\sum_{j=1}^{n} \frac{(a_j - r)\det(\mathbf{C}_j)}{b_j} > 0$, it is easily seen that equation [9.14] has a unique solution $\hat{R} \in (0, R_\infty)$. Indeed, $h(R)$ is increasing, convex and continuous on $[0, R_\infty)$ with $h(0) = 0$ and $\lim_{R \uparrow R_\infty} h(R) = +\infty$. The linear function of R on the right-hand side of [9.14] is increasing with positive value at $R = 0$. Therefore, there is a unique solution $\hat{R} \in (0, R_\infty)$ to [9.14].

To prove theorem 9.3, we will need the following lemma.

LEMMA 9.1.– Let the conditions of theorem 9.3 hold. Then, the process $\left(V_t^{(\hat{K}^1, \hat{K}^2, \ldots, \hat{K}^n)}(x, \hat{R})\right)_{t \geq 0}$ is an $(\mathfrak{F}_t^n)$-supermartingale.

PROOF (Proof of lemma 9.1).– Since $c(u)$ is a Lipschitz continuous function on $\mathbb{R}$, equation [9.4] has a unique strong solution up to the time τ for any admissible strategy $(K_t^1, K_t^2, \ldots, K_t^n)$ by theorem 9.1. Therefore, by theorem 9.2, in order to prove that $\left(V_t^{(\hat{K}^1, \hat{K}^2, \ldots, \hat{K}^n)}(x, \hat{R})\right)_{t \geq 0}$ is an $(\mathfrak{F}_t^n)$-supermartingale, it is sufficient to show that [9.5] holds with $\hat{R}$ and the constant strategy $(\hat{K}^1, \hat{K}^2, \ldots, \hat{K}^n)$ given in the assertion of theorem 9.3. This means that it is sufficient to show that $g(\hat{K}^1, \hat{K}^2, \ldots, \hat{K}^n) \leq 0$ for $R = \hat{R}$. In fact, we now prove that $g(\hat{K}^1, \hat{K}^2, \ldots, \hat{K}^n) = 0$ for $R = \hat{R}$.

By the definition of $g(K_t^1, K_t^2, \ldots, K_t^n)$ and the strategy $(\hat{K}^1, \hat{K}^2, \ldots, \hat{K}^n)$, we have

$$
\begin{aligned}
& g(\hat{K}^1, \hat{K}^2, \ldots, \hat{K}^n) \\
&\quad = g\left(\frac{\det(\mathbf{C}_1)}{b_1 \hat{R} \det(\mathbf{C})}, \frac{\det(\mathbf{C}_2)}{b_2 \hat{R} \det(\mathbf{C})}, \ldots, \frac{\det(\mathbf{C}_n)}{b_n \hat{R} \det(\mathbf{C})}\right) \\
&\quad = \frac{1}{(\det(\mathbf{C}))^2}\left(\frac{1}{2}\sum_{j=1}^{n}(\det(\mathbf{C}_j))^2 + \sum_{j=1}^{n-1}\sum_{k=j+1}^{n} \rho_{jk}\det(\mathbf{C}_j)\det(\mathbf{C}_k)\right) \\
&\quad = \frac{1}{\det(\mathbf{C})}\sum_{j=1}^{n}\frac{(a_j - r)\det(\mathbf{C}_j)}{b_j} - \hat{R}\inf_{u \in \mathbb{R}_+}(c(u) + ru) + \lambda h(\hat{R}).
\end{aligned}
\tag{9.16}
$$

Moreover, note that [9.11] is true with $K_t^j = \hat{K}^j$, $1 \le j \le n$. This yields

$$\det(\mathbf{C}_j) + \sum_{\substack{1\le k\le n \\ k\neq j}} \rho_{jk} \det(\mathbf{C}_k) = \frac{(a_j - r)\det(\mathbf{C})}{b_j}, \quad 1 \le j \le n. \qquad [9.17]$$

Multiplying [9.17] by $\det(\mathbf{C}_j)$ and summation over all $1 \le j \le n$ give

$$\sum_{j=1}^{n} (\det(\mathbf{C}_j))^2 + \sum_{j=1}^{n} \det(\mathbf{C}_j) \sum_{\substack{1\le k\le n \\ k\neq j}} \rho_{jk} \det(\mathbf{C}_k)$$
$$= \det(\mathbf{C}) \sum_{j=1}^{n} \frac{(a_j - r)\det(\mathbf{C}_j)}{b_j},$$

which is equivalent to

$$\sum_{j=1}^{n} (\det(\mathbf{C}_j))^2 + 2\sum_{j=1}^{n-1} \sum_{k=j+1}^{n} \rho_{jk} \det(\mathbf{C}_j)\det(\mathbf{C}_k)$$
$$= \det(\mathbf{C}) \sum_{j=1}^{n} \frac{(a_j - r)\det(\mathbf{C}_j)}{b_j}. \qquad [9.18]$$

By [9.18], we can rewrite [9.16] as

$$g(\hat{K}^1, \hat{K}^2, \ldots, \hat{K}^n) = -\frac{1}{2\det(\mathbf{C})} \sum_{j=1}^{n} \frac{(a_j - r)\det(\mathbf{C}_j)}{b_j}$$
$$- \hat{R} \inf_{u\in\mathbb{R}_+} (c(u) + ru) + \lambda h(\hat{R}). \qquad [9.19]$$

By [9.14], equality [9.19] implies that $g(\hat{K}^1, \hat{K}^2, \ldots, \hat{K}^n) = 0$, which is the desired conclusion. □

PROOF (Proof of theorem 9.3).– Since $\big(V_t^{(\hat{K}^1,\hat{K}^2,\ldots,\hat{K}^n)}(x,\hat{R})\big)_{t\ge 0}$ is an $(\mathfrak{F}_t^n)$-supermartingale by lemma 9.1, for all $t \ge 0$, we have

$$\mathrm{e}^{-\hat{R}x} = V_0^{(\hat{K}^1,\hat{K}^2,\ldots,\hat{K}^n)}(x,\hat{R}) \ge \mathbb{E}\big[V_t^{(\hat{K}^1,\hat{K}^2,\ldots,\hat{K}^n)}(x,\hat{R}) \,/\, \mathfrak{F}_0\big]$$
$$= \mathbb{E}\Big[\exp\Big(-\hat{R}X_{t\wedge\tau}^{(\hat{K}^1,\hat{K}^2,\ldots,\hat{K}^n)}(x)\Big)\Big]$$

$$= \mathbb{E}\Big[\exp\Big(-\hat{R}X_\tau^{(\hat{K}^1,\hat{K}^2,\ldots,\hat{K}^n)}(x)\Big)\ \mathbb{1}_{\{\tau<t\}}\Big]$$

$$+\,\mathbb{E}\Big[\exp\Big(-\hat{R}X_t^{(\hat{K}^1,\hat{K}^2,\ldots,\hat{K}^n)}(x)\Big)\ \mathbb{1}_{\{\tau\geq t\}}\Big] \qquad [9.20]$$

$$\geq \mathbb{E}\Big[\exp\Big(-\hat{R}X_\tau^{(\hat{K}^1,\hat{K}^2,\ldots,\hat{K}^n)}(x)\Big)\ \mathbb{1}_{\{\tau<t\}}\Big].$$

Letting $t \to \infty$ in [9.20] yields

$$\mathbb{E}\Big[\exp\Big(-\hat{R}X_\tau^{(\hat{K}^1,\hat{K}^2,\ldots,\hat{K}^n)}(x)\Big)\ \mathbb{1}_{\{\tau<\infty\}}\Big] \leq \mathrm{e}^{-\hat{R}x}. \qquad [9.21]$$

Furthermore,

$$\mathbb{E}\Big[\exp\Big(-\hat{R}X_\tau^{(\hat{K}^1,\hat{K}^2,\ldots,\hat{K}^n)}(x)\Big)\ \mathbb{1}_{\{\tau<\infty\}}\Big]$$

$$= \mathbb{E}\Big[\exp\Big(-\hat{R}X_\tau^{(\hat{K}^1,\hat{K}^2,\ldots,\hat{K}^n)}(x)\Big)\ /\,\tau<\infty\Big]\ \mathbb{P}\Big[\tau^{(\hat{K}^1,\hat{K}^2,\ldots,\hat{K}^n)}(x)<\infty\Big],$$

and

$$\mathbb{E}\Big[\exp\Big(-\hat{R}X_\tau^{(\hat{K}^1,\hat{K}^2,\ldots,\hat{K}^n)}(x)\Big)\ /\,\tau<\infty\Big] \geq 1$$

by the definition of the ruin time. Thus, from [9.21] we conclude that

$$\mathbb{P}\Big[\tau^{(\hat{K}^1,\hat{K}^2,\ldots,\hat{K}^n)}(x)<\infty\Big]$$

$$\leq \frac{\mathrm{e}^{-\hat{R}x}}{\mathbb{E}\Big[\exp\Big(-\hat{R}X_\tau^{(\hat{K}^1,\hat{K}^2,\ldots,\hat{K}^n)}(x)\Big)\ /\,\tau<\infty\Big]} \leq \mathrm{e}^{-\hat{R}x},$$

which gives [9.15]. □

REMARK 9.4.– In theorem 9.3, we took $\hat{R} \in (0, R_\infty)$, such that $g(\hat{K}^1, \hat{K}^2, \ldots, \hat{K}^n) = 0$ for $R = \hat{R}$ to get the better bound in [9.15].

REMARK 9.5.– If $a_j > r$, $1 \leq j \leq n$, then $\sum_{j=1}^n \frac{(a_j-r)\det(\mathbf{C}_j)}{b_j} > 0$ in conditions of theorem 9.3.

REMARK 9.6.– If $c(u)$ is constant, then by [9.14] and the fact that $\sum_{j=1}^n \frac{(a_j-r)\det(\mathbf{C}_j)}{b_j} > 0$, it is easily seen that the bound in theorem 9.3 is better than the exponential bound in the corresponding classical risk model without any investments. However, this does not always imply that the ruin probability in the model considered in this chapter is less than the ruin probability in the corresponding classical risk model without investments.

9.5. Case of one risky asset

We now deal with the case $n = 1$. This means that the *surplus process* $(X_t)_{t\geq 0}$ follows the equation

$$X_t = x + \int_0^t \Big(c(X_s) + rX_s + (a_1 - r)K_s^1 \Big)\,\mathrm{d}s + b_1 \int_0^t K_s^1\,\mathrm{d}W_s^1 - \sum_{i=1}^{N_t} Y_i, \quad t \geq 0. \qquad [9.22]$$

It is easy to check that theorems 9.1 and 9.2 are true for $n = 1$. In this case, [9.5] takes the form

$$\mathbb{P}\left[\frac{\hat{R}^2 b_1^2 (\hat{K}_t^1)^2}{2} - \hat{R}\Big(\inf_{u\in\mathbb{R}_+} (c(u) + ru) + (a_1 - r)\hat{K}_t^1 \Big) + \lambda h(\hat{R}) \leq 0 \right] = 1 \quad \text{for all} \quad t \geq 0. \qquad [9.23]$$

The following theorem gives an upper exponential bound for the ruin probability.

THEOREM 9.4.– Let $a_1 > r$, $c(u)$ be a Lipschitz continuous function on $\mathbb{R}$ and the surplus process $\big(X_t^{(K_t^1)}(x)\big)_{t\geq 0}$ follow [9.22] under the above assumptions with $K_t^1 = \hat{K}^1$ for all $t \geq 0$, where $\hat{K}^1 = \frac{a_1 - r}{\hat{R} b_1^2}$ and $\hat{R} \in (0, R_\infty)$ is the unique solution to the equation

$$\lambda h(R) = \frac{(a_1 - r)^2}{2b_1^2} + R \inf_{u\in\mathbb{R}_+} (c(u) + ru). \qquad [9.24]$$

Then, for all $x \geq 0$, we have

$$\psi^{(\hat{K}^1)}(x) \leq \mathrm{e}^{-\hat{R}x}.$$

PROOF.– The proof of the theorem is similar to the proof of theorem 9.3. We only show that the process $\big(V_t^{(\hat{K}^1)}(x, \hat{R})\big)_{t\geq 0}$ is an $(\mathfrak{F}_t^1)$-supermartingale, i.e. [9.23] holds with $\hat{R}$ and the constant strategy $(\hat{K}^1)$ given in the assertion of the theorem.

The inequality

$$\frac{\hat{R}^2 b_1^2 (\hat{K}_t^1)^2}{2} - \hat{R}\left(\inf_{u \in \mathbb{R}_+} (c(u) + ru) + (a_1 - r)\hat{K}_t^1\right) + \lambda h(\hat{R}) \leq 0 \qquad [9.25]$$

holds for some $\hat{K}_t^1$ provided that the discriminant of the corresponding quadratic w.r.t. $\hat{K}_t^1$ equation is non-negative, i.e.

$$2\hat{R}^2 b_1^2 \left(\frac{(a_1 - r)^2}{2b_1^2} + \hat{R} \inf_{u \in \mathbb{R}_+} (c(u) + ru) - \lambda h(\hat{R}) \right) \geq 0. \qquad [9.26]$$

By [9.24], equality holds in [9.26]. Hence, there is only one point $\hat{K}_t^1 = \frac{a_1 - r}{\hat{R} b_1^2}$ for all $t \geq 0$ where [9.25] is true. In fact, equality holds in [9.25] at this point. Thus, [9.23] is true, which completes the proof. □

REMARK 9.7.– Since $a_1 > r$, equation [9.24] has a unique solution $\hat{R} \in (0, R_\infty)$. The explanation of this fact is the same as in remark 9.3. Moreover, we took $\hat{R} \in (0, R_\infty)$, such that equality holds in [9.26] to get the better exponential bound.

REMARK 9.8.– If $c(u)$ is constant, our exponential bound coincides with that obtained in [GAI 03].

9.6. Examples

We now consider a few examples. In all examples below we deal with exponentially distributed claim sizes and constant $c(u)$, i.e. $h(R) = \frac{\mu R}{1 - \mu R}$ for all $R \in [0, 1/\mu)$ and $c(u) = c > 0$ for all $u \in \mathbb{R}$. In this case,

$$\hat{R} = \frac{c - A\mu - \lambda\mu + \sqrt{(A\mu + \lambda\mu - c)^2 + 4Ac\mu}}{2c\mu} \qquad [9.27]$$

is the unique solution to [9.14] and [9.24], such that $\hat{R} \in (0, 1/\mu)$. Here

$$A = \begin{cases} \dfrac{(a_1 - r)^2}{2b_1^2} & \text{if} \quad n = 1, \\[2ex] \dfrac{1}{2\det(\mathbf{C})} \displaystyle\sum_{j=1}^{n} \dfrac{(a_j - r)\det(\mathbf{C}_j)}{b_j} & \text{if} \quad n \geq 2. \end{cases}$$

Note that $A > 0$ by the conditions of theorems 9.3 and 9.4.

Since

$$\sqrt{(A\mu + \lambda\mu - c)^2 + 4Ac\mu} > A\mu + \lambda\mu - c,$$

it is obvious that $\hat{R} > 0$.

To show that $\hat{R} < 1/\mu$, it is enough to establish that

$$c - A\mu - \lambda\mu + \sqrt{(A\mu + \lambda\mu - c)^2 + 4Ac\mu} < 2c. \qquad [9.28]$$

We can rewrite [9.28] as

$$\sqrt{(A\mu + \lambda\mu - c)^2 + 4Ac\mu} < A\mu + \lambda\mu + c,$$

which is equivalent to

$$4Ac\mu < (A\mu + \lambda\mu + c)^2 - (A\mu + \lambda\mu - c)^2. \qquad [9.29]$$

Simplifying [9.29] gives $\lambda > 0$, which is true. Hence, [9.28] is also true and $\hat{R} < 1/\mu$.

In addition, recall that in the classical risk model without any investments,

$$\psi(x) = \frac{\lambda\mu}{c} \exp\left(-\frac{(c - \lambda\mu)x}{c\mu}\right) \qquad [9.30]$$

for all $x \geq 0$ provided that $c > \lambda\mu$ (see section 1.1.6 and [ASM 10, BEA 84, BÜH 70, DIC 10, GER 79, GRA 91, ROL 99]).

EXAMPLE 9.1.– Let $n = 1$, $c = 6$, $\lambda = 1$, $\mu = 5$, $r = 0.001$, $a_1 = 0.0015$ and $b_1 = 0.0005$. Then, by theorem 9.4 and [9.27], we have

$$\psi^{(\hat{K}^1)}(x) \leq \mathrm{e}^{-0.106498\,x}$$

for all $x \geq 0$ with $\hat{K}^1 = 18\,779.73$. Note that by [9.30], in the classical risk model without any investments,

$$\psi(x) \approx 0.833333\,\mathrm{e}^{-0.033333\,x}$$

for all $x \geq 0$. Thus, $\psi^{(\hat{K}^1)}(x) < \psi(x)$ for large enough x.

EXAMPLE 9.2.– Let $n = 2$. Hence,

$$\mathbf{C} = \begin{pmatrix} 1 & \rho_{12} \\ \rho_{12} & 1 \end{pmatrix}$$

and we have

$$\det(\mathbf{C}) = 1 - \rho_{12}^2,$$

$$\det(\mathbf{C}_1) = \frac{b_2(a_1 - r) - b_1\rho_{12}(a_2 - r)}{b_1 b_2},$$

$$\det(\mathbf{C}_2) = \frac{b_1(a_2 - r) - b_2\rho_{12}(a_1 - r)}{b_1 b_2},$$

$$\sum_{j=1}^{2} \frac{(a_j - r)\det(\mathbf{C}_j)}{b_j}$$

$$= \frac{b_2^2(a_1 - r)^2 - 2b_1 b_2 \rho_{12}(a_1 - r)(a_2 - r) + b_1^2(a_2 - r)^2}{b_1^2 b_2^2}.$$

The matrix $\mathbf{C}$ is positive definite if and only if $\rho_{12} \in (-1, 1)$. By theorem 9.3, if

$$b_2(a_1 - r) - b_1\rho_{12}(a_2 - r) \geq 0,$$

$$b_1(a_2 - r) - b_2\rho_{12}(a_1 - r) \geq 0$$

and

$$b_2^2(a_1 - r)^2 - 2b_1 b_2 \rho_{12}(a_1 - r)(a_2 - r) + b_1^2(a_2 - r)^2 > 0,$$

then [9.15] holds with

$$\hat{K}^1 = \frac{b_2(a_1 - r) - b_1\rho_{12}(a_2 - r)}{b_1^2 b_2 \hat{R}(1 - \rho_{12}^2)},$$

$$\hat{K}^2 = \frac{b_1(a_2 - r) - b_2\rho_{12}(a_1 - r)}{b_1 b_2^2 \hat{R}(1 - \rho_{12}^2)}$$

and $\hat{R}$ given by [9.27].

Set $c = 6$, $\lambda = 1$, $\mu = 5$, $r = 0.001$, $a_1 = 0.0015$, $b_1 = 0.0005$, $a_2 = 0.002$, $b_2 = 0.001$ and

$$\mathbf{C} = \begin{pmatrix} 1 & 0.4 \\ 0.4 & 1 \end{pmatrix}.$$

Then, we get $\det(\mathbf{C}) = 0.84$, $\det(\mathbf{C}_1) = 0.6$ and $\det(\mathbf{C}_2) = 0.6$. By theorem 9.3 and [9.27], we have

$$\psi^{(\hat{K}^1,\hat{K}^2)}(x) \leq \mathrm{e}^{-0.117287\, x}$$

for all $x \geq 0$ with $\hat{K}^1 = 12\,180.10$ and $\hat{K}^2 = 6\,090.05$.

EXAMPLE 9.3.– Let $n = 3$. Hence,

$$\mathbf{C} = \begin{pmatrix} 1 & \rho_{12} & \rho_{13} \\ \rho_{12} & 1 & \rho_{23} \\ \rho_{13} & \rho_{23} & 1 \end{pmatrix}$$

and we have

$$\det(\mathbf{C}) = 1 - \rho_{12}^2 - \rho_{13}^2 - \rho_{23}^2 + 2\rho_{12}\rho_{13}\rho_{23},$$

$$\begin{aligned}
\det(\mathbf{C}_1) = {} & \frac{1}{b_1 b_2 b_3}\big(b_2 b_3(1-\rho_{23}^2)(a_1 - r) \\
& + b_1 b_3(\rho_{13}\rho_{23} - \rho_{12})(a_2 - r) + b_1 b_2(\rho_{12}\rho_{23} - \rho_{13})(a_3 - r)\big),
\end{aligned}$$

$$\begin{aligned}
\det(\mathbf{C}_2) = {} & \frac{1}{b_1 b_2 b_3}\big(b_2 b_3(\rho_{13}\rho_{23} - \rho_{12})(a_1 - r) \\
& + b_1 b_3(1-\rho_{13}^2)(a_2 - r) + b_1 b_2(\rho_{12}\rho_{13} - \rho_{23})(a_3 - r)\big),
\end{aligned}$$

$$\begin{aligned}
\det(\mathbf{C}_3) = {} & \frac{1}{b_1 b_2 b_3}\big(b_2 b_3(\rho_{12}\rho_{23} - \rho_{13})(a_1 - r) \\
& + b_1 b_3(\rho_{12}\rho_{13} - \rho_{23})(a_2 - r) + b_1 b_2(1-\rho_{12}^2)(a_3 - r)\big),
\end{aligned}$$

$$\begin{aligned}
\sum_{j=1}^{3} \frac{(a_j - r)\det(\mathbf{C}_j)}{b_j} = {} & \frac{1}{b_1^2 b_2^2 b_3^2}\big(b_2^2 b_3^2(1-\rho_{23}^2)(a_1 - r)^2 \\
& + b_1^2 b_3^2(1-\rho_{13}^2)(a_2 - r)^2 + b_1^2 b_2^2(1-\rho_{12}^2)(a_3 - r)^2 \\
& + 2 b_1 b_2 b_3^2(\rho_{13}\rho_{23} - \rho_{12})(a_1 - r)(a_2 - r) \\
& + 2 b_1 b_2^2 b_3(\rho_{12}\rho_{23} - \rho_{13})(a_1 - r)(a_3 - r) \\
& + 2 b_1^2 b_2 b_3(\rho_{12}\rho_{13} - \rho_{23})(a_2 - r)(a_3 - r)\big).
\end{aligned}$$

The matrix $\mathbf{C}$ is positive definite if and only if $\rho_{12} \in (-1, 1)$ and

$$1 - \rho_{12}^2 - \rho_{13}^2 - \rho_{23}^2 + 2\rho_{12}\rho_{13}\rho_{23} > 0.$$

Moreover, $\rho_{13} \in (-1, 1)$ and $\rho_{23} \in (-1, 1)$ by our assumption.

By theorem 9.3, if $\det(\mathbf{C}_j) \geq 0$, $1 \leq j \leq 3$, and $\sum_{j=1}^{3} \frac{(a_j - r)\det(\mathbf{C}_j)}{b_j} > 0$, then [9.15] holds with

$$\begin{aligned}
\hat{K}^1 &= \frac{1}{b_1^2 b_2 b_3 \hat{R}(1 - \rho_{12}^2 - \rho_{13}^2 - \rho_{23}^2 + 2\rho_{12}\rho_{13}\rho_{23})} \\
&\quad \times \big(b_2 b_3(1 - \rho_{23}^2)(a_1 - r) + b_1 b_3(\rho_{13}\rho_{23} - \rho_{12})(a_2 - r) \\
&\quad + b_1 b_2(\rho_{12}\rho_{23} - \rho_{13})(a_3 - r)\big), \\
\hat{K}^2 &= \frac{1}{b_1 b_2^2 b_3 \hat{R}(1 - \rho_{12}^2 - \rho_{13}^2 - \rho_{23}^2 + 2\rho_{12}\rho_{13}\rho_{23})} \\
&\quad \times \big(b_2 b_3(\rho_{13}\rho_{23} - \rho_{12})(a_1 - r) + b_1 b_3(1 - \rho_{13}^2)(a_2 - r) \\
&\quad + b_1 b_2(\rho_{12}\rho_{13} - \rho_{23})(a_3 - r)\big), \\
\hat{K}^3 &= \frac{1}{b_1 b_2 b_3^2 \hat{R}(1 - \rho_{12}^2 - \rho_{13}^2 - \rho_{23}^2 + 2\rho_{12}\rho_{13}\rho_{23})} \\
&\quad \times \big(b_2 b_3(\rho_{12}\rho_{23} - \rho_{13})(a_1 - r) + b_1 b_3(\rho_{12}\rho_{13} - \rho_{23})(a_2 - r) \\
&\quad + b_1 b_2(1 - \rho_{12}^2)(a_3 - r)\big)
\end{aligned}$$

and $\hat{R}$ given by [9.27].

Set $c = 6$, $\lambda = 1$, $\mu = 5$, $r = 0.001$, $a_1 = 0.0015$, $b_1 = 0.0005$, $a_2 = 0.002$, $b_2 = 0.001$, $a_3 = 0.0025$, $b_3 = 0.0015$ and

$$\mathbf{C} = \begin{pmatrix} 1 & 0.4 & 0.2 \\ 0.4 & 1 & -0.1 \\ 0.2 & -0.1 & 1 \end{pmatrix}.$$

Then, we get $\det(\mathbf{C}) = 0.774$, $\det(\mathbf{C}_1) = 0.33$, $\det(\mathbf{C}_2) = 0.72$ and $\det(\mathbf{C}_3) = 0.78$. By theorem 9.3 and [9.27], we have

$$\psi^{(\hat{K}^1, \hat{K}^2, \hat{K}^3)}(x) \leq \mathrm{e}^{-0.132873\, x}$$

for all $x \geq 0$ with $\hat{K}^1 = 6\,417.53$, $\hat{K}^2 = 7\,000.94$ and $\hat{K}^3 = 5\,056.23$.

Thus, in the numerical examples considered above we get a better exponential bound if we have a greater number of risky assets available for investments.

Appendix

Mathematical Background

This chapter gives some well-known mathematical results which are used in Chapters 2, 3 and 6–9.

A.1. Differentiability of integrals

We now formulate some classical theorems of calculus concerning differentiability of integrals. All the integrals in theorems A.1–A.5 are assumed to be the Riemann integrals. For details, see, for example [COU 88, FIC 90].

Theorem A.1 establishes continuity and differentiability of an integral with variable upper limit of integration.

THEOREM A.1.–

1) Let the function $g(y)\colon [a,b] \to \mathbb{R}$ be integrable on $[a,b]$, where $-\infty < a < b < +\infty$. Then the function

$$G(x) = \int_a^x g(y)\,\mathrm{d}y$$

is continuous on $[a,b]$.

2) If, moreover, $g(y)$ is continuous at the point $y \in (a,b)$, then $G(x)$ has a derivative at the point $x = y$ and

$$G'(x) = g(x).$$

REMARK A.1.– If we define

$$G(x) = \int_x^b g(y)\,\mathrm{d}y$$

in theorem A.1, it remains true with

$$G'(x) = -g(x).$$

REMARK A.2.– If in theorem A.1 the function $g(y)$ is only left-continuous at the point $y \in (a, b]$ (or right-continuous at the point $y \in [a, b)$), then $G(x)$ has a left (or right) derivative at the point $x = y$ and

$$G'_-(x) = g(x) \quad (\text{or } G'_+(x) = g(x)),$$

where $G'_-(x)$ and $G'_+(x)$ are the left and right derivatives of $G(x)$, respectively.

Theorems A.2–A.5 below establish differentiability of integrals w.r.t. a parameter.

THEOREM A.2.– Let the function $g(x, y)\colon [a_1, b_1] \times [a_2, b_2] \to \mathbb{R}$ be continuous w.r.t. y on $[a_2, b_2]$ for every fixed $x \in [a_1, b_1]$ and the function $G(x)$ be defined by

$$G(x) = \int_{a_2}^{b_2} g(x, y)\,\mathrm{d}y$$

on $[a_1, b_1]$, where $-\infty < a_1 < b_1 < +\infty$ and $-\infty < a_2 < b_2 < +\infty$. Moreover, let $g(x, y)$ have a partial derivative w.r.t. x, which is continuous as a function of two variables on $[a_1, b_1] \times [a_2, b_2]$. Then, $G(x)$ is differentiable on $[a_1, b_1]$ and

$$G'(x) = \int_{a_2}^{b_2} \frac{\partial g(x, y)}{\partial x}\,\mathrm{d}y.$$

Theorem A.3 is a generalization of theorem A.2.

THEOREM A.3.– Let the function $g(x, y)\colon [a_1, b_1] \times [a_2, b_2] \to \mathbb{R}$ be integrable w.r.t. y on $[a_2, b_2]$ for every fixed $x \in [a_1, b_1]$ and the function $G(x)$ be defined by

$$G(x) = \int_{a_2}^{b_2} g(x, y)\,\mathrm{d}y$$

on $[a_1, b_1]$, where $-\infty < a_1 < b_1 < +\infty$ and $-\infty < a_2 < b_2 < +\infty$. Moreover, let $g(x, y)$ have a partial derivative w.r.t. x, which is also integrable w.r.t. y on $[a_2, b_2]$ for every fixed $x \in [a_1, b_1]$ and bounded for all $(x, y) \in [a_1, b_1] \times [a_2, b_2]$. Then, $G(x)$ is differentiable on $[a_1, b_1]$ and

$$G'(x) = \int_{a_2}^{b_2} \frac{\partial g(x,y)}{\partial x}\, \mathrm{d}y.$$

Theorem A.4 is an extension of theorem A.2 to the case of an integral over an infinite interval.

THEOREM A.4.– Let the function $g(x, y)\colon [a_1, b_1] \times [a_2, +\infty] \to \mathbb{R}$ be continuous w.r.t. y on $[a_2, +\infty]$ for every fixed $x \in [a_1, b_1]$ and have a partial derivative w.r.t. x, which is continuous as a function of two variables on $[a_1, b_1] \times [a_2, +\infty]$, where $-\infty < a_1 < b_1 < +\infty$ and $-\infty < a_2 < +\infty$. Moreover, let the integral

$$G(x) = \int_{a_2}^{+\infty} g(x, y)\, \mathrm{d}y$$

be convergent for all $x \in [a_1, b_1]$ and the integral

$$\int_{a_2}^{+\infty} \frac{\partial g(x,y)}{\partial x}\, \mathrm{d}y$$

be uniformly convergent for all $x \in [a_1, b_1]$. Then, $G(x)$ is differentiable on $[a_1, b_1]$ and

$$G'(x) = \int_{a_2}^{+\infty} \frac{\partial g(x,y)}{\partial x}\, \mathrm{d}y.$$

In theorem A.5, both limits of integration, as well as an integrand, are dependent on a parameter.

THEOREM A.5.– Let the function $g(x, y)\colon [a_1, b_1] \times [a_2, b_2] \to \mathbb{R}$ be continuous as a function of two variables, the functions $\gamma_1(x)\colon [a_1, b_1] \to [a_2, b_2]$ and $\gamma_2(x)\colon [a_1, b_1] \to [a_2, b_2]$ be continuous and the function $G(x)$ be defined by

$$G(x) = \int_{\gamma_1(x)}^{\gamma_2(x)} g(x, y)\, \mathrm{d}y$$

on $[a_1, b_1]$, where $-\infty < a_1 < b_1 < +\infty$ and $-\infty < a_2 < b_2 < +\infty$. Moreover, let $g(x, y)$ have a partial derivative w.r.t. x, which is continuous as a function of two variables on $[a_1, b_1] \times [a_2, b_2]$ and the functions $\gamma_1(x)$ and $\gamma_2(x)$ have derivatives on $[a_1, b_1]$. Then, $G(x)$ is differentiable on $[a_1, b_1]$ and

$$G'(x) = \int_{\gamma_1(x)}^{\gamma_2(x)} \frac{\partial g(x, y)}{\partial x} \, \mathrm{d}y + \gamma_2'(x)\, g\big(x, \gamma_2(x)\big) - \gamma_1'(x)\, g\big(x, \gamma_1(x)\big).$$

A.2. Hoeffding's inequality

We now give a simple version of *Hoeffding's inequality* which is used in section 6.4. For details and more general versions of this inequality, see [HOE 63].

THEOREM A.6.– Let Y_i, $1 \le i \le n$, be independent r.v.'s, such that $\mathbb{P}\big[0 \le Y_i \le 1\big] = 1$. Then, for any $\varepsilon > 0$, we have

$$\mathbb{P}\left[\frac{Y_1 + Y_2 + \ldots + Y_n}{n} - \frac{\mathbb{E}[Y_1 + Y_2 + \ldots + Y_n]}{n} \ge \varepsilon\right] \le \mathrm{e}^{-2n\varepsilon^2}.$$

A.3. Some results on one-dimensional homogeneous stochastic differential equations

We now give some classical results on one-dimensional homogeneous stochastic differential equations. For details and possible generalizations of these results, see, for example [CHE 05, IKE 89, KAR 91, ØKS 03].

Consider the following stochastic differential equation:

$$X_t = x + \int_0^t q(X_s)\, \mathrm{d}s + \int_0^t b(X_s)\, \mathrm{d}W_s, \quad t \ge 0, \qquad \text{[A.1]}$$

where $x \in \mathbb{R}$, $(W_t)_{t \ge 0}$ is a standard Brownian motion, $q \colon \mathbb{R} \to \mathbb{R}$ and $b \colon \mathbb{R} \to \mathbb{R}$.

In what follows, we imply the pathwise uniqueness of solutions.

A.3.1. *Existence and uniqueness theorems*

THEOREM A.7.– If the functions $q(u)$ and $b(u)$ are Lipschitz continuous on $\mathbb{R}$, then [A.1] has a unique strong solution on $\mathbb{R}_+$.

THEOREM A.8.– If the functions $q(u)$ and $b(u)$ are locally Lipschitz continuous on $\mathbb{R}$, then [A.1] has a unique strong solution up to the time $t^* = \inf\{t \geq 0\colon X_t \notin (-\infty, +\infty)\}$.

A.3.2. *Comparison theorem*

In addition to [A.1], consider the stochastic differential equation

$$\bar{X}_t = \bar{x} + \int_0^t \bar{q}(\bar{X}_s)\,\mathrm{d}s + \int_0^t b(\bar{X}_s)\,\mathrm{d}W_s, \quad t \geq 0, \qquad \text{[A.2]}$$

where $\bar{x} \in \mathbb{R}$, $\bar{q}\colon \mathbb{R} \to \mathbb{R}$ and $b\colon \mathbb{R} \to \mathbb{R}$ is the same as in [A.1].

THEOREM A.9.– Let $(X_t)_{t\geq 0}$ and $(\bar{X}_t)_{t\geq 0}$ be strong unique solutions to [A.1] and [A.2], respectively, $x \leq \bar{x}$ and $q(u) \leq \bar{q}(u)$ for all $u \in \mathbb{R}$. Then, $X_t \leq \bar{X}_t$ a.s. for all $t \geq 0$.

A.3.3. *Theorem on the first exit time from an interval*

THEOREM A.10.– Let $(X_t)_{t\geq 0}$ be a strong unique solution to [A.1], the functions $q(u)$ and $b(u)$ be continuously differentiable on (l_1, l_2) and $b^2(u) > 0$ on (l_1, l_2), where l_1 and l_2 are such that $-\infty \leq l_1 < x < l_2 \leq +\infty$. Moreover, let $I(l)$ be defined by

$$I(l) = \int_x^l \exp\left(-\int_x^v \frac{2q(u)}{b^2(u)}\,\mathrm{d}u\right)\mathrm{d}v$$

on (l_1, l_2), $I(l_1) = \lim_{l\downarrow l_1} I(l)$, $I(l_2) = \lim_{l\uparrow l_2} I(l)$ and $t^*_{(l_1,l_2)} = \inf\{t \geq 0\colon X_t \notin (l_1, l_2)\}$.

i) If $I(l_1) = -\infty$ and $I(l_2) = +\infty$, then

$$\mathbb{P}\big[t^*_{(l_1,l_2)} = \infty\big] = \mathbb{P}\big[\limsup_{t\uparrow\infty} X_t = l_2\big] = \mathbb{P}\big[\liminf_{t\uparrow\infty} X_t = l_1\big] = 1.$$

ii) If $I(l_1) > -\infty$ and $I(l_2) = +\infty$, then $\lim_{t \uparrow t^*_{(l_1,l_2)}} X_t$ exists a.s. and

$$\mathbb{P}\big[\lim_{t \uparrow t^*_{(l_1,l_2)}} X_t = l_1\big] = \mathbb{P}\big[\sup_{t < t^*_{(l_1,l_2)}} X_t < l_2\big] = 1.$$

iii) If $I(l_1) = -\infty$ and $I(l_2) < +\infty$, then $\lim_{t \uparrow t^*_{(l_1,l_2)}} X_t$ exists a.s. and

$$\mathbb{P}\big[\lim_{t \uparrow t^*_{(l_1,l_2)}} X_t = l_2\big] = \mathbb{P}\big[\inf_{t < t^*_{(l_1,l_2)}} X_t > l_1\big] = 1.$$

iv) If $I(l_1) > -\infty$ and $I(l_2) < +\infty$, then $\lim_{t \uparrow t^*_{(l_1,l_2)}} X_t$ exists a.s. and

$$\mathbb{P}\big[\lim_{t \uparrow t^*_{(l_1,l_2)}} X_t = l_1\big] = 1 - \mathbb{P}\big[\lim_{t \uparrow t^*_{(l_1,l_2)}} X_t = l_2\big] = \frac{I(l_2)}{I(l_2) - I(l_1)}.$$

A.4. Itô's formula for semimartingales

We now give Itô's formula for semimartingales which is used in sections 7.4, 8.4 and 9.3. For details, see, for example, Chapter 12 in [ELL 82].

A real valued $(\mathfrak{F}_t)$-adapted process $(X_t)_{t \geq 0}$ defined on the filtered probability space $\big(\Omega, \mathfrak{F}, (\mathfrak{F}_t)_{t \geq 0}, \mathbb{P}\big)$ is called a *semimartingale* if it has a decomposition of the form

$$X_t = X_0 + M_t + A_t, \quad t \geq 0, \qquad \text{[A.3]}$$

where $(M_t)_{t \geq 0}$ is a local $(\mathfrak{F}_t)$-martingale with $M_0 = 0$ a.s. and $(A_t)_{t \geq 0}$ is an $(\mathfrak{F}_t)$-adapted càdlàg process of locally bounded variation with $A_0 = 0$ a.s.

In general, decomposition [A.3] is not unique.

Let $(X^{\mathrm{c}}_t)_{t \geq 0}$ be the continuous part of the local martingale $(M_t)_{t \geq 0}$ in the decomposition of $(X_t)_{t \geq 0}$. Note that $(X^{\mathrm{c}}_t)_{t \geq 0}$ is independent of decomposition [A.3].

We denote by $\big(\langle X^{\mathrm{c}}, X^{\mathrm{c}} \rangle_t\big)_{t \geq 0}$ the predictable quadratic variation of $(X^{\mathrm{c}}_t)_{t \geq 0}$.

THEOREM A.11.– Let $(X_t)_{t\geq 0}$ be a semimartingale and $g\colon \mathbb{R} \to \mathbb{R}$ be a twice continuously differentiable function. Then, $\big(g(X_t)\big)_{t\geq 0}$ is also a semimartingale and

$$g(X_t) = g(X_0) + \int_{0_+}^{t} g'(X_{s-})\,\mathrm{d}X_s + \frac{1}{2}\int_{0_+}^{t} g''(X_{s-})\,\mathrm{d}\langle X^{\mathrm{c}}, X^{\mathrm{c}}\rangle_s$$

$$+ \sum_{0<s\leq t} \big(g(X_s) - g(X_{s-}) - g'(X_{s-})(X_s - X_{s-})\big), \quad t \geq 0.$$

Bibliography

[ABA 92] ABATE J., WHITT W., "The Fourier-series method for inverting transforms of probability distributions", *Queueing Systems*, vol. 10, no. 1, pp. 5–88, 1992.

[ABA 95] ABATE J., WHITT W., "Numerical inversion of Laplace transforms of probability distributions", *ORSA Journal on Computing*, vol. 7, no. 1, pp. 36–43, 1995.

[AGA 08] AGARWAL R.P., O'REGAN D., *An Introduction to Ordinary Differential Equations*, Springer, New York, 2008.

[ALB 01] ALBRECHER H., TEUGELS J.L., TICHY R.F., "On a gamma series expansion for the time-dependent probability of collective ruin", *Insurance: Mathematics and Economics*, vol. 29, no. 3, pp. 345–355, 2001.

[ALB 10a] ALBRECHER H., AVRAM F., KORTSCHAK D., "On the efficient evaluation of ruin probabilities for completely monotone claim size distributions", *Journal of Computational and Applied Mathematics*, vol. 233, no. 10, pp. 2724–2736, 2010.

[ALB 10b] ALBRECHER H., HIPP C., KORTSCHAK D., "Higher-order expansions for compound distributions and ruin probabilities with subexponential claims", *Scandinavian Actuarial Journal*, vol. 2010, no. 2, pp. 105–135, 2010.

[AND 07a] ANDROSHCHUK M., MISHURA Y., "Another approach to the problem of the ruin probability estimate for risk process with investments", *Theory of Stochastic Processes*, vol. 13(29), no. 4, pp. 1–18, 2007.

[AND 07b] ANDROSHCHUK M., MISHURA Y.S., "Estimation of the ruin probability of an insurance company operating on a BS-market", *Ukrainian Mathematical Journal*, vol. 59, no. 11, pp. 1617–1631, 2007.

[AND 08] ANDROSHCHUK M., MISHURA Y., "Optimal investment for an insurance company into risky and non-risky assets in the model with variable premium income process", *Journal of Numerical and Applied Mathematics*, vol. 1, no. 96, pp. 11–27, 2008.

[ARF 55] ARFWEDSON G., "Research in collective risk theory. Part II", *Scandinavian Actuarial Journal*, nos 1–2, pp. 53–100, 1955.

[ASM 84] ASMUSSEN S., "Approximations for the probability of ruin within finite time", *Scandinavian Actuarial Journal*, vol. 1984, no. 1, pp. 31–57, 1984.

[ASM 85] ASMUSSEN S., "Conjugate processes and the simulation of ruin problems", *Stochastic Processes and their Applications*, vol. 20, no. 2, pp. 213–229, 1985.

[ASM 97] ASMUSSEN S., BINSWANGER K., "Simulation of ruin probabilities for subexponential claims", *ASTIN Bulletin*, vol. 27, no. 2, pp. 297–318, 1997.

[ASM 98] ASMUSSEN S., "Subexponential asymptotics for stochastic processes: extremal behaviour, stationary distributions and first passage probabilities", *The Annals of Applied Probability*, vol. 8, no. 2, pp. 354–374, 1998.

[ASM 00] ASMUSSEN S., BINSWANGER K., HØJGAARD B., "Rare events simulation for heavy-tailed distributions", *Bernoulli*, vol. 6, no. 2, pp. 303–322, 2000.

[ASM 03] ASMUSSEN S., *Applied Probability and Queues*, Springer-Verlag, New York, 2003.

[ASM 06] ASMUSSEN S., KROESE D.P., "Improved algorithms for rare event simulation with heavy tails", *Advances in Applied Probabability*, vol. 38, no. 2, pp. 545–558, 2006.

[ASM 07] ASMUSSEN S., GLYNN P.W., *Stochastic Simulation: Algorithms and Analysis*, Springer-Verlag, New York, 2007.

[ASM 10] ASMUSSEN S., ALBRECHER H., *Ruin Probabilities*, World Scientific, Singapore, 2010.

[AZC 09] AZCUE P., MULER N., "Optimal investment strategy to minimize the ruin probability of an insurance company under borrowing constraints", *Insurance: Mathematics and Economics*, vol. 44, no. 1, pp. 26–34, 2009.

[AZC 14] AZCUE P., MULER N., *Stochastic Optimization in Insurance*, Springer-Verlag, New York, 2014.

[BAI 08] BAI L., GUO J., "Optimal proportional reinsurance and investment with multiple risky assets and no-shorting constraint", *Insurance: Mathematics and Economics*, vol. 42, no. 3, pp. 968–975, 2008.

[BAL 99] BALTRNAS A., "Second order behaviour of ruin probabilities", *Scandinavian Actuarial Journal*, vol. 1999, no. 2, pp. 120–133, 1999.

[BAL 05] BALTRNAS A., "Second order behaviour of ruin probabilities in the case of large claims", *Insurance: Mathematics and Economics*, vol. 36, no. 3, pp. 485–498, 2005.

[BAR 95] BARNDORFF-NIELSEN O.E., SCHMIDLI H., "Saddlepoint approximations for the probability of ruin in finite time", *Scandinavian Actuarial Journal*, vol. 1995, no. 2, pp. 169–186, 1995.

[BEA 84] BEARD R.E., PENTIKÄINEN T., PESONEN E., *Risk Theory*, Chapman and Hall, London, 1984.

[BEE 69] BEEKMAN J.A., "A ruin function approximation", *Transactions of the Society of Actuaries*, vol. 21, no. 59 AB, pp. 41–48, 275–279, 1969.

[BEE 85] BEEKMAN J.A., "A series for infinite time ruin probabilities", *Insurance: Mathematics and Economics*, vol. 4, no. 2, pp. 129–134, 1985.

[BEN 03] BENING V.E., KOROLEV V.Y., "Nonparametric estimation of the ruin probability for generalized risk processes", *Theory of Probability and Its Applications*, vol. 47, no. 1, pp. 1–16, 2003.

[BIR 78] BIRKHOFF G., ROTA G.-C., *Ordinary Differential Equations*, John Wiley & Sons, New York, 1978.

[BLA 08] BLANCHET J., GLYNN P., "Efficient rare-event simulation for the maximum of heavy-tailed random walks", *The Annals of Applied Probability*, vol. 18, no. 4, pp. 1351–1378, 2008.

[BOI 03] BOIKOV A.V., "The Cramér–Lundberg model with stochastic premium process", *Theory of Probability and Its Applications*, vol. 47, no. 3, pp. 489–493, 2003.

[BON 12] BONDAREV B.V., RAGULINA E.Y., "On the finite-time nonruin probability of an insurance company with investments in the financial (B,S)-market", *Cybernetics and Systems Analysis*, vol. 48, no. 5, pp. 736–748, 2012.

[BOO 87] BOOGAERT P., CRIJNS V., "Upperbounds on ruin probabilities in case of negative loadings and positive interest rates", *Insurance: Mathematics and Economics*, vol. 6, no. 3, pp. 221–232, 1987.

[BOO 88] BOOGAERT P., HAEZENDONCK J., DELBAEN F., "Limit theorems for the present value of the surplus of an insurance portfolio", *Insurance: Mathematics and Economics*, vol. 7, no. 2, pp. 131–138, 1988.

[BRE 01] BREKELMANS R., DE WAEGENAERE A., "Approximating the finite-time ruin probability under interest force", *Insurance: Mathematics and Economics*, vol. 29, no. 2, pp. 217–229, 2001.

[BRO 95] BROWNE S., "Optimal investment policies for a firm with a random risk process: exponential utility and minimizing the probability of ruin", *Mathematics of Operations Research*, vol. 20, no. 4, pp. 937–958, 1995.

[BÜH 70] BÜHLMANN H., *Mathematical Methods in Risk Theory*, Springer-Verlag, Berlin Heidelberg, 1970.

[BÜH 84] BÜHLMANN H., "The general economic premium principle", *ASTIN Bulletin*, vol. 14, no. 1, pp. 13–21, 1984.

[CAI 99] CAI J., GARRIDO J., "Two-sided bounds for ruin probabilities when the adjustment coefficient does not exist", *Scandinavian Actuarial Journal*, vol. 1999, no. 1, pp. 80–92, 1999.

[CAI 03] CAI J., DICKSON D.C.M., "Upper bounds for ultimate ruin probabilities in the Sparre Andersen model with interest", *Insurance: Mathematics and Economics*, vol. 32, no. 1, pp. 61–71, 2003.

[CAI 04] CAI J., "Ruin probabilities and penalty functions with stochastic rates of interest", *Stochastic Processes and their Applications*, vol. 112, no. 1, pp. 53–78, 2004.

[CAI 05] CAI J., YANG H., "Ruin in the perturbed compound Poisson risk process under interest force", *Advances in Applied Probabability*, vol. 37, no. 3, pp. 819–835, 2005.

[CAI 06] CAI J., XU C., "On the decomposition of the ruin probability for a jump-diffusion surplus process compounded by a geometric Brownian motion", *North American Actuarial Journal*, vol. 10, no. 1, pp. 120–129, 2006.

[CAR 03] CARDOSO R.M.R., WATERS H.R., "Recursive calculation of finite time ruin probabilities under interest force", *Insurance: Mathematics and Economics*, vol. 33, no. 3, pp. 659–676, 2003.

[CAR 05] CARDOSO R.M.R., WATERS H.R., "Calculation of finite time ruin probabilities for some risk models", *Insurance: Mathematics and Economics*, vol. 37, no. 2, pp. 197–215, 2005.

[CHE 05] CHERNY A.S., ENGELBERT H.-J., *Singular Stochastic Differential Equations*, Springer-Verlag, Berlin Heidelberg, 2005.

[CHO 10] CHOI S.K., CHOI M.H., LEE H.S. *et al.*, "New approximations of ruin probability in a risk process", *Quality Technology and Quantitative Management*, vol. 7, no. 4, pp. 377–383, 2010.

[CON 04] CONT R., TANKOV P., *Financial Modelling with Jump Processes*, Chapman & Hall/CRC, Boca Raton, 2004.

[COU 88] COURANT R., *Differential and Integral Calculus, I, II*, John Wiley & Sons, New York, 1988.

[CRA 30] CRAMÉR H., *On the Mathematical Theory of Risk*, Skandia Jubilee Volume, Stockholm, 1930.

[CRA 55] CRAMÉR H., *Collective Risk Theory*, Skandia Jubilee Volume, Stockholm, 1955.

[CSÖ 90] CSÖRG M., TEUGELS J.L., "Empirical Laplace transform and approximation of compound distributions", *Journal of Applied Probability*, vol. 27, no. 1, pp. 88–101, 1990.

[CSÖ 91] CSÖRG M., STEINEBACH J., "On the estimation of the adjustment coefficient in risk theory via intermediate order statistics", *Insurance: Mathematics and Economics*, vol. 10, no. 1, pp. 37–50, 1991.

[DAV 93] DAVIS M., *Markov Models and Optimization*, Chapman & Hall/CRC, Boca Raton, 1993.

[DEH 90] DEHEUVELS P., STEINEBACH J., "On some alternative estimates of the adjustment coefficient in risk theory", *Scandinavian Actuarial Journal*, vol. 1990, nos 2–3, pp. 135–159, 1990.

[DEL 87] DELBAEN F., HAEZENDONCK J., "Classical risk theory in an economic environment", *Insurance: Mathematics and Economics*, vol. 6, no. 2, pp. 85–116, 1987.

[DEN 90] DENNEBERG D., "Premium calculation: why standard deviation should be replaced by absolute deviation", *ASTIN Bulletin*, vol. 20, no. 2, pp. 181–190, 1990.

[DEV 78] DE VYLDER F., "A practical solution to the problem of ultimate ruin probability", *Scandinavian Actuarial Journal*, vol. 1978, no. 2, pp. 114–119, 1978.

[DEV 84] DE VYLDER F., GOOVAERTS M., "Bounds for classical ruin probabilities", *Insurance: Mathematics and Economics*, vol. 3, no. 2, pp. 121–131, 1984.

[DIC 91] DICKSON D.C.M., WATERS H.R., "Recursive calculation of survival probabilities", *ASTIN Bulletin*, vol. 21, no. 2, pp. 199–221, 1991.

[DIC 05] DICKSON D.C.M., WILLMOT G.E., "The density of the time to ruin in the classical Poisson risk model", *ASTIN Bulletin*, vol. 35, no. 1, pp. 45–60, 2005.

[DIC 10] DICKSON D.C.M., *Insurance Risk and Ruin*, Cambridge University Press, Cambridge, 2010.

[DOO 53] DOOB J.L., *Stochastic Processes*, John Wiley & Sons, New York, 1953.

[DOT 90] DOTHAN M.U., *Prices in Financial Markets*, Oxford University Press, 1990.

[DUF 88] DUFRESNE F., GERBER H.U., "The probability and severity of ruin for combinations of exponential claim amount distributions and their translations", *Insurance: Mathematics and Economics*, vol. 7, no. 2, pp. 75–80, 1988.

[DUF 89] DUFRESNE F., GERBER H.U., "Three methods to calculate the probability of ruin", *ASTIN Bulletin*, vol. 19, no. 1, pp. 71–90, 1989.

[DUP 07] DUPUIS P., LEDER K., WANG H., "Importance sampling for sums of random variables with regularly varying tails", *ACM Transactions on Modeling and Computer Simulation*, vol. 17, no. 3, article no. 14, 2007.

[EDA 13] EDALATI A., HIPP C., "Solving a Hamilton–Jacobi–Bellman equation with constraints", *Stochastics*, vol. 85, no. 4, pp. 637–651, 2013.

[ELL 82] ELLIOTT R.J., *Stochastic Calculus and Applications*, Springer-Verlag, New York, 1982.

[EMA 75] EMANUEL D.C., HARRISON J.M., TAYLOR A.J., "A diffusion approximation for the ruin function of a risk process with compounding assets", *Scandinavian Actuarial Journal*, vol. 1975, no. 4, pp. 240–247, 1975.

[EMB 82] EMBRECHTS P., VERAVERBEKE N., "Estimates for the probability of ruin with special emphasis on the possibility of large claims", *Insurance: Mathematics and Economics*, vol. 1, no. 1, pp. 55–72, 1982.

[EMB 88] EMBRECHTS P., VILLASENOR J.A., "Ruin estimates for large claims", *Insurance: Mathematics and Economics*, vol. 7, no. 4, pp. 269–274, 1988.

[EMB 91] EMBRECHTS P., MIKOSCH T., "A bootstrap procedure for estimating the adjustment coefficient", *Insurance: Mathematics and Economics*, vol. 10, no. 3, pp. 181–190, 1991.

[EMB 93] EMBRECHTS P., GRÜBEL R., PITTS S.M., "Some applications of the fast Fourier transform in insurance mathematics", *Statistica Neerlandica*, vol. 47, no. 1, pp. 59–75, 1993.

[EMB 94] EMBRECHTS P., SCHMIDLI H., "Ruin estimation for a general insurance risk model", *Advances in Applied Probability*, vol. 26, no. 2, pp. 404–422, 1994.

[FEL 71] FELLER W., *An Introduction to Probability Theory and Its Applications*, John Wiley & Sons, New York, 1971.

[FIC 90] FICHTENHOLZ G.M., *Differential- und Integralrechnung, II*, VEB Deutscher Verlag der Wissenschaften, Berlin, 1990.

[FIS 96] FISHMAN G.S., *Monte Carlo*, Springer-Verlag, New York, 1996.

[FLE 06] FLEMING W.H., SONER H.M., *Controlled Markov Processes and Viscosity Solutions*, Springer-Verlag, New York, 2006.

[FRA 82] FRANKEN P., KÖNIG D., ARNDT U. *et al.*, *Queues and Point Processes*, John Wiley & Sons, New York, 1982.

[FRO 02] FROLOVA A., KABANOV Y., PERGAMENSHCHIKOV S., "In the insurance business risky invesments are dangerous", *Finance and Stochastics*, vol. 6, no. 2, pp. 227–235, 2002.

[GAI 02] GAIER J., GRANDITS P., "Ruin probabilities in the presence of regularly varying tails and optimal investment", *Insurance: Mathematics and Economics*, vol. 30, no. 2, pp. 211–217, 2002.

[GAI 03] GAIER J., GRANDITS P., SCHACHERMAYER W., "Asymptotic ruin probabilities and optimal investment", *The Annals of Applied Probability*, vol. 13, no. 3, pp. 1054–1076, 2003.

[GAI 04] GAIER J., GRANDITS P., "Ruin probabilities and investment under interest force in the presence of regularly varying tails", *Scandinavian Actuarial Journal*, vol. 2004, no. 4, pp. 256–278, 2004.

[GAR 05] GARCIA J.M.A., "Explicit solutions for survival probabilities in the classical risk model", *ASTIN Bulletin*, vol. 35, no. 1, pp. 113–130, 2005.

[GEM 02] GEMAN H., "Pure jump Lévy processes for asset price modelling", *Journal of Banking & Finance*, vol. 26, no. 7, pp. 1297–1316, 2002.

[GER 71] GERBER H.U., "Der Einfluss von Zins auf die Ruinwahrscheinlichkeit", *Mitteilungen der Schweizerischen Vereinigung fur Versicherungsmathematiker*, vol. 71, no. 1, pp. 63–70, 1971.

[GER 73] GERBER H.U., "Martingales in risk theory", *Mitteilungen der Schweizerischen Vereinigung fur Versicherungsmathematiker*, vol. 73, no. 2, pp. 205–216, 1973.

[GER 79] GERBER H.U., *An Introduction to Mathematical Risk Theory*, S.S. Huebner Foundation for Insurance Education, University of Pennsylvania, Philadelphia, 1979.

[GER 07] GERBER H.U., YANG H., "Absolute ruin probabilities in a jump diffusion risk model with investment", *North American Actuarial Journal*, vol. 11, no. 3, pp. 159–169, 2007.

[GĪH 72] GĪHMAN Ī.Ī., SKOROHOD A.V., *Stochastic Differential Equations*, Springer-Verlag, New York-Heidelberg, 1972.

[GIL 91] GILBERT G.T., "Positive definite matrices and Sylvester's criterion", *The American Mathematical Monthly*, vol. 98, no. 1, pp. 44–46, 1991.

[GLY 89] GLYNN P.W., IGLEHART D.L., "Importance sampling for stochastic simulations", *Management Science*, vol. 35, no. 11, pp. 1367–1392, 1989.

[GOO 82] GOOVAERTS M., DE VYLDER F., HAEZENDONCK J., "Ordering of risks: a review", *Insurance: Mathematics and Economics*, vol. 1, no. 2, pp. 131–161, 1982.

[GOO 84a] GOOVAERTS M., DE VYLDER F., "A stable recursive algorithm for evaluation of ultimate ruin probabilities", *ASTIN Bulletin*, vol. 14, no. 1, pp. 53–59, 1984.

[GOO 84b] GOOVAERTS M., DE VYLDER F., HAEZENDONCK J., *Insurance Premiums*, North-Holland, Amsterdam, 1984.

[GRA 77] GRANDELL J., "A class of approximations of ruin probabilities", *Scandinavian Actuarial Journal*, vol. 1977, suppl. 1, pp. 37–52, 1977.

[GRA 78] GRANDELL J., "A remark on 'A class of approximations of ruin probabilities'", *Scandinavian Actuarial Journal*, vol. 1978, no. 2, pp. 77–78, 1978.

[GRA 79] GRANDELL J., "Empirical bounds for ruin probabilities", *Stochastic Processes and their Applications*, vol. 8, no. 3, pp. 243–255, 1979.

[GRA 91] GRANDELL J., *Aspects of Risk Theory*, Springer-Verlag, New York, 1991.

[GRA 00] GRANDELL J., "Simple approximations of ruin probabilities", *Insurance: Mathematics and Economics*, vol. 26, nos 2–3, pp. 157–173, 2000.

[GRA 04a] GRANDITS P., "An analogue of the Cramér-Lundberg approximation in the optimal investment case", *Applied Mathematics and Optimization*, vol. 50, no. 1, pp. 1–20, 2004.

[GRA 04b] GRANDITS P., "A Karamata-type theorem and ruin probabilities for an insurer investing proportionally in the stock market", *Insurance: Mathematics and Economics*, vol. 34, no. 2, pp. 297–305, 2004.

[GRA 05] GRANDITS P., "Minimal ruin probabilities and investment under interest force for a class of subexponential distributions", *Scandinavian Actuarial Journal*, vol. 2005, no. 6, pp. 401–416, 2005.

[GUS 05a] GUSAK D.V., "Risk processes with stochastic premiums and distributions of their functionals", *Theory of Stochastic Processes*, vol. 11(27), nos 1–2, pp. 29–39, 2005.

[GUS 05b] GUSAK D.V., KARNAUKH E.V., "On the exit from a finite interval for the risk processes with stochastic premiums", *Theory of Stochastic Processes*, vol. 11(27), nos 3–4, pp. 71–81, 2005.

[HAL 04] HALD M., SCHMIDLI H., "On the maximisation of the adjustment coefficient under proportional reinsurance", *ASTIN Bulletin*, vol. 34, no. 1, pp. 75–83, 2004.

[HAR 77] HARRISON J.M., "Ruin problems with compounding assets", *Stochastic Processes and their Applications*, vol. 5, no. 1, pp. 67–79, 1977.

[HER 86] HERKENRATH U., "On the estimation of the adjustment coefficient in risk theory by means of stochastic approximation procedures", *Insurance: Mathematics and Economics*, vol. 5, no. 4, pp. 305–313, 1986.

[HIP 00a] HIPP C., PLUM M., "Optimal investment for insurers", *Insurance: Mathematics and Economics*, vol. 27, no. 2, pp. 215–228, 2000.

[HIP 00b] HIPP C., TAKSAR M., "Stochastic control for optimal new business", *Insurance: Mathematics and Economics*, vol. 26, nos 2–3, pp. 185–192, 2000.

[HIP 03a] HIPP C., PLUM M., "Optimal investment for investors with state dependent income, and for insurers", *Finance and Stochastics*, vol. 7, no. 3, pp. 299–321, 2003.

[HIP 03b] HIPP C., VOGT M., "Optimal dynamic XL reinsurance", *ASTIN Bulletin*, vol. 33, no. 2, pp. 193–207, 2003.

[HIP 04a] HIPP C., "Asymptotics of ruin probabilities for controlled risk processes in the small claims case", *Scandinavian Actuarial Journal*, vol. 2004, no. 5, pp. 321–335, 2004.

[HIP 04b] HIPP C., "Stochastic control with application in insurance", in FRITTELLI M., RUNGGALDIER W.J., (eds), *Stochastic Methods in Finance*, Springer, Berlin Heidelberg, 2004.

[HIP 10] HIPP C., TAKSAR M., "Optimal non-proportional reinsurance control", *Insurance: Mathematics and Economics*, vol. 47, no. 2, pp. 246–254, 2010.

[HOE 63] HOEFFDING W., "Probability inequalities for sums of bounded random variables", *Journal of the American Statistical Association*, vol. 58, no. 301, pp. 13–30, 1963.

[HOG 84] HOGG R.V., KLUGMAN S.A., *Loss Distributions*, John Wiley & Sons, New York, 1984.

[HOR 85] HORN R.A., JOHNSON C.R., *Matrix Analysis*, Cambridge University Press, Cambridge, 1985.

[HUS 07] HUSAK D.V., "Behavior of risk processes with random premiums after ruin and a multivariate ruin function", *Ukrainian Mathematical Journal*, vol. 59, no. 11, pp. 1653–1667, 2007.

[IGL 69] IGLEHART D.L., "Diffusion approximations in collective risk theory", *Journal of Applied Probabability*, vol. 6, no. 2, pp. 285–292, 1969.

[IKE 89] IKEDA N., WATANABE S., *Stochastic Differential Equations and Diffusion Processes*, North-Holland Publishing Co., Kodansha Ltd., Amsterdam, Tokyo, 1989.

[JAC 06] JACOBSEN M., *Point Process Theory and Applications: Marked Point and Piecewise Deterministic Processes*, Birkhäuser, Boston, 2006.

[JAG 78] JAGERMAN D.L., "An inversion technique for the Laplace transform with application to approximation", *Bell System Technical Journal*, vol. 57, no. 3, pp. 669–710, 1978.

[JAG 82] JAGERMAN D.L., "An inversion technique for the Laplace transform", *Bell System Technical Journal*, vol. 61, no. 8, pp. 1995–2002, 1982.

[JUN 02] JUNEJA S., SHAHABUDDIN P., "Simulating heavy-tailed processes using delayed hazard rate twisting", *ACM Transactions on Modeling and Computer Simulation*, vol. 12, no. 2, pp. 94–118, 2002.

[JUN 07] JUNEJA S., "Estimating tail probabilities of heavy tailed distributions with asymptotically zero relative error", *Queueing Systems*, vol. 57, no. 2, pp. 115–127, 2007.

[KAA 91] KAAS R., "The Schmitter problem and a related problem: a partial solution", *ASTIN Bulletin*, vol. 21, no. 1, pp. 133–146, 1991.

[KAA 08] KAAS R., GOOVAERTS M., DHAENE J. *et al.*, *Modern Actuarial Risk Theory*, Springer-Verlag, Berlin Heidelberg, 2008.

[KAL 96] KALASHNIKOV V., "Two-sided bounds of ruin probabilities", *Scandinavian Actuarial Journal*, vol. 1996, no. 1, pp. 1–18, 1996.

[KAL 99] KALASHNIKOV V., "Bounds for ruin probabilities in the presence of large claims and their comparison", *North American Actuarial Journal*, vol. 3, no. 2, pp. 116–128, 1999.

[KAL 00] KALASHNIKOV V., KONSTANTINIDES D., "Ruin under interest force and subexponential claims: a simple treatment", *Insurance: Mathematics and Economics*, vol. 27, no. 1, pp. 145–149, 2000.

[KAL 02] KALASHNIKOV V., NORBERG R., "Power tailed ruin probabilities in the presence of risky investments", *Stochastic Processes and their Applications*, vol. 98, no. 2, pp. 211–228, 2002.

[KAR 91] KARATZAS I., SHREVE S.E., *Brownian Motion and Stochastic Calculus*, Springer-Verlag, New York, 1991.

[KAR 98] KARATZAS I., SHREVE S.E., *Methods of Mathematical Finance*, Springer-Verlag, New York, 1998.

[KAS 05] KASOZI J., PAULSEN J., "Numerical ultimate ruin probabilities under interest force", *Journal of Mathematics and Statistics*, vol. 1, no. 3, pp. 246–251, 2005.

[KIN 64] KINGMAN J.F.C., "A martingale inequality in the theory of queues", *Mathematical Proceedings of the Cambridge Philosophical Society*, vol. 60, no. 2, pp. 359-361, 1964.

[KLI 91] KLING B.M., GOOVAERTS M.J., "A recursive evaluation of the finite time ruin probability based on an equation of Seal", *Insurance: Mathematics and Economics*, vol. 10, no. 2, pp. 93–97, 1991.

[KLO 92] KLOEDEN P.E., PLATEN E., *Numerical Solution of Stochastic Differential Equations*, Springer-Verlag, Berlin Heidelberg, 1992.

[KLÜ 89] KLÜPPELBERG C., "Estimation of ruin probabilities by means of hazard rates", *Insurance: Mathematics and Economics*, vol. 8, no. 4, pp. 279–285, 1989.

[KLÜ 98] KLÜPPELBERG C., STADTMÜLLER U., "Ruin probabilities in the presence of heavy-tails and interest rates", *Scandinavian Actuarial Journal*, vol. 1998, no. 1, pp. 49–58, 1998.

[KNE 96] KNESSL C., PETERS C.S., "Exact and asymptotic solutions for the time-dependent problem of collective ruin II", *SIAM Journal on Applied Mathematics*, vol. 56, no. 5, pp. 1471–1521, 1996.

[KON 02] KONSTANTINIDES D., TANG Q., TSITSIASHVILI G., "Estimates for the ruin probability in the classical risk model with constant interest force in the presence of heavy tails", *Insurance: Mathematics and Economics*, vol. 31, no. 3, pp. 447-460, 2002.

[KON 04] KONSTANTINIDES D., TANG Q.H., TSITSIASHVILI G.S., "Two-sided bounds for ruin probability under constant interest force", *Journal of Mathematical Sciences*, vol. 123, no. 1, pp. 3824–3833, 2004.

[KOR 10] KORN R., KORN E., KROISANDT G., *Monte Carlo Methods and Models in Finance and Insurance*, Chapman & Hall/CRC, Boca Raton, 2010.

[LEH 92] LEHTONEN T., NYRHINEN H., "Simulating level-crossing probabilities by importance sampling", *Advances in Applied Probability*, vol. 24, no. 4, pp. 858–874, 1992.

[LIU 04] LIU C.S., YANG H., "Optimal investment for an insurer to minimize its probability of ruin", *North American Actuarial Journal*, vol. 8, no. 2, pp. 11–31, 2004.

[LIV 99] LIVSHITS K.I., "Probability of ruin of an insurance company for the Poisson model", *Russian Physics Journal*, vol. 42, no. 4, pp. 394–399, 1999.

[LUN 26] LUNDBERG F., *Försäkringsteknisk Riskutjämning*, F. Englunds Boktryckeri AB, Stockholm, 1926.

[LUN 03] LUNDBERG F., *I. Approximerad Framställning av Sannolikhetsfunktionen, II. Återförsäkring av Kollektivrisker*, Almqvist & Wiksell, Uppsala, 1903.

[LUO 08] LUO S., "Ruin minimization for insurers with borrowing constraints", *North American Actuarial Journal*, vol. 12, no. 2, pp. 143–174, 2008.

[LUO 11] LUO S., TAKSAR M., "On absolute ruin minimization under a diffusion approximation model", *Insurance: Mathematics and Economics*, vol. 48, no. 1, pp. 123–133, 2011.

[MA 03] MA J., SUN X., "Ruin probabilities for insurance models involving investments", *Scandinavian Actuarial Journal*, vol. 2003, no. 3, pp. 217–237, 2003.

[MAL 94] MALINOVSKII V.K., "Corrected normal approximation for the probability of ruin within finite time", *Scandinavian Actuarial Journal*, vol. 1994, no. 2, pp. 161–174, 1994.

[MAM 86] MAMMITZSCH V., "A note on the adjustment coefficient in ruin theory", *Insurance: Mathematics and Economics*, vol. 5, no. 2, pp. 147–149, 1986.

[MER 76] MERTON R.C., "Option pricing when underlying stock returns are discontinuous", *Journal of Financial Economics*, vol. 3, nos 1–2, pp. 125–144, 1976.

[MEY 87] MEYERS G., BEEKMAN J.A., "An improvement to the convolution method of calculating $\psi(u)$", *Insurance: Mathematics and Economics*, vol. 6, no. 4, pp. 267–274, 1987.

[MIJ 10] MIJATOVIĆ A., "Local time and the pricing of time-dependent barrier options", *Finance and Stochastics*, vol. 14, no. 1, pp. 13–48, 2010.

[MIK 01] MIKOSCH T., NAGAEV A., "Rates in approximations to ruin probabilities for heavy-tailed distributions", *Extremes*, vol. 4, no. 1, pp. 67–78, 2001.

[MIS 06] MISHURA Y., "An estimate of ruin probabilities for long range dependence models", *Theory of Probabability and Mathematical Statistics*, vol. 72, pp. 103–111, 2006.

[MIS 15] MISHURA Y., PERESTYUK M., RAGULINA O., "Ruin probability in a risk model with variable premium intensity and risky investments", *Opuscula Mathematica*, vol. 35, no. 3, pp. 333–352, 2015.

[NOR 06] NORKIN B.V., "The method of successive approximation applied to find the probability for an insurance company with random premiums", *Cybernetics and Systems Analysis*, vol. 42, no. 1, pp. 98–110, 2006.

[ØKS 03] ØKSENDAL B., *Stochastic Differential Equations*, Springer-Verlag, Berlin Heidelberg, 2003.

[ØKS 07] ØKSENDAL B., SULEM A., *Applied Stochastic Control of Jump Diffusions*, Springer-Verlag, Berlin Heidelberg, 2007.

[PAN 86] PANJER H.H., "Direct calculation of ruin probabilities", *Journal of Risk and Insurance*, vol. 53, no. 3, pp. 521–529, 1986.

[PAU 93] PAULSEN J., "Risk theory in a stochastic economic environment", *Stochastic Processes and their Applications*, vol. 46, no. 2, pp. 327–361, 1993.

[PAU 97] PAULSEN J., GJESSING H.K., "Ruin theory with stochastic return on investments", *Advances in Applied Probability*, vol. 29, no. 4, pp. 965–985, 1997.

[PAU 98] PAULSEN J., "Sharp conditions for certain ruin in a risk process with stochastic return on investments", *Stochastic Processes and their Applications*, vol. 75, no. 1, pp. 135–148, 1998.

[PAU 02] PAULSEN J., "On Cramér-like asymptotics for risk processes with stochastic return on investments", *The Annals of Applied Probability*, vol. 12, no. 4, pp. 1247–1260, 2002.

[PAU 05] PAULSEN J., KASOZI J., STEIGEN A., "A numerical method to find the probability of ultimate ruin in the classical risk model with stochastic return on investments", *Insurance: Mathematics and Economics*, vol. 36, no. 3, pp. 399–420, 2005.

[PAU 08] PAULSEN J., "Ruin models with investment income", *Probability Surveys*, vol. 5, pp. 416–434, 2008.

[PER 98] PERVOZVANSKY JR. A.A., "Equation for survival probability in a finite time interval in case of non-zero real interest force", *Insurance: Mathematics and Economics*, vol. 23, no. 3, pp. 287–295, 1998.

[PER 06] PERGAMENSHCHIKOV S., ZEITOUNY O., "Ruin probability in the presence of risky investments", *Stochastic Processes and their Applications*, vol. 116, no. 2, pp. 267–278, 2006.

[PIC 97] PICARD P., LEFÈVRE C., "The probability of ruin in finite time with discrete claim size distribution", *Scandinavian Actuarial Journal*, vol. 1997, no. 1, pp. 58–69, 1997.

[PIT 96] PITTS S.M., GRÜBEL R., EMBRECHTS P., "Confidence bounds for the adjustment coefficient", *Advances in Applied Probability*, vol. 28, no. 3, pp. 802–827, 1996.

[PIT 14] PITACCO E., *Health Insurance*, Springer International Publishing, Switzerland, 2014.

[PRA 61] PRABHU N.U., "On the ruin problem of collective risk theory", *The Annals of Mathematical Statistics*, vol. 32, no. 3, pp. 757–764, 1961.

[PRO 05] PROMISLOW S.D., YOUNG V.R., "Minimizing the probability of ruin when claims follow Brownian motion with drift", *North American Actuarial Journal*, vol. 9, no. 3, pp. 109–128, 2005.

[RAG 10] RAGULINA O., "On differentiability of the non-ruin probability of an insurance company in models with constant interest rate (in Ukrainian)", *Applied Statistics. Actuarial and Financial Mathematics*, vol. 2010, nos 1–2, pp. 82–116, 2010.

[RAG 11] RAGULINA O., "On the non-ruin probability of an insurance company in the classical risk model using conditional franchise and limit of responsibility", *Applied Statistics. Actuarial and Financial Mathematics*, vol. 2011, nos 1–2, pp. 27–46, 2011.

[RAG 12] RAGULINA O., "Estimates and properties of the survival probability of an insurance company in the classical risk model with investments to the financial (B,S)-market (in Ukrainian)", *Bulletin of Taras Shevchenko National University of Kyiv, Series: Physics & Mathematics*, vol. 2012, no. 2, pp. 23–30, 2012.

[RAG 14] RAGULINA O., "Maximization of the survival probability by franchise and deductible amounts in the classical risk model", in KOROLYUK V., LIMNIOS N., MISHURA Y. *et al.* (eds), *Modern Stochastics and Applications*, Springer International Publishing, Switzerland, 2014.

[RAG 15] RAGULINA O., "Continuity and differentiability properties of the survival probabilities in risk models with investments and their applications", in VANMAELE M., DEELSTRA G., DE SCHEPPER A. *et al.* (eds), *Actuarial and Financial Mathematics Conference, Interplay between Finance and Insurance*, available at: http://www.afmathconf.ugent.be/FormerEditions/Proceedings2015.pdf, 2015.

[REI 86] REICH A., "Properties of premium calculation principles", *Insurance: Mathematics and Economics*, vol. 5, no. 1, pp. 97–101, 1986.

[RIP 87] RIPLEY B.D., *Stochastic Simulation*, John Wiley & Sons, New York, 1987.

[ROB 04] ROBINSON J.C., *An Introduction to Ordinary Differential Equations*, Cambridge University Press, 2004.

[ROG 00] ROGERS L.C.G., WILLIAMS D., *Diffusions, Markov Processes, and Martingales, I, II*, Cambridge University Press, 2000.

[ROL 99] ROLSKI T., SCHMIDLI H., SCHMIDT V. *et al.*, *Stochastic Processes for Insurance and Finance*, John Wiley & Sons, Chichester, 1999.

[ROU 94] ROUSSEEUW P.J., MOLENBERGHS G., "The shape of correlation matrices", *The American Statistician*, vol. 48, no. 4, pp. 276–279, 1994.

[RUB 08] RUBINSTEIN R.Y., KROESE D.P., *Simulation and the Monte Carlo Method*, John Wiley & Sons, Hoboken, 2008.

[RUL 04] RULLIÈRE D., LOISEL S., "Another look at the Picard-Lefèvre formula for finite-time ruin probabilities", *Insurance: Mathematics and Economics*, vol. 35, no. 2, pp. 187–203, 2004.

[SAM 65] SAMUELSON P.A., "Rational theory of warrant pricing", *Industrial Management Review*, vol. 6, no. 2, pp. 13–31, 1965.

[SCH 90] SCHMITTER H., "The ruin probability of a discrete claims distribution with a finite number of steps", *XXII ASTIN-colloquium*, Montreux, Switzerland, 1990.

[SCH 94] SCHMIDLI H., "Diffusion approximations for a risk process with the possibility of borrowing and investment", *Communications in Statistics: Stochastic Models*, vol. 10, no. 2, pp. 365–388, 1994.

[SCH 97] SCHMIDLI H., "Estimation of the Lundberg coefficient for a Markov modulated risk model", *Scandinavian Actuarial Journal*, vol. 1997, no. 1, pp. 48–57, 1997.

[SCH 01] SCHMIDLI H., "Optimal proportional reinsurance policies in a dynamic setting", *Scandinavian Actuarial Journal*, vol. 2001, no. 1, pp. 55–68, 2001.

[SCH 02a] SCHMIDLI H., Asymptotics of ruin probabilities for risk processes under optimal reinsurance policies: the small claim case, Working paper 180, Laboratory of Actuarial Mathematics, University of Copenhagen, 2002.

[SCH 02b] SCHMIDLI H., "On minimizing the ruin probability by investment and reinsurance", *The Annals of Applied Probability*, vol. 12, no. 3, pp. 890–907, 2002.

[SCH 04a] SCHMIDLI H., "Asymptotics of ruin probabilities for risk processes under optimal reinsurance and investment policies: the large claim case", *Queueing Systems*, vol. 46, no. 1, pp. 149–157, 2004.

[SCH 04b] SCHMIDLI H., On Cramér–Lundberg approximations for ruin probabilities under optimal excess of loss reinsurance, Working paper 193, Laboratory of Actuarial Mathematics, University of Copenhagen, 2004.

[SCH 05] SCHMIDLI H., "On optimal investment and subexponential claims", *Insurance: Mathematics and Economics*, vol. 36, no. 1, pp. 25–35, 2005.

[SCH 08a] SCHMIDLI H., "On Cramér–Lundberg approximations for ruin probabilities under optimal excess of loss reinsurance", *Journal of Numerical and Applied Mathematics*, vol. 1, no. 96, pp. 198–205, 2008.

[SCH 08b] SCHMIDLI H., *Stochastic Control in Insurance*, Springer-Verlag, London, 2008.

[SEA 69] SEAL H.L., *Stochastic Theory of a Risk Business*, John Wiley & Sons, New York, 1969.

[SEA 74] SEAL H.L., "The numerical calculation of $U(\omega, t)$, the probability of non-ruin in an interval $(0, t)$", *Scandinavian Actuarial Journal*, vol. 1974, no. 3, pp. 121–139, 1974.

[SEG 55] SEGERDAHL C.-O., "When does ruin occur in the collective theory of risk?", *Scandinavian Actuarial Journal*, vol. 1955, nos 1 2, pp. 22 36, 1955.

[SHI 88] SHIU E.S.W., "Calculation of the probability of eventual ruin by Beekman's convolution series", *Insurance: Mathematics and Economics*, vol. 7, no. 1, pp. 41–47, 1988.

[SHI 89] SHIU E.S.W., "Ruin probability by operational calculus", *Insurance: Mathematics and Economics*, vol. 8, no. 3, pp. 243–249, 1989.

[SHR 04] SHREVE S.E., *Stochastic Calculus for Finance, I, II*, Springer-Verlag, New York, 2004.

[SØR 96] SØRENSEN M., "Semimartingale approach to some problems in risk theory", *ASTIN Bulletin*, vol. 26, no. 1, pp. 15–23, 1996.

[STR 88] STRANG G., *Linear Algebra and Its Applications*, Harcourt Brace Jovanovich, Inc., San Diego, 1988.

[SUN 95] SUNDT B., TEUGELS J.L., "Ruin estimates under interest force", *Insurance: Mathematics and Economics*, vol. 16, no. 1, pp. 7–22, 1995.

[SUN 97] SUNDT B., TEUGELS J.L., "The adjustment function in ruin estimates under interest force", *Insurance: Mathematics and Economics*, vol. 19, no. 2, pp. 85–94, 1997.

[TAK 03] TAKSAR M.I., MARKUSSEN C., "Optimal dynamic reinsurance policies for large insurance portfolios", *Finance and Stochastics*, vol. 7, no. 1, pp. 97–121, 2003.

[TAN 05] TANG Q., "The finite-time ruin probability of the compound Poisson model with constant interest force", *Journal of Applied Probability*, vol. 42, no. 3, pp. 608–619, 2005.

[TAY 76] TAYLOR G.C., "Use of differential and integral inequalities to bound ruin and queuing probabilities", *Scandinavian Actuarial Journal*, no. 4, pp. 197–208, 1976.

[TEU 73] TEUGELS J.L., VERAVERBEKE N., Cramér-type estimates for the probability of ruin, CORE discussion paper No. 7316, 1973.

[THO 73a] THORIN O., "The ruin problem in case the tail of the claim distribution is completely monotone", *Scandinavian Actuarial Journal*, vol. 1973, no. 2, pp. 100–119, 1973.

[THO 73b] THORIN O., WIKSTAD N., "Numerical evaluation of ruin probabilities for a finite period", *ASTIN Bulletin*, vol. 7, no. 2, pp. 137–153, 1973.

[THO 77a] THORIN O., "Ruin probabilities prepared for numerical calculation", *Scandinavian Actuarial Journal*, vol. 1977, suppl. 1, pp. 7–17, 1977.

[THO 77b] THORIN O., WIKSTAD N., "Calculation of ruin probabilities when the claim distribution is lognormal", *ASTIN Bulletin*, vol. 9, no. 1–2, pp. 231–246, 1977.

[THO 86] THORIN O., Ruin probabilities when the claim amounts are gamma distributed, Forsakringstekniska forskningsnamnden. Meddelande nr 69, 1986.

[VIN 99] VINOGRADOV O.P., "The Cramér condition is necessary and sufficient for asymptotically exponential decrease of ruin probability", *Journal of Mathematical Sciences*, vol. 93, no. 4, pp. 609–611, 1999.

[VIT 87] VITTAL P.R., VASUDEVAN R., "Some first passage time problems with restricted reserve and two components of income", *Scandinavian Actuarial Journal*, vol. 1987, nos 3-4, pp. 198–210, 1987.

[VON 75] VON BAHR B., "Asymptotic ruin probabilities when exponential moments do not exist", *Scandinavian Actuarial Journal*, vol. 1975, no. 1, pp. 6–10, 1975.

[WAN 01] WANG G., WU R., "Distributions for the risk process with a stochastic return on investments", *Stochastic Processes and their Applications*, vol. 95, no. 2, pp. 329–341, 2001.

[WAT 79] WATERS H.R., "Excess of loss reinsurance limits", *Scandinavian Actuarial Journal*, vol. 1979, no. 1, pp. 37–43, 1979.

[WAT 83] WATERS H.R., "Some mathematical aspects of reinsurance", *Insurance: Mathematics and Economics*, vol. 2, no. 1, pp. 17–26, 1983.

[YUE 04] YUEN K.C., WANG G., NG K.W., "Ruin probabilities for a risk process with stochastic return on investments", *Stochastic Processes and their Applications*, vol. 110, no. 2, pp. 259–274, 2004.

[YUE 06] YUEN K.C., WANG G., WU R., "On the renewal risk process with stochastic interest", *Stochastic Processes and their Applications*, vol. 116, no. 10, pp. 1496–1510, 2006.

[ZIN 08] ZINCHENKO N., ANDRUSIV A., "Risk process with stochastic premiums", *Theory of Stochastic Processes*, vol. 14(30), nos. 3–4, pp. 189–208, 2008.

Abbreviations and Notation

Abbreviations

a.s.	almost surely
c.d.f.	cumulative distribution function
i.i.d.	independent identically distributed
p.d.f.	probability density function
r.v.	random variable
w.r.t.	with respect to

Mathematical notation

$\approx$	approximately equal
$\sim$	asymptotically equivalent
$\lceil a \rceil$	ceiling function, i.e. the smallest integer not less than a
$\lfloor a \rfloor$	floor function, i.e. the largest integer not greater than a
$a \vee b$	maximum value of a and b, i.e. $a \vee b = \max\{a, b\}$
$a \wedge b$	minimum value of a and b, i.e. $a \wedge b = \min\{a, b\}$
$\mathbb{E}$	expectation
$\mathbb{1}_{\{E\}}$	indicator function of the event E
$\mathbb{P}$	probability measure
$\mathbb{R}$	set of real numbers, i.e. $\mathbb{R} = (-\infty, +\infty)$
$\mathbb{R}_+$	set of non-negative real numbers, i.e. $\mathbb{R} = [0, +\infty)$
$\mathbb{R}_+^2$	Cartesian square of $\mathbb{R}_+$, i.e. $\mathbb{R}_+^2 = [0, +\infty) \times [0, +\infty)$

Principal notation for the risk models

$(B_t)_{t\geq 0}$	price of the risk-free asset
c	constant premium intensity
$c(u)$	variable premium intensity
d	franchise amount
$(d_t)_{t\geq 0}$, (d_t)	admissible strategy of the franchise amount choice
(d_t^*)	optimal strategy of the franchise amount choice
$(\bar{d}_t)_{t\geq 0}$, $(\bar{d}_t)$	admissible strategy of the deductible amount choice
$(\bar{d}_t^*)$	optimal strategy of the deductible amount choice
$F(y)$	c.d.f. of Y_i
$\bar{F}(y)$	c.d.f. of $\bar{Y}_i$
$F^{\mathrm{c}}(y)$	continuous component of $F(y)$
$F^{\mathrm{d}}(y)$	discrete component of $F(y)$
$f(y)$	p.d.f. of Y_i
$\bar{f}(y)$	p.d.f. of $\bar{Y}_i$
$h(R)$	shifted moment generating function of Y_i
L	liability limit amount
$(N_t)_{t\geq 0}$	number of claims on $[0,t]$, a homogeneous Poisson process
$(\bar{N}_t)_{t\geq 0}$	number of premiums on $[0,t]$, a homogeneous Poisson process
r	risk-free interest rate
$(S_t)_{t\geq 0}$	price of the risky asset
$(S_t^j)_{t\geq 0}$, $1\leq j\leq n$	price of the jth risky asset
$(W_t)_{t\geq 0}$	standard Brownian motion
x	initial surplus
$\big(X_t(x)\big)_{t\geq 0}$, $(X_t)_{t\geq 0}$	surplus process
Y_i, $i\geq 1$	claim sizes, i.i.d. r.v.'s
$\bar{Y}_i$, $i\geq 1$	premium sizes, i.i.d. r.v.'s
θ	safety loading
λ	intensity of $(N_t)_{t\geq 0}$
$\bar{\lambda}$	intensity of $(\bar{N}_t)_{t\geq 0}$
μ	expectation of Y_i
$\bar{\mu}$	expectation of $\bar{Y}_i$

$\tau(x), \tau$	ruin time
$\tau_i, i \geq 1$	time when the ith claim arrives
$\bar{\tau}_i, i \geq 1$	time when the ith premium arrives
$\varphi(x)$	infinite-horizon survival probability
$\varphi'_-(x)$	left derivative of $\varphi(x)$
$\varphi'_+(x)$	right derivative of $\varphi(x)$
$\varphi(x, t)$	finite-horizon survival probability
$\psi(x)$	infinite-horizon ruin probability
$\psi(x, t)$	finite-horizon ruin probability
$(\Omega, \mathfrak{F}, \mathbb{P})$	probability space

Index

A, B, C, D

E, F, G, I

L, M, N, P

R, S, T

FSC
www.fsc.org
MIX
Papier | Fördert
gute Waldnutzung
FSC® C083411

Kolibri 360 GmbH
Ferdinand-Jühlke-Straße 7
99095 Erfurt, Deutschland
produktsicherheit@kolibri360.de